EXPLAINABLE ARTIFICIAL INTELLIGENCE FOR SMART CITIES

EXPLAINABLE ARTIFICIAL INTELLIGENCE FOR SMART CITIES

Edited by
Mohamed Lahby
Utku Kose
Akash Kumar Bhoi

CRC Press
Taylor & Francis Group
Boca Raton London New York

CRC Press is an imprint of the
Taylor & Francis Group, an **informa** business

First Edition published 2022
by CRC Press
6000 Broken Sound Parkway NW, Suite 300, Boca Raton, FL 33487-2742

and by CRC Press
2 Park Square, Milton Park, Abingdon, Oxon, OX14 4RN

© 2022 selection and editorial matter, Mohamed Lahby, Utku Kose, and Akash Kumar Bhoi; individual chapters, the contributors.

CRC Press is an imprint of Taylor & Francis Group, LLC

ISBN: 978-1-032-00112-8 (hbk)
ISBN: 978-1-032-00113-5 (pbk)
ISBN: 978-1-003-17277-2 (ebk)

DOI: 10.1201/9781003172772

Typeset in Times
by MPS Limited, Dehradun

Contents

Contributors

W. Abbass
National Institute of Posts and
  Telecommunication 'INPT'
Madinat Al Irfane
Rabat, Morocco

M. Afif
Laboratory of Electronics and
  Microelectronics (EµE), Faculty of Sciences of
  Monastir
University of Monastir
Monastir, Tunisia

B. Agarwal
Indian Institute of Information Technology
Kota, India

Vishal Ahuja
Himachal Pradesh University
Himachal Pradesh, India

Muna Al-Kharraz
Computer Science Department, Faculty of
  Computing and Information Technology
King Abdulaziz University
Jeddah, Saudi Arabia

Wadee Alhalabi
Computer Science Department, Faculty of
  Computing and Information Technology
King Abdulaziz University
Jeddah, Saudi Arabia
and
Computer Science Department, College of
  Engineering
Effat University
Jeddah, Saudi Arabia

Rangel Arthur
Faculty of Technology (FT)
State University of Campinas (UNICAMP)
Limeira, São Paulo, Brazil

M. Atri
College of Computer Science
King Khalid University
Abha, Saudi Arabia

R. Ayachi
Laboratory of Electronics and Microelectronics
  (EµE), Faculty of Sciences of Monastir
University of Monastir
Monastir, Tunisia

A. Baina
National Institute of Posts and Telecommunication
  'INPT'
Madinat Al Irfane
Rabat, Morocco

Reem Bashmail
Computer Science Department, Faculty of
  Computing and Information Technology
King Abdulaziz University
Jeddah, Saudi Arabia

M. Bellafkih
National Institute of Posts and Telecommunication
  'INPT'
Madinat Al Irfane
Rabat, Morocco

Siham Benhadou
National Higher School of Electricity and
  Mechanics (ENSEM)
Hassan II University
Casablanca, Morocco

A. Biswas
Chittagong University of Engineering and
  Technology
Chittagong, Bangladesh

Zineb Boudanga
National Higher School of Electricity and
    Mechanics (ENSEM)
Hassan II University
Casablanca, Morocco
and
Foundation for Development and Innovation in
    Science and Engineering (FRDISI)
Casablanca, Morocco

Shree Charran R.
Indian Institute of Science
Bengaluru, India

Thomas M. Chen
City, University of London
London, UK

A. Daaif
ENSET Mohammedia
Hassan II University of Casablanca
Casablanca, Morocco

R. Dhaya
Department of Computer Science
King Khalid University-Sarat Abidha Campus
Abha, Saudi Arabia

Rahul Kumar Dubey
Robert Bosch Engineering and Business Solutions
    Private Limited
Bengaluru, India

M. El Khaili
ENSET Mohammedia
Hassan II University of Casablanca
Casablanca, Morocco

S. El Motaki
University Sidi Mohamed Ben Abdellah
Fez, Morocco

A. El-Fengour
University Ibn Tofail
Kenitra, Morocco
and
University of Castilla-La Mancha
Ciudad Real, Spain

Lamiaa A. Elrefaei
Computer Science Department, Faculty of
    Computing and Information Technology
King Abdulaziz University
Jeddah, Saudi Arabia
and
Electrical Engineering Department, Faculty of
    Engineering at Shoubra
Benha University
Cairo, Egypt

Ossama Embarak
Higher Colleges of Technology
Abu Dhabi, UAE

Emre Erturk
Eastern Institute of Technology
Napier, New Zealand

Mai Fadel
Computer Science Department, Faculty of
    Computing and Information Technology
King Abdulaziz University
Jeddah, Saudi Arabia

Reinaldo Padilha França
School of Electrical Engineering and Computing
    (FEEC)
State University of Campinas (UNICAMP)
Campinas, São Paulo, Brazil

P. Harjule
Department to Mathematics
MNIT Jaipur
Jaipur, India

Yuzo Iano
School of Electrical Engineering and Computing
    (FEEC)
State University of Campinas (UNICAMP)
Campinas, São Paulo, Brazil

M.S. Islam
Chittagong University of Engineering and
    Technology
Chittagong, Bangladesh

A. Jain
Indian Institute of Information Technology
Kota, India

M. Jain
Indian Institute of Information Technology
Kota, India

R. Kanthavel
Department of Computer Engineering
King Khalid University
Abha, Saudi Arabia

Jaspreet Kaur
PG Department of Commerce & Management
Hans Raj Mahila Maha Vidyalaya
Jalandhar, India

Nabeel Khan
Department of Information Technology, College
of Computer
Qassim University
Buraydah, Saudi Arabia

A. Kumar
Indian Institute of Information Technology
Kota, India

Jean Philippe Leroy
Liser IPI Paris, Group IGS
Paris, France

Dobrila Lopez
Eastern Institute of Technology
Napier, New Zealand

Ana Carolina Borges Monteiro
School of Electrical Engineering and Computing
(FEEC)
State University of Campinas (UNICAMP)
Campinas, São Paulo, Brazil

Lakshika S. Nawarathna
Department of Statistics and Computer Science
University of Peradeniya
Peradeniya, Sri Lanka

H. Ouajji
ENSET Mohammedia
Hassan II University of Casablanca
Casablanca, Morocco

Andria Procopiou
City, University of London
London, UK

Mohamed Uwaiz Fathima Rushda
Postgraduate Institute of Science
University of Peradeniya
Peradeniya, Sri Lanka

Harpreet Singh
PG Department of Bioinformatics
Hans Raj Mahila Maha Vidyalaya
Jalandhar, India

L. Terrada
ENSET Mohammedia
Hassan II University of Casablanca
Casablanca, Morocco

R.A. Verma
Indian Institute of Information Technology
Kota, India

C.Y. Yong
University College of Technology Sarawak
Sarawak, Malaysia

Weiyang Yu
Eastern Institute of Technology
Napier, New Zealand

An Overview of Explainable Artificial Intelligence (XAI) from a Modern Perspective

1

Ana Carolina Borges Monteiro[1],
Reinaldo Padilha França[1], Rangel Arthur[2],
and Yuzo Iano[1]

[1]*School of Electrical Engineering and Computing (FEEC) – State University of Campinas (UNICAMP), Campinas, São Paulo, Brazil*
[2]*Faculty of Technology (FT) – State University of Campinas (UNICAMP), Limeira, São Paulo, Brazil*

Contents

1.1 INTRODUCTION

Similar to disruptive all-purpose technologies, Artificial Intelligence (AI) will move from its current perspective-related experimental state to be incorporated into the fabric of most modern businesses, considering that smart technology does things and does it well, from smart smartphone assistants to the autopilot that controls the plane most of the time of your flight. Or even pondering insurance companies that use machine learning to motorize, industrialize, condition, robotize, that is, automate and improve/customize/personalize customer support; commercial companies optimizing their business with neural networks (Figure 1.1) and even AI for the automation of medical diagnosis. There is a part of a society

DOI: 10.1201/9781003172772-1

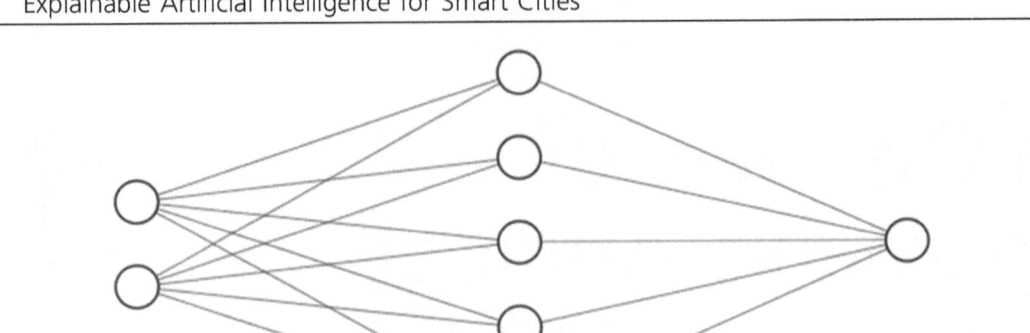

Input Layer $\in \mathbb{R}^2$ Hidden Layer $\in \mathbb{R}^5$ Output Layer $\in \mathbb{R}^1$

FIGURE 1.1   Neural Networks.

where intelligent robots have already taken over manual labour, still reflecting the potential of AI that as time goes by becomes more intelligent taking on analytical tasks (Došilović et al., 2018).

In a smart city, it is possible to provide interaction and connection with the citizens of smart cities through safer, building this concept that derives from innovative and even resilient technological solutions that are inclusive and accessible to citizens. Inclusive sustainable urban planning for a deep involvement of residents, companies, and governments improves collaboration, transparency, and sustainability with safer and more compatible tools. The transformation that leads a city to be identified as 'smart' takes into account the characteristics and demands of citizens, improving government services through the application of data analysis to enable city leaders and employees to make actionable and informed decisions. Even providing better digital services to residents, incorporating intelligent solutions that provide answers to specific questions, as well as offering technology applications and services in a safe, helping to disseminate digital technologies that can bring benefits to people and the city where they operate (França et al., 2021).

Due to the advances in computational power that have become more accessible and the large volume of data generated daily, Machine Learning (ML) models have become more complex, reaching increasingly impressive levels of performance. However, the increase in performance of ML models over time has not always been accompanied by an advance in their transparency, that is, intelligent models did not obtain an exact explanation of how this came to their conclusions (França et al., 2021).

Whereas the field of AI has developed computational devices with properties and characteristics that allow drive cars, fold proteins, synthesize chemical compounds, and even detect and identify high-energy particles at a superhuman level (high computational processing), that is, which specialist professionals would not be able to achieve with such precision and efficiency. However, these AI algorithms cannot explain the thinking procedure behind their digital decisions. Since a computer (computational device) has the capacity and properties to dominate protein folding (a chemical process in which the structure of a protein assumes its functional configuration with importance for cellular metabolism and related problems) and also informs about the rule and properties of Science Biology, reflecting that it is more useful, advantageous and even convenient than a computational device to just fold proteins without the need for explanation than a human being (Haenlein & Kaplan, 2019; Holzinger et al., 2019).

The lack of explainability becomes a problem in the exemplary panorama of a well-respected and recognized teacher by her students and colleagues, generating motivation for her students and sharing her techniques with other teachers. And be conceptualized minimally by an AI algorithm that aims to improve the quality of teaching by evaluating the performance of teachers, however without guidance and explanation of what may have impacted the algorithm by conceptualizing in this way. This molds a possible problem of not being able to explain the decisions made by an AI algorithm, when it evaluates a person's performance, considering that it is important that employees and workers who are evaluated by intelligent technology have access to the factors that can lead to this digital judgement (AI-oriented decision making) (Putnik et al., 2018).

In that way, that person could challenge the decision made by the algorithm or work to improve these factors. It is in this context that the importance of the XAI study comes in, addressing the need to be able to interpret an ML model. This arises because it is common for the formulation of problems addressed by ML to be incomplete, considering that usually, insight is not enough to deal with a problem, considering the importance of more than just 'what', but also 'why', and even the 'how', that is, reason for an improvement to be achieved (Emmert-Streib et al., 2020).

In this sense, XAI is the field of research dedicated to the study of methods so that AI applications produce solutions that can be explained to human beings, appearing as a counterpoint to the development of completely black-box models, that is, opaque models in which not even developers know how decisions are made. However, one of the great challenges of XAI is the difficulty of reaching a concise definition of what a sufficiently explainable model is (Gunning, 2017).

Also considering the transformation of machine learning that has been occurring over time, pondering the recent past that aims to obtain intelligent model the digital decision system, digital behavior, and your reactions (answers to the problem presented). These results obtained in the field of ML have led to fast growth in the elaboration and realization of AI. Reflecting that the conception of Deep Learning (DL) (Figure 1.2) is grounded on the past, there are many techniques such as consisting of convolutional neural networks (CNN), recursive neural networks (RNN), reinforcement learning, and even contentious networks demonstrating remarkable success. Although these successful achievements were obtained, it is difficult to clarify and even elucidate the digital decisions and actions of these intelligent systems related to users (França, Monteiro, et al., 2020; Yan et al., 2015).

Considering that these DL models projected with hundreds of millions of layered artificial neural networks (ANN) are not foolproof, given the disadvantage when it is simply duped, relating case of a pixel attack. With respect to the pixel attack, this is similar to the association inference attack, related to extracting images from facial recognition systems, just by having API access to the ML model, this can be performed as a series of progressive queries using an image digital base, even considering if you know anything about the target (such as age, sex, or even race), you can try to choose a digital image closer to the likeness of this person. It then executes a series of queries using the flip attack, changing the pixels to increase the accuracy or reliability of the ML smart system. At some point, high confidence is achieved, which produces a digital image similar to the original, which, although not perfect, is very close to the person's appearance. Thus, the complexity of this type of advanced applications-oriented in AI increases with the difficulty related to successes and the ability to explain. Thus, XAI aims to explain the reasons for the new ML/DL systems, understand, and determine how this algorithm behaves in the future, producing better definable intelligent models (França, Monteiro, et al., 2020; Yan et al., 2015).

These XAI-oriented models are planned to be associated with an interactive human-computer interface, that is, HMI techniques, which allows converting AI models into useful and understandable

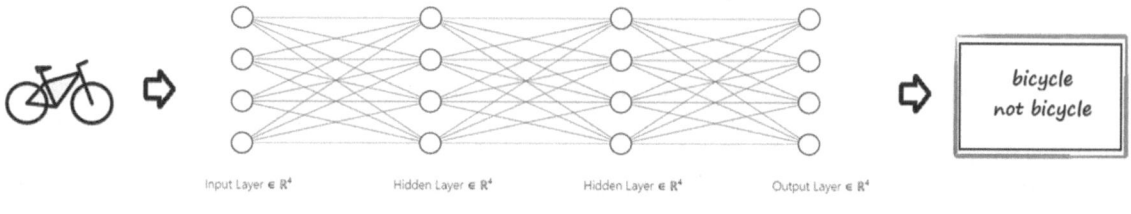

Input   Feature extraction + Classification   Output

FIGURE 1.2 Deep Learning.

explanation dialogue for the end-user. Consider three basic expectations: addressing the explanation of the purpose behind how the parts (algorithm modules) that design and employ the system are influenced; or even how data sources (dataset) and results (inferences) are utilized; and finally, how the inputs of an AI model drive to outputs (outcome) (Gunning et al., 2019).

It is worthwhile to exemplify XAI from medical practice, concerning after examining patient data, the physician must comprehend and explain to the patient the medical treatment/therapy proposed to the patient based on the recommendation of the AI-oriented medical decision support system. In this regard, first, what medical data is assessed is another fundamental criterion, it is also essential to identify and recognize what medical data is required and what needs to be done for a suitable medical assessment (Samek & Müller, 2019).

The XAI, in this sense, emerges as a new field of research, which seeks more transparent approaches, especially for deep neural networks. Therefore, this chapter aims to provide an updated overview of the XAI and your technologies, showing the fundamentals of this disruptive technology, demonstrating a landscape view to the applied aspect, as also key concerns, and challenges, with a concise bibliographic background, featuring the potential of technologies (Islam et al., 2021).

1.2 EXPLAINABLE ARTIFICIAL INTELLIGENCE (XAI) CONCEPT

Explainable Artificial Intelligence (XAI) is artificial intelligence programmed to explain its own purpose, and rationalization of the decision process, in a way that can be understood by the average user. Either way, XAI offers important information about how an Artificial Intelligence program makes decisions, by listing the program's strengths and weaknesses; the specific criteria used by the program to achieve a result; the appropriate level of confidence for different types of decisions; the reason why a decision was made, at the expense of other options; what errors the intelligent system in question is vulnerable to; and even how these errors can be corrected (Longo et al., 2020).

Considering an important objective of XAI, it is to offer open algorithms, breaking the paradigm acting as 'black boxes', about the algorithms used to reach a given decision were not understood, that is, skepticism in the face of inexplicable responses given by AI mechanisms and machine learning. Still pondering that as the use of artificial intelligence spreads, it becomes increasingly important to pay attention to issues of digital trust and data protection (Arrieta et al., 2020; Lin et al., 2020).

The term 'black-box' is associated with the underlying AI process, that is, basically the data is introduced in an AI-oriented process to train the AI model. This AI model then learns patterns from the training data (through a learning machine) and employs that to predict patterns (insights) in volume data, or even perform recommendations (inference) based on digital learning. Thus, underlying digital thinking behind the decision that the AI/ML system affords is shrouded in mystery (i.e., unexplained AI) (Arrieta et al., 2020; Lin et al., 2020).

In this context, it is worth mentioning five metrics that often appear as desirable in XAI models as Justice related to the AI model predictions containing some bias implicitly or explicitly discriminating against a minority group, still considering the imbalance of data or inappropriate content. Privacy related to the sensitive data used in the model to be really protected, and even digital reliability with respect to small changes in the model input, which can cause large unexpected changes in the output. Causality is associated with the explanation of a decision made by the AI model, which can be explained by a causal relationship that is interpretable by a human being. And even digital trust associated with human beings relying on the AI model, which is a black-box model of the who in a transparent model (Duval, 2019; Vilone and Longo, 2020).

AI approaches differ for insurance companies, banks, healthcare providers, and other distinct sectors, as this is due to the targeted AI models for these sectors bring different ethical requirements and even legal regulations. In this case, when implementing an AI system that is explained in the following condition, it will be necessary to replace it with the simplest one that is not very strong digitally, for now. Just as it doesn't matter which sector and who is used by AI systems, considering that there is a relationship between clarity and precision that an exchange is inevitable and tends to extend (Das & Rad, 2020).

Local accountability is a simple, visual audit trail to determine how an AI model decision was made and should always be offered when feasible. It is associated with the factor (reason) that an AI model suggested the doubling of units (spare parts or specific raw material) in stock, or even the reduction of the maintenance periodicity in a certain class of assets in the oil field. Since in some cases of inherent explainable AI models, it relies on visually intuitive methods for local explainability, such as decision trees, which can be used to solve a problem, and this choice alone facilitates explainability (Dağlarli, 2020).

Global Explainability consists of the need for a broader view of how the AI-oriented model really works; this tries to understand at a higher level the reasoning of a model, instead of focusing on the steps that lead to a specific decision, among other characteristics, to ensure that the AI model is not biased in the data. Exemplifying a context in which a biased AI model penalizes certain ethnic groups or social groups when recommending the granting or not of loans (Mata et al., 2018).

In explaining how AI makes decisions, there are two methods, that are possibly used on corporate software, depending on the use case. One of these methods, the Shapley Additive (SHAP) explanation structure, is based on the work of Lloyd Shapley's game theory and allows to 'reverse engineer' the output of a predictive algorithm and understand which variables contributed most to it. Likewise, Local Interpretable Model-Agnostic Explanations (LIME) can help determine the importance and contribution of the resource, but it actually considers fewer variables than SHAP and can therefore be less accurate, but what is lacking in consistency is compensated for speed. Considering that both SHAP and LIME are global, robust, and sophisticated methods, which can combine the interpretability of statistical models and the flexibility of ML models (Nohara et al., 2019).

If the model is simple enough and the use case is not very sensitive, LIME may be sufficient, but a complex model in a highly regulated industry may require more effort and resources to provide the appropriate level of perception, associating better SHAP, in both scenarios these methods are used to provide explanations for AI processes to end-users. Still, from a cost-benefit point of view, ML models are generally very little biased, since this commonly has few prerequisites, and even depending on the applied context few assumptions about the data; however, the cost of this is in the absence of interpretability (Kumarakulasinghe et al., 2020).

Transparency is considered essential for effective deployment of intelligent systems in the real world, such as machine learning; this need to understand and present transparent displays of ML models has led to the growth of explainable AI, through inexplicable responses given by AI and learning mechanisms machine, mainly to ensure that the models work as planned. But to generate a transformative impact on organizations, AI needs to be reliable, comprising a mechanism to justify its conclusions in an intelligible way (explainable rationalization), with an interactive way to explore and validate the conclusions, with the maximum possible simplicity, helping the human being to better understand their data (Wang & Siau, 2019).

1.3 NEED FOR XAI IN NEURAL NETWORK-ORIENTED APPLICATIONS

Considering that an artificial neural network (ANN) works with a system of synapses and weight adjustments and that, although very efficient, there is no way to objectively understand its logic. Although

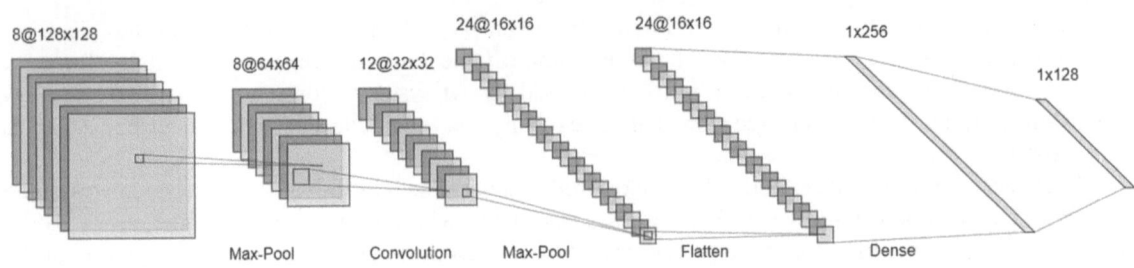

FIGURE 1.3   Convolutional Neural Networks.

there are classifiers that have an understandable decision process, such as Decision Trees and Bayesian Networks, there is no way of knowing with certainty if applying any of them in a given context, the performance obtained with the neural network will be achieved (Walczak, 2019).

Or even reflecting on a classifier (ML algorithm) in which the computer is able to learn through data processing, with different types of classifiers, but one of the most important ones is the convolutional neural networks (CNN) (Figure 1.3), functioning by a system of neuron devices connected through synapses and weights, which are adjusted during learning, which analyzes the existence of a spatial relationship between close pixels in a digital image, in which operations performed on an image channel can be performed in parallel for all channels (weight can be unique). Thus, the activations in the image can be located in a region, in search of features that define the region, still considering the possibility of learning several features, generating a volume that represents the activations of neurons (weights), since a feature is also a matrix, and activation can then be achieved by a convolution operation (which preserves or not the original size). Thus, the process consists of encoding the digital image in a smaller number of neurons that preserve the previous spatial relationships (França et al., 2021; Negrete et al., 2021).

Although RNA and CNN are the best of AI, one of the most important and disruptive technologies of the century, it is subject to bias; it is possible to explain how it works, but understanding your decision-making process after being trained is extremely complex and not at all intuitive. Pondering that both RNA and CNN are the most important pieces of AI and, at the same time, one of the most complex and unexplainable. Still pointing out that a CNN can present a high precision in the recognition of patterns, and then discover that the high accuracy was the result of an object that was always present when photographing the different classes, and not due to the characteristics of the objects themselves. Or even, the AI used in sentiment analysis (combining natural language processing (PLN) and ML techniques to assign weighted sentiment scores to sentences) specifically with the pragmatic competence of communication, that is, it interprets what is written, but it does not necessarily need what the user actually meant (Negrete 2021).

As the models become better, considering DL, and the volume of data becomes bigger, consequently the DL algorithms become more and more complex, with deep biological neural connections, with hundreds of deep layers and thousands of nodes. The cost for such complexity is the lack of interpretability, through the response to making certain decisions, considering that DL technology is optimal in capturing data trends and using them to promote their predictions (Monteiro et al., 2020).

In this context, it is necessary to understand the AI in the context in which it is today, already part of the lives of people and the daily lives of organizations, going beyond that, increasingly responsible for human lives, whether with the control of an autonomous vehicle without any human action, or even in the form of intelligent assistants, as also applied either in the fight against hunger and epidemics or in medical diagnosis, making instant decisions about the future of people's lives. In other words, the more AI grows, both in terms of application and performance, the more it will be necessary to understand it, trust its decisions; however, trusting a machine is easy if it is possible to understand that its predictions are correct, and in this respect, the XAI is a promising science in which it focuses on techniques such as showing the relevance of attributes (Wang & Siau, 2019).

1.4 DISCUSSION

XAI moves the employment of AI technology with the autonomy of computational devices to increase the efficiency of human workers where this technology is deployed, focusing on the analysis of intelligence and processes of interest in autonomous systems and the learning of classification and reinforcement (two ML approaches), generating as a product end a set of tools allowing the growth of a user-friendly AI. Considering that AI technology is feasible in view of digital viability, allowing use with added value through more transparent approaches of artificial intelligence and, especially, for deep neural networks. Seeking that professional human experts teach these machines to explain themselves in a way that is understandable, towards a desirable future state, considering it destined to be the next broad application technology of the era of modern society.

Reinforcement Learning (RL) techniques explain how computers artificially learn from their own experiences, receiving negative or positive feedback associated with their actions. This ML approach allows intelligent algorithms from independently artificial learning to play chess on a level of excellence to proving mathematical theorems without any human interference.

Through RL, AI independently learns to resolve problems that even humans would have difficulties solving, this associates XAI's premise of being able to teach people how to resolve it too, that is, digital capacity related to humans can learn from AI. To that end, there was research developing XAI exemplifying the digital solution that an AI found for a Rubik's Cube, that is, a three-dimensional puzzle, used as a source of analysis. This problem (Rubik's cube) involves a path discovery problem, associating a path from point A (scrambled cube) until reaching point B (cube solved).

Whereas the solutions for the Rubik's cube can be divided into generic steps for example, 'forming a kind of cross', or 'putting the corner pieces in place'; although this mathematical three-dimensional puzzle has possible combinations of 10^{19}, it is possible to develop a generalized step-by-step guide, and through this logic, it will be applicable in many different scenarios, or even with an AI algorithm solving the Rubik's cube. And in that sense, addressing a mathematic three-dimensional puzzle problem (Rubik's Cube) at this level by dividing it into stages is frequently the standard way in which people explain things to one another, as it also fits into the step-by-step structure of an algorithm, which allows being an initial example of how XAI can really work.

Although many times, the use of XAI tools has led to incorrect assumptions about AI data and models, which proves that interpretability is not a rigid and singular concept. And although most people agree that interpretability should describe a human understanding of an AI system.

'Explainability' is important for people to understand what AI is doing and why, so it can do its own job better. Relating the business management software companies that add some tools (corporate software) with business logic oriented to the events in the ERP (Enterprise Resource Planning), it is necessary to explain to auditors, regulators, or even litigants how the decisions or determinations were made. Whereas even though the intelligence is artificial, human understanding and criteria must be organic, still making explainable AI omnipresent in the system, reducing the time to obtain value and increasing the ability to explain.

ERP and other business systems can drive intelligent process automation (IPA), using the underlying business logic encompassing very clear processes and procedures, collecting and analyzing the processes and results that this produces. In this approach, ML models can, through sufficient transaction history, improve these automated processes, requiring business rules that define how decisions are made, as well as that these rules change based on the breakdown of information.

It is in this criterion that ML can evaluate the results of rule-based decisions and revise business rules in an explainable way, since ultimately ML may not only be able to augment existing rules, but also suggest new ones. Considering XAI, the advantage of the 'ability to explain' of this intelligent approach can be even better understood when looking at the two levels of interpretation, that is, local and global 'explicability'.

However, it is important to note that to understand how to interpret ML algorithms, we first need to understand what an ML algorithm is, considering this in a very general way, it is a paradigm shift from 'traditional programming' where is necessary to explicitly pass all the heuristics to a new concept, where instead of writing every action that the algorithm must perform, just pass several examples and let ML (AI) learn what the 'best' (lowest cost) decisions are. That is, ML algorithms receive inputs and are able to transform them into outputs without being explicitly programmable, so interpreting ML algorithms is to understand how this can transform inputs into outputs.

Thus, there is a very expressive paradox related to explainability in relation to the trade-off associated with bias, that is, for ML algorithms to be interpretable, these models need to be Simple, however, if the models are simple, this is highly biased. Associating that bias has different interpretations in different contexts, meaning the simplifications of the model, the hypotheses that are assumed for the model to work. And in this criterion, simpler ML models, by definition, are more skewed models, with stronger prerequisites, that is, the bias decreases with increasing complexity.

It is also worth noting that research on explainable artificial intelligence is being increasingly expanded, due to the need to obtain results that are accepted and understood by users, especially those related to the analysis of medical images to explain the decisions made by machine learning algorithms. But, as DL is integrated into more critical areas of society, such as medical diagnostics, it is only fair that it is possible to learn more about what happens under the hood. Thus, through the creation of AI algorithms with this ability (explainable), a wide niche of results will be opened for humans to understand, in the best and quickest way, how to solve problems and better understand the responses of these results.

1.5 TRENDS

The integration of AI with Blockchain and the IoT, as autonomous cars may not make much sense without IoT sensors working with AI, considering that IoT technology activates and regulates the sensors of the car, which collect data in real-time, while the AI models act on the decision-making part of the vehicles. Likewise, Blockchain technology can work in conjunction with AI to address security, scalability, and trust issues (de Sá et al., 2019; França, Borges, et al., 2020).

Cyber attacks are increasing rather quickly, overcoming existing defensive measures, overcoming human capacity in detecting and preventing today's cybercrime. As such, AI systems tend to continue to play a significant role in controlling these digital attacks, using ML to detect these security breaches acting on the cybersecurity of institutions issues (de Sá et al., 2019; França, Borges, et al., 2020).

Today, most real-time marketing activities are limited to automatic responses only, but if it is driven by AI, organizations will manage interactions with customers in real time across all channels. In addition, they will be able to use AI marketing to improve customer retention, finding new audiences on social media and other platforms (Chaib-Draa et al., 1992; Jordan & Mitchell, 2015).

However, if data filtering and the necessary digital care are not used, the machines can perform coding prejudices, through the data collected through the web. Leading to the phenomenon of 'redlining', that is, automatic denial of medical assistance and other financial services to people living in certain areas associated with ethnic minorities or a certain race (Chaib-Draa et al., 1992; Jordan & Mitchell, 2015).

In this regard, equitable AI (transparent AI) can act by verifying the data used for each forecast, by providing evidence that its predictions are diligently correct and fair, and can be changed immediately if it believes it is doing some forecast unfairly, by combining ML technology with interpretation and explanation techniques, such as the LIME model or global substitutes. Considering that the history of mankind is one of automation, however, trust must not be automated, this is acquired through equity and transparency. And through XAI it is possible to understand the process confirming whether or not this is impartial and correct (Haenlein & Kaplan, 2019).

Companies can create and enforce formal digital governance policies, as well as processes and controls around AI technology including designing and implementing standard procedures around AI in different areas, including monitoring and managing risks, performance, and value while maintaining appropriate levels of digital trust and digital transparency (Cheng et al., 2016; Lee et al., 2018; Ustundag & Cevikcan, 2017).

'As-a-Service' (SaaS) is emerging around AI, that is (AIaaS), including APIs with pre-trained machine learning models, meaning artificial intelligence as a service, giving companies more options to access the resources of smart technology through managed services, and microservices and bot shops, among other aspects (Cheng et al., 2016; Lee et al., 2018; Ustundag & Cevikcan, 2017).

1.6 CONCLUSIONS

Considering also that for a deep understanding of the impacts that ML models generate in society and to ensure that it is in fact accompanied by an ethical discussion, more than just the optimization of metrics is necessary, that is, to achieve explicability in AI models, it is necessary to study more than how to optimize metrics. For this reason, research aimed at qualitative studies also offers an essential complement to the field of XAI. Thus, the study of XAI is necessary, so that wrong decisions in AI-oriented models can be contested and corrected, and so that the models can evolve together with the society that it affects.

It does not necessarily mean that all ML models need to be explainable since when working with well-documented problems that have been addressed in the field for many years, it is not necessary, for the algorithm to be explainable. And it is not even necessary to explain to an AI model when the impact of your incorrect decisions is low, for example, an AI capable of learning to dance. But in the case of ML models that have a direct impact on people's lives, like those algorithms used to fire a person, explainability is important.

Or even mentioning the lack of digital transparency that can cause discomfort for the doctor or even the patient, relating hospitals and clinics that employ advanced digital intelligence, in some cases where an AI-oriented medical decision support system recommends surgery instead of chemotherapy, representing a recommendation can be counterintuitive, that is, AI tools today can suggest and show things, but it cannot say why behind those things.

Reflecting that AI related to the engineering and science of developing and produce intelligent machines (computational devices), especially intelligent software, simply exposes solutions without giving reasons for the result (insights), exemplifying, a computer trained to recognize animals learning about diverse and distinct types of eyes and even ears, gathering this information (learned from the dataset) to correctly identify, recognize and classify the animal. This type of approach determines and shows that several AI algorithms artificially think in ways similar to human beings, but it does not confirm this effect.

XAI has enabled AI-oriented systems to provide an understanding of the context in which intelligent devices operate, building underlying explanatory AI-oriented models allowing to describe real-world procedures and processes. At last, it is not consistent to transform AI technology into a divine power that people will seek without establishing a cause-and-effect relationship. In contrast, it is not feasible to ignore the digital insight that this provides for society. Basically, it is necessary to create interpretable as also flexible AI models that enable working together with the experts with knowledge and academic level in different sectors in which society operates and disciplines.

Thus, XAI will be essential for future operators to properly understand, manage, and even trust an emerging generation of AI-oriented machines, considering the ability to explain their logic, artificial reasoning, method set, among other characteristics, and convey a suitable understanding of how the algorithms behave.

REFERENCES

Arrieta, A. B., Díaz-Rodríguez, N., Del Ser, J., Bennetot, A., Tabik, S., Barbado, A., Garcia, S., Gil-Lopez, S., Molina, D., Benjamins, R., Chatila, R., & Herrera, F. (2020). Explainable artificial intelligence (XAI): Concepts, taxonomies, opportunities and challenges toward responsible AI. *Information Fusion, 58*, 82–115.

Chaib-Draa, B., Moulin, B., Mandiau, R., & Millot, P. (1992). Trends in distributed artificial intelligence. *Artificial Intelligence Review, 6*(1), 35–66.

Cheng, G. J., Liu, L. T., Qiang, X. J., & Liu, Y. (2016, June). *Industry 4.0 development and application of intelligent manufacturing.* 2016 International Conference on Information System and Artificial Intelligence (ISAI) (pp. 407–410), Hong Kong. IEEE.

Dağlarli, E. (2020). Explainable artificial intelligence (XAI) approaches and deep meta-learning models. In *Advances in deep learning.* IntechOpen.

Das, A., & Rad, P. (2020). Opportunities and challenges in explainable artificial intelligence (XAI): A survey. *arXiv preprint arXiv:2006.11371.*

de Sá, L. A. R., Iano, Y., de Oliveira, G. G., Pajuelo, D., Monteiro, A. C. B., & França, R. P. (2019, October). *An insight into applications of internet of things security from a blockchain perspective.* Brazilian Technology Symposium (pp. 143–152), Campinas, Brazil. Springer.

Došilović, F. K., Brčić, M., & Hlupić, N. (2018, May). *Explainable artificial intelligence: A survey.* 2018 41st International Convention on Information and Communication Technology, Electronics and Microelectronics (MIPRO) (pp. 0210–0215), Opatija, Croatie. IEEE.

Duval, A. (2019). *Explainable artificial intelligence (XAI).* MA4K9 Scholarly Report. Mathematics Institute, The University of Warwick.

Emmert-Streib, F., Yli-Harja, O., & Dehmer, M. (2020). Explainable artificial intelligence and machine learning: A reality rooted perspective. *Wiley Interdisciplinary Reviews: Data Mining and Knowledge Discovery, 10*(6), e1368.

França, R. P., Borges, A. C., Monteiro, R. A., & Iano, Y. (2020). An overview of blockchain and its applications in the modern digital age. In *Security and trust issues in internet of things: Blockchain to the rescue* (p. 185–207), CRC Press.

França, R. P., Monteiro, A. C. B., Arthur, R., & Iano, Y. (2020). An overview of deep learning in big data, image, and signal processing in the modern digital age. *Trends in Deep Learning Methodologies: Algorithms, Applications, and Systems, 4*, 63.

França, R. P., Monteiro, A. C. B., Arthur, R., & Iano, Y. (2021). An overview of the machine learning applied in smart cities. In *Smart cities: A data analytics perspective* (pp. 91–111), Springer.

Gunning, D. (2017). *Explainable artificial intelligence (XAI)* (Vol. 2, Iss. 2). Defense Advanced Research Projects Agency (DARPA), and Web.

Gunning, D., Stefik, M., Choi, J., Miller, T., Stumpf, S., & Yang, G. Z. (2019). XAI—Explainable artificial intelligence. *Science Robotics, 4*(37), pp. 1–2.

Haenlein, M., & Kaplan, A. (2019). A brief history of artificial intelligence: On the past, present, and future of artificial intelligence. *California Management Review, 61*(4), 5–14.

Holzinger, A., Langs, G., Denk, H., Zatloukal, K., & Müller, H. (2019). Causability and explainability of artificial intelligence in medicine. *Wiley Interdisciplinary Reviews: Data Mining and Knowledge Discovery, 9*(4), e1312.

Islam, S. R., Eberle, W., Ghafoor, S. K., & Ahmed, M. (2021). Explainable artificial intelligence approaches: A survey. *arXiv preprint arXiv:2101.09429.*

Jordan, M. I., & Mitchell, T. M. (2015). Machine learning: Trends, perspectives, and prospects. *Science*, *349*(6245), 255–260.

Kumarakulasinghe, N. B., Blomberg, T., Liu, J., Leao, A. S., & Papapetrou, P. (2020, July). *Evaluating local interpretable model-agnostic explanations on clinical machine learning classification models*. 2020 IEEE 33rd International Symposium on Computer-Based Medical Systems (CBMS) (pp. 7–12), Minnesota, USA. IEEE.

Lee, J., Davari, H., Singh, J., & Pandhare, V. (2018). Industrial Artificial Intelligence for Industry 4.0-based manufacturing systems. *Manufacturing Letters*, *18*, 20–23.

Lin, Y. S., Lee, W. C., & Celik, Z. B. (2020). What do you see? Evaluation of explainable artificial intelligence (XAI) interpretability through neural backdoors. *arXiv preprint arXiv:2009.10639*.

Longo, L., Goebel, R., Lecue, F., Kieseberg, P., & Holzinger, A. (2020, August). *Explainable artificial intelligence: Concepts, applications, research challenges and visions*. International Cross-Domain Conference for Machine Learning and Knowledge Extraction (pp. 1–16), Dublin, Ireland. Springer.

Mata, J., de Miguel, I., Duran, R. J., Merayo, N., Singh, S. K., Jukan, A., & Chamania, M. (2018). Artificial intelligence (AI) methods in optical networks: A comprehensive survey. *Optical Switching and Networking*, *28*, 43–57.

Monteiro, A. C. B., Iano, Y., França, R. P., & Arthur, R. (2020). Deep learning methodology proposal for the classification of erythrocytes and leukocytes. *Trends in Deep Learning Methodologies: Algorithms, Applications, and Systems* (pp. 129–156).

Negrete, P. D. M., Iano, Y., Monteiro, A. C. B., França, R. P., de Oliveira, G. G., & Pajuelo, D. (2021). *Classification of dermoscopy skin images with the application of deep learning techniques*. Proceedings of the 5th Brazilian Technology Symposium (pp. 73–81), Campinas, Brazil. Springer.

Nohara, Y., Matsumoto, K., Soejima, H., & Nakashima, N. (2019, September). *Explanation of machine learning models using improved Shapley Additive Explanation*. Proceedings of the 10th ACM International Conference on Bioinformatics, Computational Biology and Health Informatics (pp. 546–557), Niagara Falls, NY, USA.

Putnik, G.D., Varela, L., & Modrák, V. (2018). Intelligent collaborative decision-making models, methods, and tools. *Mathematical Problems in Engineering*, *2018*, 9627917.

Samek, W., & Müller, K. R. (2019). Towards explainable artificial intelligence. In *Explainable AI: Interpreting, explaining and visualizing deep learning* (pp. 5–22). Springer.

Ustundag, A., & Cevikcan, E. (2017). *Industry 4.0: Managing the digital transformation*. Springer.

Vilone, G., & Longo, L. (2020). Explainable artificial intelligence: A systematic review. *arXiv preprint arXiv:2006.00093*.

Walczak, S. (2019). Artificial neural networks. In *Advanced methodologies and technologies in artificial intelligence, computer simulation, and human-computer interaction* (pp. 40–53). IGI Global.

Wang, W., & Siau, K. (2019). Artificial intelligence, machine learning, automation, robotics, future of work and future of humanity: A review and research agenda. *Journal of Database Management*, *30*(1), 61–79.

Yan, L., Yoshua, B., & Geoffrey, H. (2015). Deep learning. *Nature*, *521*(7553), 436–444.

Explainable Artificial Intelligence for Services Exchange in Smart Cities

2

Ossama Embarak

Higher Colleges of Technology, Abu Dhabi, UAE

Contents

2.1 INTRODUCTION

Smart urban research has become a hot area for research and application where the physical infrastructure, human and social factors of Smart City are part of it (Kumar et al., 2020). Digitalization aims to link all aspects of urban life to create smart city services through the fourth Industrial Revolution in AI, ML, big data, advanced modelling, autonomous machines, additive manufacturing, cybersecurity, cloud computing, internet of things (IoT), virtual reality (VR), and universal convergence (Allam, 2021). Core aspects of these technologies are the adoption and advancement of AI techniques that have shown their ability to process large amounts

DOI: 10.1201/9781003172772-2

of data to develop rules for learning, such as through ML, complex associations, and predicting results of complex physical processes (Rathore et al., 2018). Several major components of smart cities, such as intelligent living, intelligent governance, smart people, smart mobility, smart economics, and intelligent infrastructure, have been established (Allam & Dhunny, 2019). ML and recently Deep Learning (DL), a traditional AI technology used in smart cities solutions, rely more on best representative data sets and engineering features and less on available field expertise. Recent developments particularly in ML have advanced significantly in smart city applications in AI (Kim et al., 2021). A critical component of Smart Cities is the smart infrastructure that has wireless sensor networks that autonomously collect, analyze, and disclose 'smart monitoring' structural data. AI algorithms allow large amounts of data to be processed that would remain undetected using traditional approaches (Du et al., 2019). Although AI algorithms are used in smart monitoring, they are still restricted because of a lack of confidence by users in AI processes unseen (Chui et al., 2018). The new paradigm called explainable artificial intelligence (XAI) will play a key role in boosting confidence in AI by technologists to understand the 'black-box' nature of AI algorithms for smart monitoring (Putnam & Conati, 2019). In Smart Monitoring, AI algorithms can be utilized for two purposes:

1. Detect patterns that would otherwise remain unrecognized represent complex physical processes. For example, multi-agent systems allow agent-based software to move autonomously from one wireless sensor node to another to analyze smart infrastructure on demand. This significantly reduces resource consumption for wireless, intelligent monitoring systems in comparison to traditional approaches.
2. Exploit large amounts of data that are only partially used in long-lasting infrastructures. For instance, distributed artificial intelligence deployed to advance various smart surveillance areas such as Dam inspection, Wind turbine observation, and Bridge Examination.

ML algorithms for smart monitoring in smart cities can be categorized as symbolic and subsymbolic, as shown in Table 2.1. Symbolic ML (SML) derives intelligent conclusions and decisions from memorized facts and rules assembled using propositional logic and predicate calculus techniques, but cannot handle complex problems as sub-symbolic algorithms (Telesko et al., 2020). In contrast, in sub-symbolic machine learning (SSML), raw data is provided to the machine, which uses that data to learn about patterns and discover its own high-dimensional representations of that raw sensory data (Cambria et al., 2020). The users of smart cities and service providers need to understand the reasoning behind ML algorithms' reasoning such that it is crucial to provide more explainable machine learning algorithms (XML) for better reasoning.

Symbolic ML algorithms rely on input features, and when specific algorithms are used, they can produce human-understandable predictions or intelligent decisions. The main issue with SSML is that no specific features are fed into the model for prediction or decision making, but the model extracts the features and makes predictions without explicitly knowing the reason for the decision, as shown in Figure 2.1.

The widespread use of machine learning algorithms in critical areas such as medicine, finance, self-driving cars, government, bioinformatics, churn prediction, and content recommendation has drawn attention to critical trust-related issues (Bechberger, n.d.). Some recruitment systems rely on deep learning to determine the best candidate, while decision makers are sometimes unaware of why one applicant is given more weightage than another, which could be due to unfavourable factors such as resume format, style, and font symmetries (Cambria et al., 2020). As a result, explainable artificial intelligence (XAI) is required to bring trust in machine learning and the decisions it generates.

TABLE 2.1 Machine learning algorithms for smart monitoring

	SYMBOLIC ALGORITHMS	SUBSYMBOLIC ALGORITHMS
Description	• Feed algorithms with human-readable information • Use explicit programming to provide its functionalities • All steps are based on the use of logic and search to symbolic readable representatives • Simulate symbolic, conscious reasoning in humans	• Feed raw information into the algorithms to apply the specific mechanism to construct its implicit knowledge • Used to analyze complex problems that involve large data sets • Simulate basic physical (neural) brain processes
Transparency	• Explicit mechanism	• Black-box internal mechanism
Usage	• Wide use	• Limited use
Functionalities	• Inference • Searching	• Focus on symbolic reasoning and logic
Algorithm type	• Classification • Linear models • Reinforcement learning • Decision trees • Rules induction	• Bayesian Learning • Deep Learning • Artificial Neural Networks (ANN) • Backpropagation NN • Support Vector Machine (SVM) • Graphical models • Boosted Trees • Random Forests
Human cognition simulation	• Knowledge representation and reasoning • Planning • Interaction via multi-agent systems	• Perception • Learning
Advantages	• The reasoning process can be easily understood • Easier to debug • Easier to control • No need for big data processing • More useful for explaining people's thought • Better for abstract problems • Robust and predictable	• More robust against noise • Better performance • Less knowledge upfront • Easier to scale up • Better for perceptual problems • Can handle complex problems
Disadvantages	• Much coding is required for the processing of rules and knowledge. • Cannot handle complex problems	• Difficult to understand how the system came to a conclusion • Untrusted by users due to its inner mechanism • Needs a big dataset for processing • Cannot make high-risk choices

Smart cities aim to provide dynamic, real-time, and flexible services to end-users by leveraging ML algorithms and using the fourth Industrial Revolution technologies.

As a result, systems, like humans, must meet various criteria to increase trust, fairness, reliability, safety, explanatory justifiability, privacy, usability, etc.

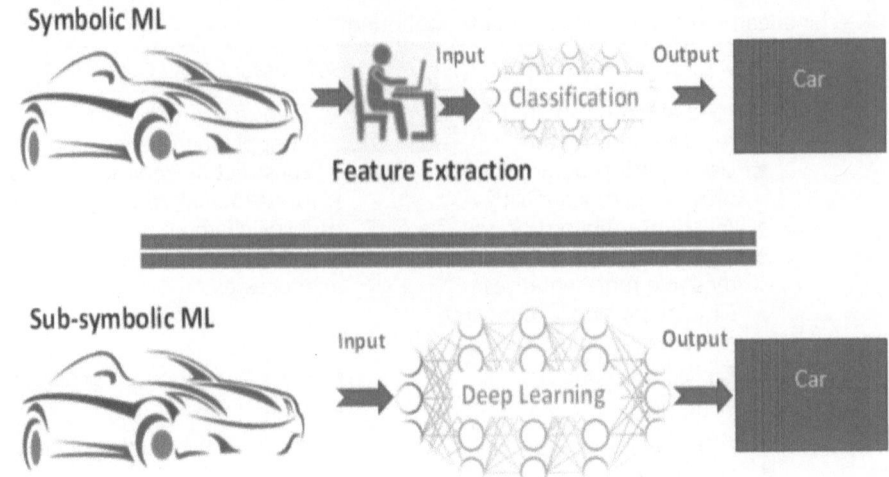

FIGURE 2.1 Symbolic and Non-Symbolic Form.

2.2 EXPLAINABLE AI AND DECISION-MAKING

Building and using explainable machine learning models for decision making in all aspects, particularly in smart cities services, is critical and must maintain the following properties (Gunning, 2016).

A. *Trustworthiness* – Considering the model's characteristics, the ML model is trustworthy to users who willingly and securely become vulnerable to use. The term trustworthiness encompasses the accuracy of outputs, privacy safeguards, and fairness.

B. *Interpretability (Comprehensibility)* – The model abstract concepts can be mapped into a domain that humans can understand.

C. *Explainability* – The model maintains a collection of interpretable domain features that have contributed to the production of a decision in a given problem.

D. *Selective* – The model focuses on one or two possible causes – rather than all possible causes – for a decision or recommendation; that is, explanations should not overwhelm the user with too much information.

E. *Social Conversation* – It is necessary to engage in interactive knowledge transfer in which information is tailored to the recipient's background and level of expertise.

F. *Interactive* – The user can interact with the ML model and feedback to the underlying model to improve its performance.

Considering the properties mentioned above, we can transition from classical ML to explainable ML, which both experts and end-users can use. Many researchers attempted to highlight the goals for explainability, as summarized in Table 2.2.

Semantic explainable ML systems can use semantic data to transmit knowledge to the users to build clear and understandable mental models (Nilsson et al., 2019). Explainable ML can also use user profiles to describe their behaviours and the reasoning behind the decision (Gunning & Aha, 2019). The use of local or global explanations using evaluation metrics of, for instance, accuracy and fidelity help users to understand model behaviour (Lundberg et al., 2019). The Hybrid ML system uses the sub-symbolic algorithms to develop predictive models through connectionist machine learning and data processing

TABLE 2.2 Explainability ML focuses

FOCUS	PRESENCES	DESCRIPTION
Knowledge transmission	Semantic Data	• For effective knowledge transmission to human users, combine semantic information with an explanation
Reasoning	Profiles	• Based on a knowledge base, symbolic reasoning systems deduce why a particular decision is made
Local or global post hoc details	Evaluation Metrics (Accuracy & Fidelity)	• Users gain an understanding of how decisions are made through local or global ad hoc explanations that employ evaluation metrics such as accuracy and fidelity
Causality	Measurements of Model Effectiveness, Efficiency, and Satisfaction	• Demonstrates how a different decision or prediction could have been obtained in a given context with a certain level of causal understanding of the relationship between inputs and outputs
Symbolic features	Domain Knowledge Features	• The models employ symbolic ML elements to explain the decisions made by the sub-symbolic ML components

(Alías & Alsina-Pagès, 2019). Simultaneously, the symbolic system contains a rich representation of domain knowledge and provides higher-level structured reasoning.

Explainable artificial intelligence (XAI), specifically ML, is essential in providing smart services to citizens and experts in smart cities. A service call passes through four phases.

1. Call for a service
   a. Business to end-user, where the end-user can be an expert or a citizen
   b. Business-to-business transactions also include requests for services from governmental to non-governmental entities
2. Select features or refeaturing; transparent features are selected or refeaturing is applied
3. Select and apply a ML model (or models), either symbolic or sub-symbolic
4. Evaluate the model using evaluation matrix with satisfaction, accuracy, and fidelity

2.3 BIG DATA AND ICT FOR XAI SERVICES IN THE CONTEXT OF SMART CITIES

The main value of big data is its huge impact on many aspects of cities and lives. Big data is projected to grow 50% in the amount of global data generated compared to only 7% of global IT expenditure (Yigitcanlar et al., 2020). In recent years, around 80% of the world's digital data has been collected (Lamba & Singh, 2017). Many policymakers use large amounts of data to promote growth and the need for smart cities. This highlighted how cities are to implement key intelligent city features. These features include durability, resilience, governance, well-being, natural resources, smart management, and urban facilities. Components of the smart city framework are still under review, namely mobility, governance, the environment and inhabitants, technologies and services such as smart education, healthcare, transport, and energy (Xie et al., 2019). Cloud computing and storage facilities are needed for smart city applications. The smart city will be dependent on cloud technologies. This creates a huge demand for smart networks to incorporate and distribute services to stakeholders intelligently. In connecting millions of computers, the internet of things (IoT) plays a crucial role too (Lv et al., 2020). The collected data will be fed into an XML model to generate predictions and make decisions. Users should be able to reselect features (re-featuring) to be used for model prediction, which will aid in revealing the black-box point in classical machine learning. Big data and cloud services maintain massive data and operations that help refeaturing data needed for explainable machine learning (XML) model for smart services.

On the other hand, the effectiveness of city operations and the quality of services provided to smart urban residents typically form an urban ecosystem that uses ICT and related technologies. Several academics and business experts describe smart cities as 'A concept is to include physical, social, commercial and ICT infrastructure to improve the intelligence of a smart city' (Kumar et al., 2020). A 'smart city' also has an extensive concept of a modern city using ICT and other technologies to increase the quality of life, productivity, and urban service efficiency. It ensures that resources are available in the social, economic, and environmental aspects of current and future generations (Serrano, 2018). The ultimate goal of smart cities is to improve the quality of urban operations by reducing the inconsequencies in different functions between demand and supply. Modern smart cities focus on sustainable and effective approaches to energy storage, transport, health care, infrastructure, etc. Applying explainable machine learning (XML) in smart cities services imposes the need to consider the following attributes in services provided.

1. *Sustainability*: It is known as sustainability that a city can maintain an ecosystem balance in all respects while serving and performing present urban activities and in the future. Within each attribute, a few sub-attributes are concerned. The attributes of sustainable development include infrastructure and management, pollution and waste, energy and climate change, social issues, economics, and health.
2. *Quality of Life (Qol)*: An improvement in the QoL indicates the emotional and financial well-being of urban citizens.
3. *Urbanization*: Urbanization focuses on the transformation aspect of the rural to an urban environment, technological, economic, and infrastructural.
4. *Smartness*: Intelligence is defined as a will to improve the town's and its inhabitants' social, environmental, and economic benchmarks.
5. *Transparency*: Transparency is where the services explicitly show the logic of the system's decision.
6. *Usability*: Utility refers to providing users with usable services, which cannot be achieved if the services are built on sub-symbolic AI models (black-box) that are not understandable by users, whether experts or unspecialized users; thus, XAI is required to achieve usability for the provided services.

Regardless of city size, location, and available resources, the concept of Smart Cities focuses on the smart city's view as an integrated live solution. Smart grids allow for electricity to flow where needed and use the technology to improve quality of life. The availability, scale, and capabilities of those services are among the challenges of developing and maintaining a smart city. The regulatory framework, which could significantly impact the probability of success, is a further obstacle. Besides, technical challenges also require highly advanced technological solutions. Big data is generated from several sources, include smartphones, computers, environmental sensors, cameras, GPS, and people. More and more diverse technologies have been added to speed up the production of data. Big data can be stored at different locations, but unused or possibly, it could be used again.

Smart city users can understand symbolic ML model predictions (e.g,. classification, regression) because they use clear features as input to the model for processing, and the logic behind the outputs clearly reflects any changes in the inputs features. On the other hand, the sub-symbolic model is difficult to grasp due to the model's ambiguous features used for prediction. For example, a set of raw data is used

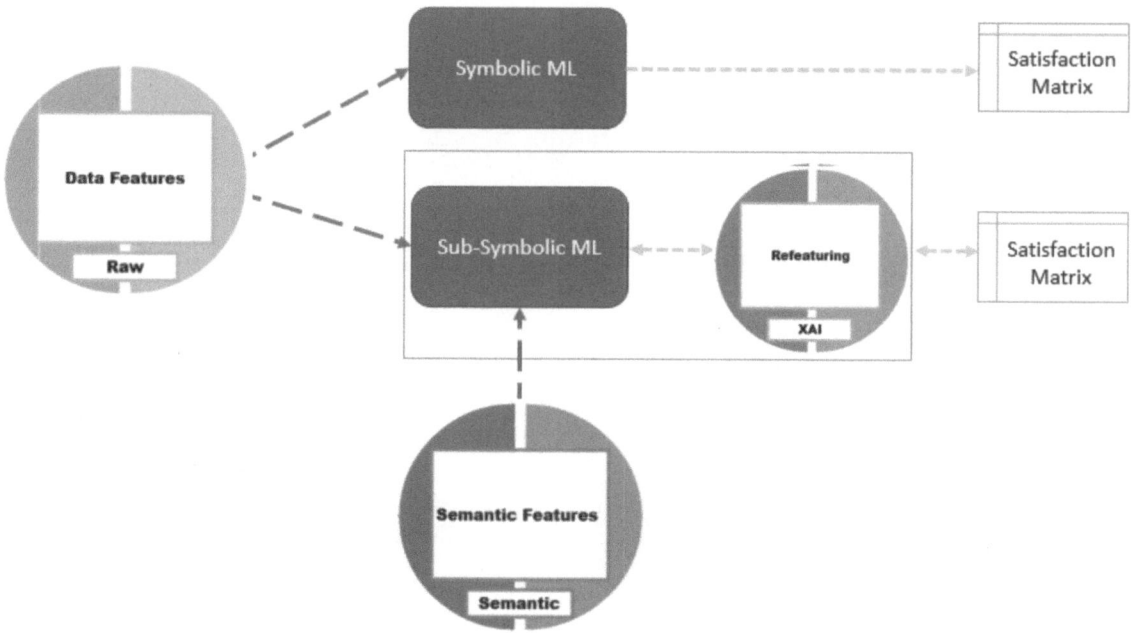

FIGURE 2.2   Data Featuring and Refeaturing in XAI.

as input to artificial Neural Networks (ANN) to make predictions, but it is unclear why ANN makes such a decision and on which specific feature(s) from the input set. As shown in Figure 2.2, we propose refeaturing the data based on user satisfaction and understanding level and taking into account the semantic features in the data set.

Refeaturing aims to explicitly reform all features used for prediction by sub-symbolic ML. Users should be able to recall the XAI model to obtain a better satisfaction matrix that includes accuracy, fidelity, and any other relevant evaluation factors. This should lead to a flashback of all patterns used for decision making or predictions.

Flashback should be conducted to declare all features involved in ML model predictions associated with its weightage.

Smart city services can be at the infrastructure or application level, and they rely on fourth industrial revolution technologies. Big data technologies are one of these core technologies that efficiently collect, process, and store information to support intelligent urban applications and produce information to improve various smart cities. Furthermore, by employing the previously mentioned refeatruing concept, comprehensive data would assist XAI modelling in providing a more precise interpretation of predictions. In this context, the following big data characteristics (Figure 2.3) should be considered for processing and time complexity (Ariyaluran 2019).

1. *Volume* refers to the overall size of the data.
2. *Velocity* describes the rate at which data is generated, stored, analyzed, and processed to enable real-time analysis.
3. *Variety* applies to the various types of data produced, structured, semi-structured, or unstructured.
4. *Variability* refers to the continuous evolution of data structure and significance, especially when dealing with data produced through natural language processing.
5. *Value* refers to big data's potential benefit for a company based on effective big data processing, management, and analysis.
6. *Volatility* refers to the structured data preservation policy from different sources.

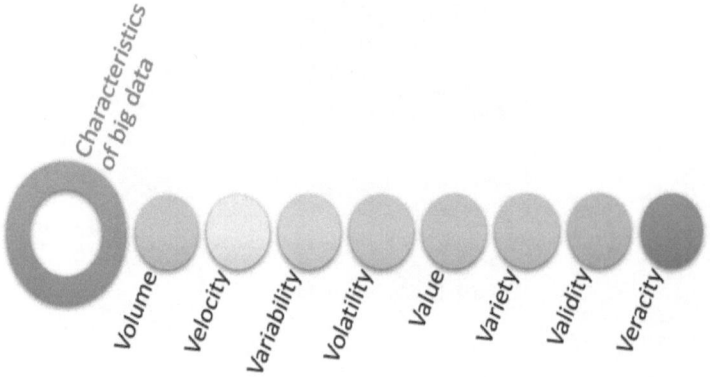

FIGURE 2.3   Characteristics of Big Data, the 8 Vs.

7. *Validity* refers to data validation, correctness, and accuracy.
8. *Veracity* refers to the coherence and integrity of the collected data and the importance of the data findings' specific problems.

2.4 THE SMART ENVIRONMENT WITH XAI

The Explainable Artificial Intelligence (XAI) refers to a new generation of existing or new artificial intelligence technology that enables users of the AI-driven system to understand, trust, and cooperate. Smart city systems services should be understandable by the end-users. However, these explanations are also affected by the domain, system usage context, governing regulations, and user role within the system under consideration. For example, some decisions may have been made based on private and sensitive data, such as health, financial, and military information, which could not be revealed and thus limited the XAI capabilities to explain the end users. Therefore, providing live and dynamic smart services based on XAI algorithms necessitates accurate and standardized data, regulations governing data exchange between entities for privilege access, and layered service production and calling. The following subsections demonstrate each with more detail.

2.4.1 Proposed Smart City Main Pillars and Architecture

The transition from heritage city to smart city requires focusing on three main pillars, as shown in Figure 2.4.

2.4.1.1 Transition Entities

Moving into a smart city requires a systemic process to examine the main factors that affect the transition to prepare and facilitate standardization. We see that major entities need to be present, including government bodies, investors in the private sector, and ad hoc research units, as shown in Figure 2.5.

Government Entities
Intelligent urban development often involves making deliberate decisions and the best way to address urban challenges by differing government authorities in various sectors, including medicine, education, agriculture, etc. For this reason, for instance, it is necessary to involve medical entities to develop an integrated smart system for health care.

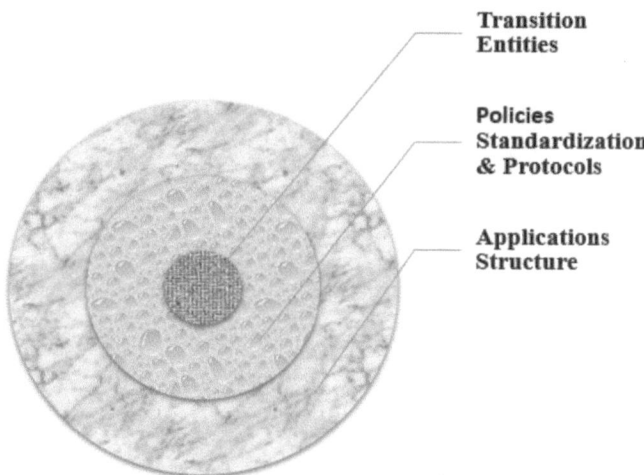

FIGURE 2.4   Smart City Transition Requirements.

FIGURE 2.5   Smart City Transition Entities.

Private Sector Investors
Most cities are provided with inadequate financial resources and staff experts. Consequently, countries must not be the principal funders and providers of all types of infrastructure and utility systems. Most intelligent city innovations are private sector companies' revenue-generating activities. Telecom providers, for example, directly use the necessary digital backbone by the public. In the development of intelligent city networks, software and high-income systems, the private sector will play a leading role. Therefore, the government's role can be to create the right regulatory environment, convene key stakeholders, provide subsidies, or amend purchasing rulings.

Ad hoc Research Units
A smart city requires research experts to develop ad hoc technologies via scientific approaches in each sector. The transition operations must therefore involve research entities and experts. This will help to maintain sustainable systems taking into account the latest findings in a certain sector.

2.4.1.2 Policies Standardization and Protocols

In Section 2.4.1.1, transitional bodies responsible for developing the necessary standards and protocols for smart cities considering the key criteria, as shown in Figure 2.6 where policies in a smart city must dominate the systems' operations.

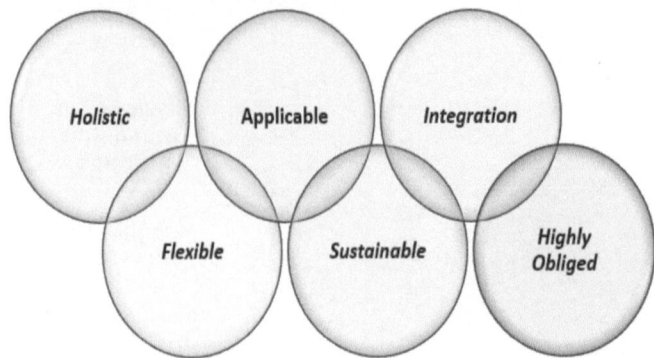

FIGURE 2.6 Standardization and Protocols Criteria.

Holistic: Strategies should control the systems as a whole instead of concentrating solely on specific features or components. For instance, robust intelligent healthcare infrastructure in a smart community should provide connected implementations.

Flexible: Policies with diversified actions should be possible to implement and remain implemented.

Applicable: Implementation strategies must be valid.

Sustainable: Policies for current implementation should be defined without any future circumstances being ignored.

Integration: Policy should be integrated and systematically designed.

Highly Obliged: In order to achieve breakthroughs, policies should take immediate action with responsibilities.

Provide sustainable services in smart cities to impose the need to regulate policies of the following:

A. Infrastructure
B. Applications
C. Services
D. Digital clouds
E. End-users and stakeholders data

2.4.1.3 Applications Layer Structure Considering XAI

Legacy communities have only two layers: the network and the user layers. Different researchers propose only three layers for a smart city which are infrastructure, services, and end user. However, we see that a real smart city framework should maintain applications and digital cloud data with many features that could serve XAI modelling. We propose three levels of structure, as shown in Figure 2.7. ML algorithms should be used at multiple layers to consider the results further and incorporate user feedback into the autonomous system. Users could use the provided interface to confirm the results, correct or refine them by providing additional information or instructions, or take another action; therefore, human-computer interaction is critical in XAI-based systems.

A. *Infrastructure Layer*

The infrastructure layer is the foundation for smart city services that interact with the physical components. The added value or resources produced by industries, education, transport, and other producer have been incorporated. This case includes communication infrastructures (satellites, landlines, etc.), networks, IoT devices, and all medium-term applications for value creation and production. There are two primary forms of infrastructure, which are the *new infrastructure and legacy infrastructure*.

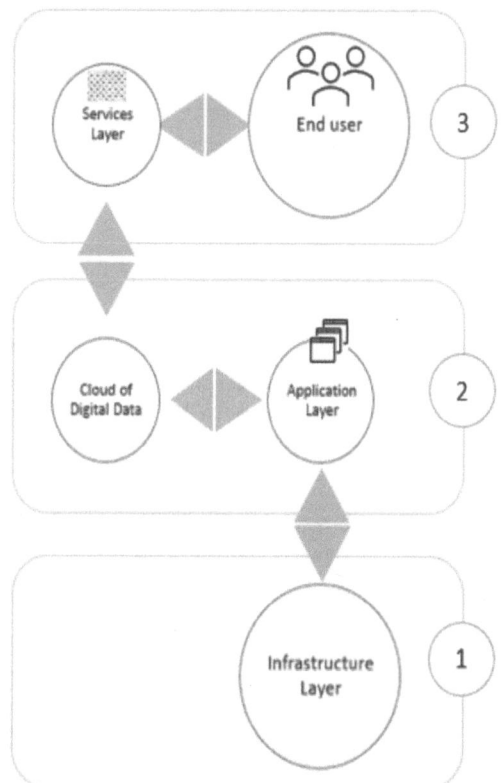

FIGURE 2.7   The Architecture for the Smart City Framework.

B. *Application Layer with XAI*

All smart city applications should interconnect and integrate with the other layers, as shown in Figure 2.8. The applications layer can communicate directly to other applications within the application layer and exchange data with the digital data cloud. Also, the application layer can exchange data with the services and Infrastructure layers as needed. The application layer receives feedback or recalls from the end-user that comes through the interaction between the services layer 3 and the application layer 2. The application layer should run XAI models based on the selected features that come from the end user at the upper layer. Meanwhile, the selected features were extracted and displayed to the end user based on their availability in the cloud data zones.

Services creation at the application layer (2) needs users/expert interaction to select the features, which could be raw features or semantic features, as shown in Figure 2.9. Users will select a ML model to be applied accordingly where the model should flashback the involved features for interpretability of the model decision. The user will reflect through the satisfaction matrix with his action: either *Satisfy* or *Recall* of the services.

Applying and utilizing semantics aids in delving deeper into the dataset and its semantic ontologies, which can be used as input for XAI processing. Form ontologies consist of *classes, individuals, attributes, and relations.*

Ontology is a vocabulary of specific knowledge terms.
Classes refer to abstract group or object types.
Individuals refer to instances from classes.
Attributes refer to properties and parameters of classes.
Relationships define how classes and individuals are related to one another.

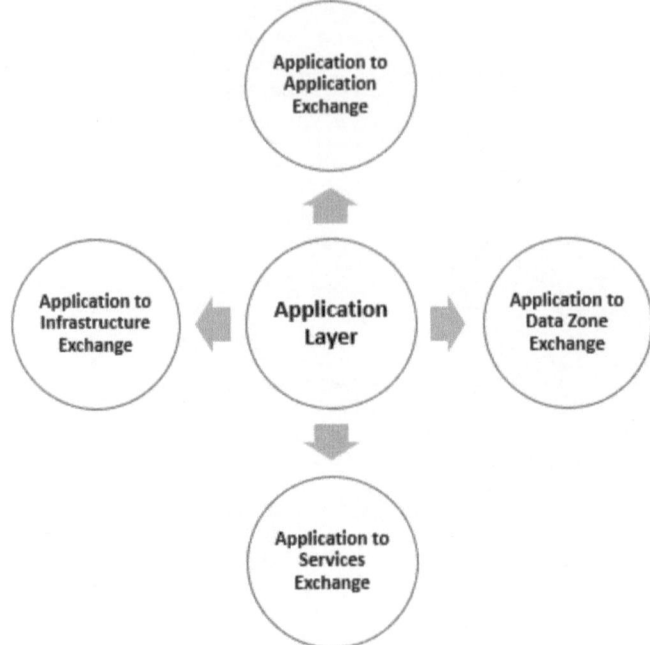

FIGURE 2.8   Application Layer Exchange with Other Layers.

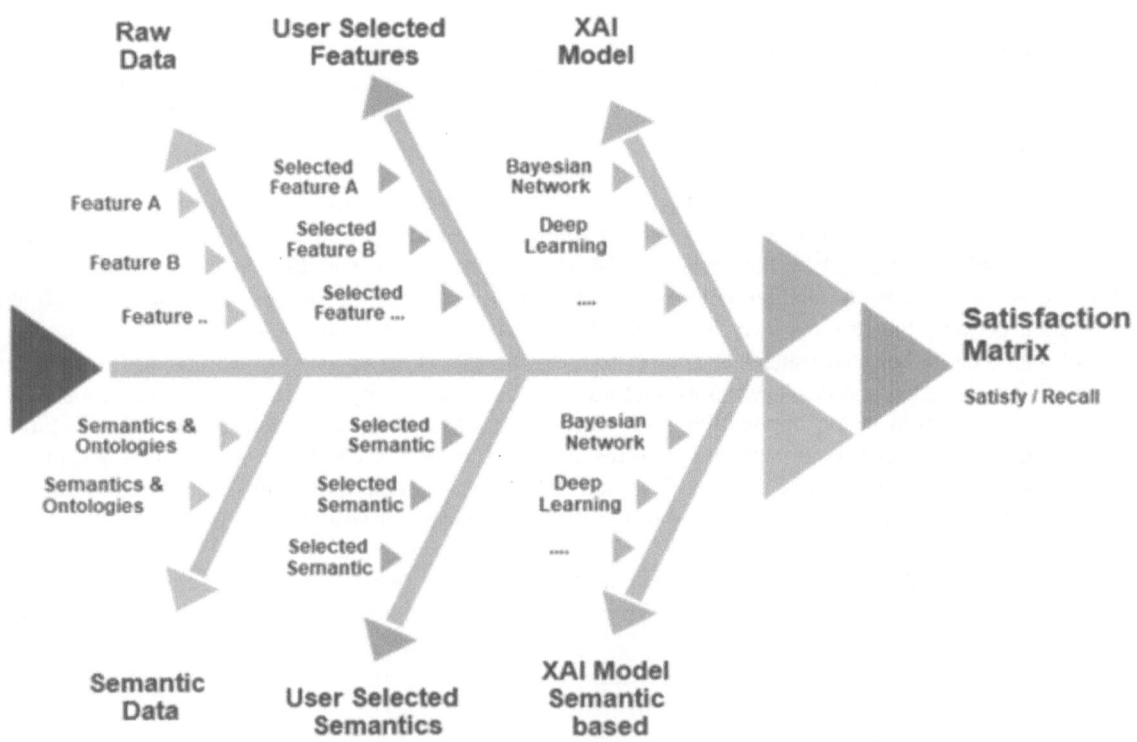

FIGURE 2.9   User Interaction with XAI Model.

Data collections at the infrastructure layer (1) can be formed in triplets of (subject, predicate, and object) and stored at the cloud zone at layer (2). While at the application layer, an inference mechanism can be applied to make predictions or decisions of new knowledge starting from the existing knowledge representation. There are numerous important factors to consider when preparing data for implementing XAI in smart cities.

1. Data preparation is crucial for XAI processing and user interaction; hence forming data features/ontologies is the preliminary phase.
   a. Raw data sets (historical/live)
   b. Semantic data sets (historical)
2. Time complexity is highly important, and hence the capacity of data involved in the processing relies on the type of services and data from for processing, whether it is raw data or semantic data.
3. The level of features of the data sets made available to users should be governed by the protocols and criteria outlined above.

C. *Cloud of Digital Data*

The data generated by the infrastructure layer should be transferred to the cloud digital data storage that the application layer can access. Data collected are expected to increase dramatically across various data collection sources in cities such as sensors, wearable technology, cameras, global positioning system (GPS), and smartphones. These data generation systems can be seen at considerably lower prices and sizes, making them very easy to use. Their storage capacity has improved significantly, resulting in an unprecedented volume of data collected from different sources such as:

- Current legacy systems
- The government, Semi-gov, and private entities
- The individual devices that generate data

D. *Services Layer*

All services received from the application layer are accommodated in the services layer. Figure 2.10 shows a sample of services produced by an application layer using smart technologies and algorithms and rely on cloud-based data. The service layer acts as an interface for the distribution and management of services which should be expandable.

The output of the services provided is supported by their respective applications' expertise on the application layer, which also depends on the quality of the cloud input data we call services tuning. However, the development of utilities in smart cities has four significant barriers.

- *Integration of services* refers to the challenges posed by collecting data from the legacy and the new platforms besides, how services are coordinated across various sectors and data integration across applications.
- *Obligation of services* refers to *Accountability* by entities that take the lead and responsibilities. *Affordability* refers to the class-based service tariff. *Accessibility* means granting access to user segmentation.
- *Quality of services* refers to quality and feedback, which implies the need to consider two issues, *Regulatory issues* to ensure that competitions are held without interruption of services – *legal issues* providing system liability for service delivery following the commitments.
- *Standardization of services* refers to the services to guarantee the design pattern and ensure that it does not contain redundant functionality.

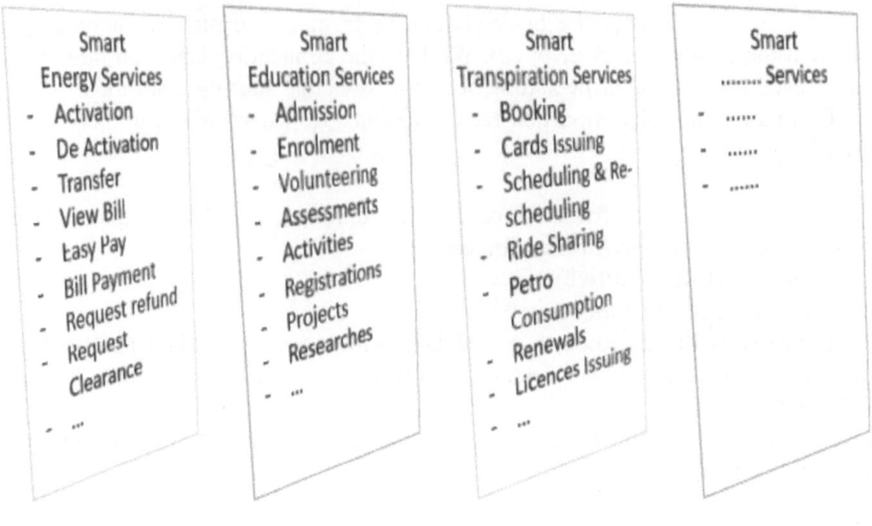

Smart Energy Services	Smart Education Services	Smart Transpiration Services	Smart .... Services
- Activation	- Admission	- Booking	- ......
- De Activation	- Enrolment	- Cards Issuing	- ......
- Transfer	- Volunteering	- Scheduling & Re-scheduling	- ......
- View Bill	- Assessments	- Ride Sharing	
- Easy Pay	- Activities	- Petro Consumption	
- Bill Payment	- Registrations	- Renewals	
- Request refund	- Projects	- Licences Issuing	
- Request Clearance	- Researches	- ...	
- ...	- ...		

FIGURE 2.10 Sample Smart Services.

- *Access rights to services* where users can revise XAI model outcomes through human-machine interactions and recall the underlying model to improve its performance based on new feature selections.

E. *End-user*

End users in smart cities can use a unique authentication to access any application which resides on a service layer. Services are developed using the latest technology, such as artificial intelligence, profound learning, virtual reality, and so forth. End-users play three leading roles:

1. Services consumer
2. Participate in service production; they must connect their smartphones, wearable devices, etc., to the infrastructure layer where data value is produced
3. The evaluation of services based on feedback leads to changes in the services provided and updates to XAI behavior

Through authentication methods controlled and overseen by authorized agencies, users can access the service grid, as shown in Figure 2.11.

2.5 BENEFITS OF USING XAI IN SMART CITIES

The change from conventional to smart cities has brought numerous economic, environmental, and social benefits. However, the transition is limited to the technologies that allow users to benefit from the services provided. Services will not be widely used unless they are easily understood by users, so XAI is required. The proposed architecture in this chapter and the concepts of refeaturing and semantic use make it easier to use ML in smart city services, resulting in numerous benefits such as given below.

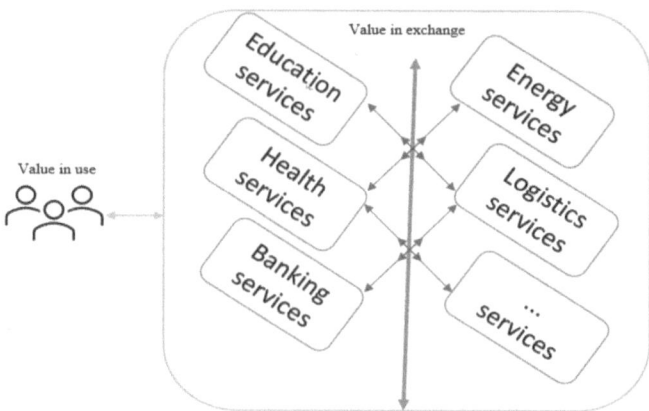

FIGURE 2.11 Service Exchange in Smart Cities.

1. *Optimize the use of ML at the infrastructure level*
   Providing durable ML methods that are understandable by users at the infrastructure level aids in the widespread adoption of these methods for processing at that lower level. This will enable quick identification of waste points and effective resource distribution while reducing costs, utilization of natural resources, and energy cost. Smart cities will create new opportunities for information sharing due to improved links between data collection and service provider applications.
2. *High standard of living*
   XAI will encourage users to use cutting-edge decision-making mechanisms instead of traditional approaches that produce traditional results. This will help to improve living/working understanding, improved transport networks, and faster availability of sufficient knowledge for informed decision making.
3. *Elimination of vagueness*
   The use of featuring/refeaturing or semantic data repositories will lead to interoperability and openness between different service providers and lead to an open ecosystem paradigm.
4. *More efficient, high-tech decision making*
   Collecting features-based (raw/semantic) data from various sources, including IoT, will aid in maintaining accurate and up-to-date datasets for better predictions and decision-making.
5. *Increased decentralized services*
   The use of XAI methods will help users to seek the services themselves anytime, anywhere. This will reduce the requests for load of services by citizens from responsible entities and give users wider flexibilities for better decision making and selections based on scientific and understandable mechanisms.
6. *Safer environments*
   Smart city users will use XAI to make decisions and select the best features that fit their objectives. Users will be given the ability to add their feedback and flashback the features impacting the developed decision by applied the ML models. They will be given the ability to add more features and provide private data if necessary, and thus their privacy controls will be preserved.
7. *New opportunities for economic growth*
   Applying explainable artificial intelligence (XAI) methods that are understandable by users speeds of user-to-user services, machine-to-user or machine-to-machine automation processing with human-like capabilities which create more opportunities for rapid economic growth.

8. *Continuous monitoring of consumptions by citizens*

   Using XAI methods, users will be able to monitor resource consumption rationally. Smart innovations help communities reduce the waste of natural capital. Smart sensors with ML monitoring allow cities to easily detect leaks in pipelines and repair fractured segments within a short period, thereby reducing wastewater. Intelligent power grids now permit two-way contact between power suppliers and customers to better identify maximum demand periods and outages.

9. *Efficient public services*

   Citizens will be able to use XAI to compare services/products from various providers and gain a better understanding of specific features for decision making. Furthermore, service providers will be eager to improve their services' effectiveness to maintain fidelity.

10. *Enhancing workforce participation*

    For a successful smart city, a highly qualified workforce is essential. By reducing the workload of employees, intelligent technology can contribute to growth. More automation, robotics, clever devices, a high-quality robotic platform designed for human (peoplebots) need more investments, and time in strategic projects throughout the city.

2.6 ADDRESSED CHALLENGES FOR XAI DATA CREATION IN SMART CITIES

Preparing data for XAI modelling in smart cities has many vital challenges which need to be examined.

- *Data sources and features* refer to the formats and characteristics of the data.
  a. Data schemes and sources refer to data sources from legacy systems and new ones in different schemes that fit AXI modelling.
  b. Data formats, data collected in different forms make it difficult to categorize and organize the unstructured nature of data (e.g., audio, video, images, tweets, logs, etc.) in an easy way to form semantic forms.
- *Data characteristics (8 Vs)*: Challenges related to the data characteristics is another issue where the smart city application should be able to cope with data Volume, Velocity, Variability, Volatility, Value, Variety, Validity, and Veracity.
- *Exchange of information and data*: Another problem is the exchange of information and data between different entities. Typical warehouses or feed storage was built for every government and municipal agency or service provider. Most of them often don't want to share their property information. Policies in smart cities must find ways to prevent or reduce smooth communication and data exchange barriers between different entities.
- *Heterogeneity and disparity*: Data collected from different archive sources are rarely stored in standard formats and stored in distinctive databases. The absence of a structure and, as a result, reliability, heterogeneity, and inconsistency problems will result in the provision of multiple suppliers with different sources and cooperation. Consequently, there is no quality assurance mechanism and universal means of automatically retrieving and converting data into a single source for useful analysis.
- *Data security and privacy*: Confidence-based data requires high security and mechanisms to prevent unauthorized access and malicious attacks. Moreover, intelligent applications integrated throughout agencies also need high safety as data movements across different

network types, some of which may be unsafe. Data encryption is necessary to protect people's privacy before employing data in smart city applications.

- *High cost*: Installing and physically deploying and testing Bigdata systems requires expensive hardware and software to develop further and monitor intelligent cities infrastructure and application. Therefore, the main question is who will pay the costs and how the service costs can be calculated, including who collects end-users' fee for these integrated and inter-connected services.
- *End-user collaboration*: Intelligent city architecture is designed explicitly for people who use loops. Therefore, policies are needed to ensure that users are involved in collecting the necessary data for the smart city applications to provide the services they need at the desired level of quality.

2.7 SUMMARY

A smart city takes into consideration allowing its citizens to have a voice. Therefore, policies are needed to ensure users are involved with the necessary data so that smart city applications can provide the services they need at the desired level of quality. ML will be at the heart of prediction and decision making as AI technologies become more prevalent in service delivery. The widespread use of ML in smart city services necessitates the need to make it understandable. SSML algorithms can produce accurate predictions, but it is not easy to understand by users or experts. First, this chapter reveals the essence of smart cities. ML algorithms for smart monitoring, symbolic and non-symbolic ML models, and their properties for smart city services, and ML explainability focus. The chapter also demonstrated the importance of big data and ICT for XAI services in smart cities, the data properties, featuring and refeaturing for XAI modelling: the smart environment with XAI, the proposed smart city main pillars and architecture, the required transition entities, policy and protocol normalization, application layer structure considering XAI, and user interaction with XAI models. Finally, the chapter also showed how XAI in smart cities is beneficial and the challenges in implementing XAI models for smart city services.

REFERENCES

Alías, F., & Alsina-Pagès, R. M. (2019). Review of wireless acoustic sensor networks for environmental noise monitoring insmart cities. *Journal of Sensors*, 2019, 1–14. 10.1155/2019/7634860

Allam, Z. (2021). Big data, artificial intelligence and the rise of autonomous smart cities. *The rise of autonomous smart cities* (pp. 7–30). Springer. 10.1007/978-3-030-59448-0_2

Allam, Z., & Dhunny, Z. A. (2019). On big data, artificial intelligence and smart cities. *Cities*, *89*, 80–91. 10.1016/j.cities.2019.01.032

Ariyaluran Habeeb, R. A., Nasaruddin, F., Gani, A., Hashem, I. A. T., Ahmed, E., & Imran, M. (2019). Real-time big data processing for anomaly detection: A survey. *International Journal of Information Management*, *45*, 289–307. 10.1016/j.ijinfomgt.2018.08.006

Bechberger, L. (n.d.). *Machine learning in conceptual spaces: Two learning processes*. Conceptuccino.Uni-Osnabrueck.De. https://www.conceptuccino.uni-osnabrueck.de/fileadmin/documents/public/Documents/CARLA-WS-18/Bechberger.pdf

Cambria, E., Li Y., Xing, F. Z., Poria, S., & Kwok, K. (2020). SenticNet 6: Ensemble application of symbolic and subsymbolic AI for sentiment analysis. *International Conference on Information and Knowledge Management, Proceedings*, *6*(October), 105–114. 10.1145/3340531.3412003

Chui, K. T., Lytras, M. D., & Visvizi, A. (2018). Energy sustainability in smart cities: Artificial intelligence, smart monitoring, and optimization of energy consumption. *Energies*, *11*(11), 2869. 10.3390/en11112869

Du, R., Santi, P., Xiao, M., Vasilakos, A. V., & Fischione, C. (2019). The sensable city: A survey on the deployment and management for smart city monitoring. *IEEE Communications Surveys and Tutorials*, *21*(2), 1533–1560. 10.1109/COMST.2018.2881008

Gunning, D. (2016). *Explainable atificial intelligence (XAI)*. Defense Advanced Research Projects Agency. http://www.darpa.mil/program/explainable-artificial-intelligence

Gunning, D., & Aha, D. W. (2019). DARPA's Explainable Artificial Intelligence Program. *AI Magazine*, *40*(2), 44–58. 10.1609/aimag.v40i2.2850

Kim, H., Choi, H., Kang, H., An, J., Yeom, S., & Hong, T. (2021). A systematic review of the smart energy conservation system: From smart homes to sustainable smart cities. *Renewable and Sustainable Energy Reviews*, *140*, 110–155. 10.1016/j.rser.2021.110755

Kumar, H., Singh, M. K., Gupta, M., & Madaan, J. (2020a.). Moving towards smart cities: Solutions that lead to the smart city transformation framework. *Technological Forecasting and Social Change*, 153. 10.1016/j.techfore.2018.04.024

Lamba, K., & Singh, S. P. (2017). Big data in operations and supply chain management: Current trends and future perspectives. *Production Planning and Control*, *28*(11–12), 877–890. 10.1080/09537287.2017.1336787

Lundberg, S. M., Erion, G., Chen, H., DeGrave, A., Prutkin, J. M., Nair, B., Katz, R., Himmelfarb, J., Bansal, N., & Lee, S. I. (2019). Explainable AI for trees: From local explanations to global understanding. *ArXiv*. https://www.nature.com/articles/s42256-019-0138-9

Lv, Z., Hu, B., & Lv, H. (2020). Infrastructure monitoring and operation for smart cities based on IoT system. *IEEE Transactions on Industrial Informatics*, *16*(3), 1957–1962. 10.1109/TII.2019.2913535

Nilsson, J., Sandin, F., & Delsing, J. (2019). Interoperability and machine-to-machine translation model with mappings to machine learning tasks. *ArXiv*. https://ieeexplore.ieee.org/abstract/document/8972085/?casa_token=HW5e7eRloowAAAAA:g7eGXBUYuaHnv9fc2E0vkC7C4x-p3j-SWPRz_KaiwhtbtKFBbs1Ho8bI5r0caSVJtN6yE9p7hAA_

Putnam, V., & Conati, C. (2019). *Exploring the need for explainable artificial intelligence (XAI) in intelligent tutoring systems (ITS)*. CEUR Workshop Proceedings. Vol. 2327. http://explainablesystems.comp.nus.edu.sg/2019/wp-content/uploads/2019/02/IUI19WS-ExSS2019-19.pdf

Rathore, M. Mazhar, A. P., Hong, W. H., Seo, H. C., Awan, I., & Saeed, S. (2018). Exploiting IoT and big data analytics: Defining smart digital city using real-time urban data. *Sustainable Cities and Society*, *40*, 600–610. 10.1016/j.scs.2017.12.022

Serrano, W. (2018). Digital systems in smart city and infrastructure: Digital as a service. *Smart Cities*, *1*(1), 134–153. 10.3390/smartcities1010008

Telesko R., Jüngling, S., & Gachnang, P. (2020). *Combining symbolic and sub-symbolic AI in the context of education and learning*. CEUR Workshop Proceedings. Vol. 2600. https://forms.gle/rj5NSqmgTth1dm2f7

Xie, J., Tang, H., Huang, T., Richard Yu, F., Xie, R., Liu, J., & Liu, Y. (2019). A survey of blockchain technology applied to smart cities: Research issues and challenges. *IEEE Communications Surveys and Tutorials*, *21*(3), 2794–2830. 10.1109/COMST.2019.2899617

Yigitcanlar, T., Kankanamge, N., & Vella, K. (2020). How are smart city concepts and technologies perceived and utilized? A systematic Geo-Twitter analysis of smart cities in Australia. *Journal of Urban Technology*, *28*, 135–154. 10.1080/10630732.2020.1753483

IoT- and XAI-Based Smart Medical Waste Management

3

Zineb Boudanga[1,2], Siham Benhadou[1], and Jean Philippe Leroy[3]

[1]*National Higher School of Electricity and Mechanics (ENSEM) Hassan II University, Casablanca, Morocco*
[2]*Foundation for Development and Innovation in Science and Engineering (FRDISI), Casablanca, Morocco*
[3]*Liser IPI Paris, Group IGS, Paris, France*

Contents

DOI: 10.1201/9781003172772-3

31

3.1 INTRODUCTION

According to world population prospects (Nygaard & David, 1994), the population size of cities is growing at a rapid rate. As a result, it will be even more difficult to live in urban areas due to limited resources and services. It is therefore important for city managers to be able to facilitate access to these services and manage them effectively. In this regard, many countries are working to develop smart city projects.

The smart city is a new concept in urban development (Boudanga et al., 2019; Kirimtat et al., 2020; Ristvej et al., 2020). It results from the adoption of new technologies such as information and communication technologies (ICT), internet of things (IoT), big data and artificial intelligence. This concept improves the efficiency of processes in all areas, such as traffic control, sustainable resource management, and safety and health of citizens, in general, the standard of living of the city's citizens.

The most developed smart applications in the context of the smart city are typically traffic management, parking management, safety and security, environmental monitoring, smart homes, lighting management, health sector management, smart government, smart grid, and waste management.

In this work, we discuss a smart solution for the management of medical waste, given its importance and the need for efficient and intelligent management in this sector. In fact, healthcare activities generate an increasing amount of medical waste (due to demographic evolution, medical technology advancements and the improvement and expansion of healthcare services). Some of this waste does not present any particular risk and can be assimilated to household waste. However, others present risks of infection, which requires a management model that takes into consideration the different types of waste with its level of risk.

In addition, according to the WHO, every year, 16 billion injections are given worldwide, but not all used needles and syringes are properly disposed of. Also, medical waste is incinerated, sometimes in the open air, and burning it can result in the release of dioxins, furans, and particulates. Moreover, healthcare waste is a reservoir of microorganisms that can infect hospital patients, healthcare workers, and the general public.

Thus, medical waste management methods can lead to a health risk if the various steps in the management process are not carried out correctly. To this end, Rolewicz-Kalińska (2016) identifies the key objectives of the medical waste management system, which can be summarized in the following points: firstly, the guarantee of the quality and safety of medical services. Second, the collection and transport of medical waste must be safe and eco-friendly. Thirdly, optimal final treatment of the waste produced in terms of environmental issues, risk reduction for humans, and economic viability.

To align with these objectives, Çetinkaya et al. (2020) propose a prediction model for accurate estimation of medical waste. This estimation can be used to plan and design medical waste management systems. They consider that sustainable medical waste management can only be implemented if an accurate waste prediction is made, so that the investment will be in the right area.

Several researchers work on the design of a reverse logistics network for efficient medical waste management (Alshraideh & Abu Qdais, 2017; Budak & Ustundag, 2017; Gergin et al., 2019; He et al., 2016; Mantzaras & Voudrias, 2017; Nolz et al., 2014; Osaba et al. 2019; Wang et al., 2019; Yu et al., 2020). Indeed, they are trying to optimize multi objectives (Minimizing costs, risks, total travel distance, emissions from transport vehicles, probability to expose to medical waste, amount of uncollected waste and impact on the environment. Finding optimal location of treatment facilities and transfer station and optimal route of transportation) by using different methods such as Inventory routing problem, Mixed-integer linear programming, Artificial bee colony, and Vehicle Routing Problem.

However, the conception of a smart solution for the optimization of the medical waste management process remains limited and represents a gap in the literature. This chapter aims to be useful in this context. In fact, our objective through this contribution is to align with the objectives of an efficient medical waste management, and to find trade-offs between economic interest and compliance with current regulations while integrating the smart concept.

In addition, this work develops a new big data analytic framework to support the decision-making process in medical waste management software platforms, and consequently to provide smarter services. The proposed analytical framework is based on explainable artificial intelligence to address the limitations of existing systems that will be discussed later.

Explained AI (XAI) is a revolution that will allow us to produce more explainable models while maintaining a high level of learning performance, and which allows us to give the human actors in our solution the ability to understand, trust appropriately, and effectively manage the smart medical waste management solution (Barredo Arrieta et al., 2020; Guo, 2020; Tjoa & Guan, 2020).

In order to achieve the objectives of our contribution, the following key research questions are formulated:

- Q1: How can explainable artificial intelligence be integrated into medical waste management platforms?
- Q2: What are the measures to be taken to optimize the medical waste management process?
- Q3: How can we make medical waste management smart?

To answer these questions, we will first explain the current waste disposal process in the Moroccan context. Next, we will propose a model for improving the current system. In a third step, we will present the architecture of our solution with a description of each layer.

3.2 RELATED WORK

The state of art includes two main areas of research: Medical waste management as an application of IoT-based smart cities, and works on XAI for smart cities.

3.2.1 Medical Waste Management as an Application of Internet of Things-Based Smart Cities

Wang et al. (2020) propose to use the IoT to monitor the level of waste at hospitals and facilitate its collection. In the same direction, Ramaa et al. (2020) propose to digitize the waste bins in order to send the unique identifier and the waste level of each bin. The collected data can then be used for route optimization and pilferage reduction. They also propose to implement a waste management platform that facilitates the tracking of waste by several parties, such as government agencies and hospitals. In addition, Brindha et al. (2020) use different types of sensors (Proximity sensor, humidity sensor, gas sensor and ultrasonic sensor, etc.) in order to detect and sort waste. Indeed, they have proposed a waste segregator that can identify the type of waste and put them in bins automatically. Similarly, Devi et al. (2020) rely on the IoT to automatically sort medical waste. To do this they have implemented a five-step system (Waste image capture, data preprocessing, median filtering, contrast enhancement and segmentation), the output data are evaluated using The Gray Level Cooccurrence Matrix (GLCM). Moreover, Chen et al. (2020) propose to equip the bins with cameras for the detection and classification of medical waste using recorded video, this system is called iWASTE.

Concerning our contribution, we have not only used IoT to sort or track waste, but rather we have used this technology to make the process, from waste generation to disposal, intelligent. Indeed, we proposed to implement smart devices at the hospital, external warehouse, transport vehicle, and waste treatment unit. In addition, we also used AI and big data to increase the efficiency of our intelligent solution. On the other hand, we proposed to establish a collaboration system between the different parties involved in this process, in order to engage them in the success of this solution.

3.2.2 Explainable Artificial Intelligence for Smart Cities

Thakker et al. (2020) use XAI with a hybrid image classification model for flood monitoring in smart cities. Indeed, this hybrid model is based on machine learning and semantic technology, which classifies drainage and gully images in cases of blockage.

Petrovic and Tosic (2020) present an approach based on XAI for autonomous cars, and also for energy efficiency in smart homes. For autonomous cars, the authors propose to provide the driver with an XAI user interface that would allow him to choose the action to be taken among the available options, in order to ensure that no damage will be caused and nobody's life will be endangered. For energy efficiency in smart homes, they propose to control the electrical devices through an XAI-based system.

Kuzlu et al. (2020) use XAI for PV solar energy forecasting. They consider that understanding the inner workings of an AI-based PV solar energy prediction model can improve PV solar energy production by highlighting relevant parameters.

In this work, we use XAI to efficiently manage decisions concerning the medical waste management process.

For the management of medical waste, it is of utmost importance to correctly identify all the risks related to this sector, as well as the probability of accidents during the transport of the waste and to act appropriately to ensure that no damage is caused and people are not in danger. Therefore, a smart solution based on XAI is crucial, so that the stakeholders trust the system and also be part of the process to ensure its reliability.

3.3 PROPOSED APPROACH

In this section, we present a smart solution to support decision making in medical waste management. To illustrate our solution, we will consider the Moroccan context as a case study.

Therefore, first, we will study the current state of waste management, highlighting their failures and areas for improvement. Then, we will try to solve these failures as part of the smart city concept.

3.3.1 The Current Medical Waste Disposal Process in Morocco

3.3.1.1 Legislative Framework for Medical Waste Management in Morocco

In the Moroccan context, the government and operators are aware of the importance of medical waste management and its consequences. Therefore, measures have been taken in this area, particularly in the legal context. Indeed, there are the following laws in action.

- (Law No. 28-00, 2006), which relates to waste management and disposal.
- (Decree No. 2-09-139, 2009) relating to the management of medical and pharmaceutical waste.
- (Decree No. 2-14-85, 2014) relating to the management of hazardous waste.
- (Decree No. 2-07-253, 2008) relating to the classification of wastes and establishing the list of hazardous wastes.

- (Decree No. 2-12-172, 2012) fixing the technical prescriptions relating to the elimination and the processes of valorization of waste by incineration.
- (Law No. 99-12, 2014) relating to the national charter of the environment and sustainable development.
- (Law No. 12-03, 2003) relating to environmental impact studies.
- (Law No. 13-03, 2015) relating to the fight against air pollution.
- (Law No. 30-05, 2011) relating to the transport by road of dangerous merchandises.

Despite all the efforts made by the government Mbarki et al. (2013), there are still some disparities between the practices put in place for medical waste management and what is recommended in the legislation.

3.3.1.2 The Current Model of Medical Waste Disposal

In the Moroccan context, the medical waste management process encompasses (see Figure 3.1):

- Sorting and packaging: Medical waste is sorted according to its category and put into bags or into resistant and watertight bins of different colours.
- Storage: Waste is stored in a secure location.
- Transport: The medical waste circuit complies with the legal regulations concerning the transport of hazardous waste.
- Treatment and disposal (Incineration or thermal and wet disinfection).
  However, this process faces several challenges:
- At the level of sorting and packaging: There are health risks to be managed, as well as the need to respect the limited storage time and to ensure that an adequate temperature is maintained in the storage facilities.
- At the transportation level: It is necessary to manage flows, balance vehicle capacity with waste generation, and manage risk on the route.

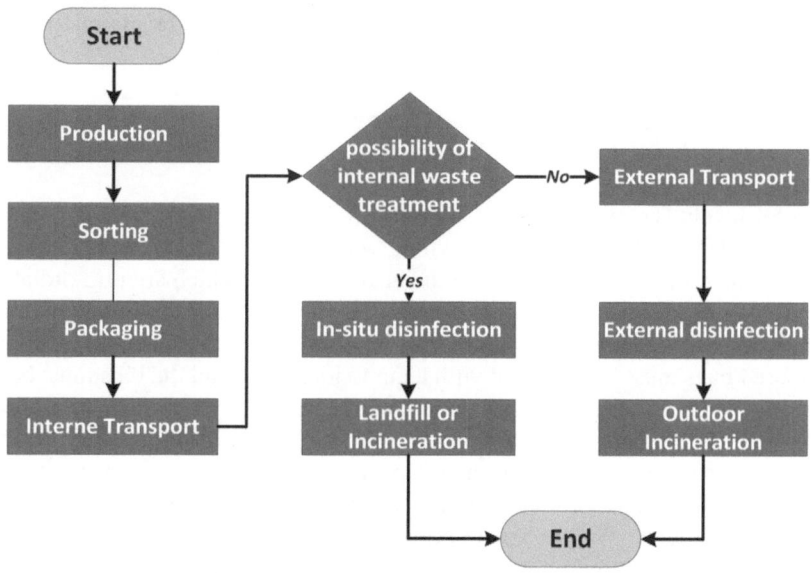

FIGURE 3.1 The Current Model of Medical Waste Disposal.

- At disposal level: It is essential to comply with the strictest regulations, control the costs of outsourcing services and, above all, reduce the negative impact on the environment.

Consequently, our proposal is oriented to respond appropriately and effectively to these issues.

3.3.2 The Proposed Smart Model for Medical Waste Management

To overcome all the above challenges, we propose a novel smart solution that encompasses the following elements (see Figure 3.2):

- Centralization of waste from hospital units in a central warehouse. This warehouse must be equipped with an intelligent belt conveyor capable of optimizing waste sorting. This centralization is motivated by the results of the work (Ahlaqqach et al., 2017). Where they proposed a model for centralizing collected waste in order to optimize collection trips, while respecting capacity constraints, time windows, and the heterogeneity of the fleet.
- Process that recycles as much waste as possible.
- Since in Morocco, there are four categories of medical waste, we take into consideration a storage capacity adapted to each type of waste (four types of bins). While integrating an intelligent solution for each bin capable of not accepting inappropriate types of waste.
- Taking into consideration that the final storage capacity of the bins must be adapted to the quantities of waste produced and the frequency of their disposal.
- Outsourcing of waste treatment in order to be able to manage the impact on the environment. In addition, in order to dedicate the work to a specialized unit capable of optimizing the process.
- Install within each treatment centre smart sensors capable of detecting air and soil quality in order to evaluate the performance of each one.
- In order to use the available infrastructure and resources, the solution integrates all existing waste treatment companies in Morocco as treatment centres. This proposal will be beneficial in the sense of homogenizing the process and strengthen efforts to optimize the process of medical waste management.

Our proposal is a collaboration between four units: hospitals, transport units, warehouses, and treatment units. The success of the process depends on all these units, so each unit has different responsibilities that are complementary to each other.

At the hospital level: The staff is responsible for collecting the waste, sorting, and storing them temporarily. We have decided to outsource the whole waste treatment in order to dedicate the work to specific units and let the hospitals concentrate on the health services which are primordial. Also, in order to centralize the waste treatment that reduces and keeps the impact on the environment in a specific area.

To succeed in these responsibilities, we proposed to implement the hospitals with smart bags and bins. Indeed, coloured bags must be equipped with a tag to identify it and the bins must be equipped with smart sensors and actuators to ensure several functions (Murugesan et al., 2021; Pardini et al., 2020). Indeed, there is a humidity sensor and a temperature sensor to control the storage conditions, a level sensor to avoid exceeding the maximum storage capacity, an actuator to close the bins automatically after filling, an object recognition sensor to control the sorting conformity.

At the transport unit level: We propose to implement a data processing software platform based on XAI in order to optimize the pick-up trip and allow the controller to understand the schedule and adapt it to the available resources and external factors. This platform must ensure that the pickup schedule is sent to the central warehouse in order to anticipate the management at its level.

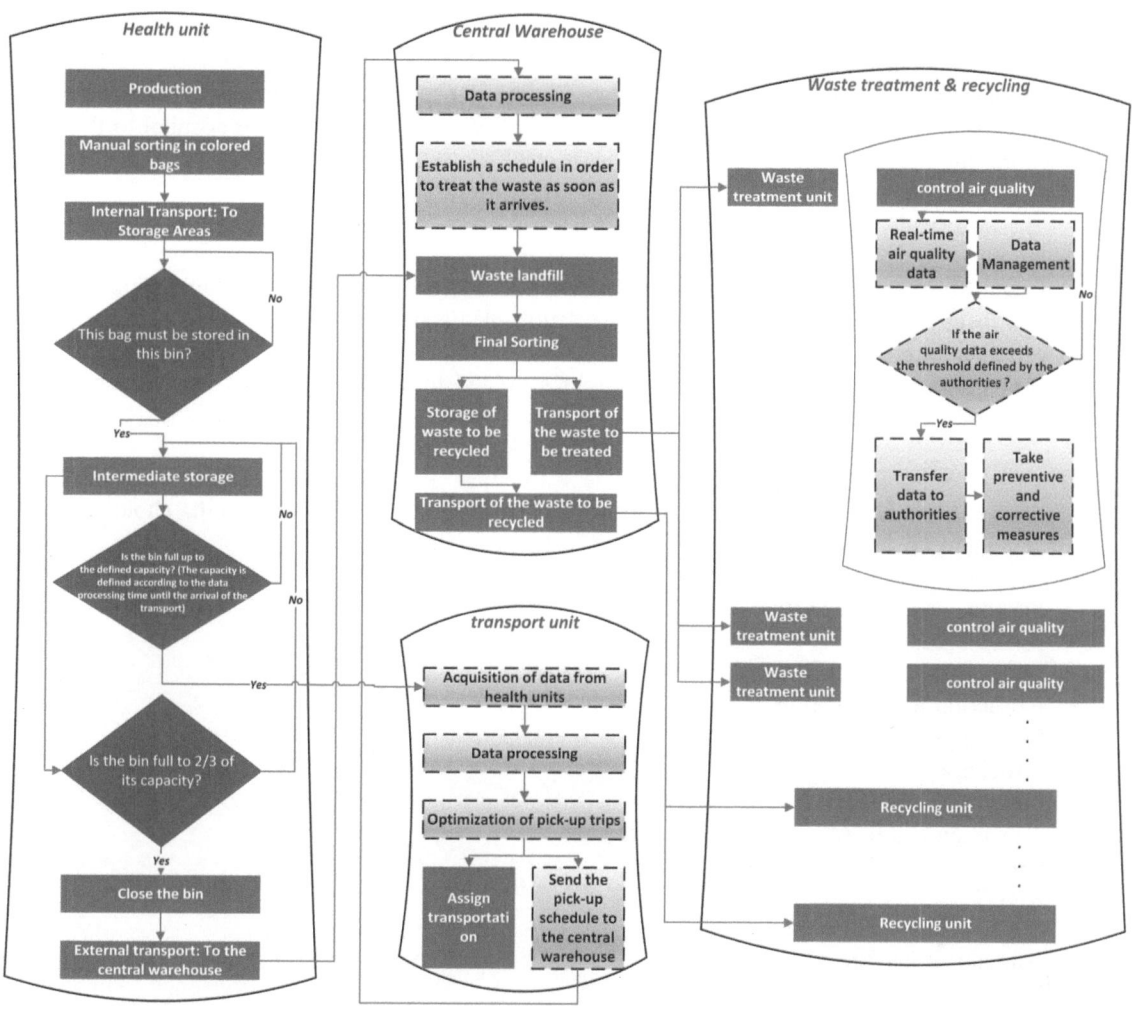

FIGURE 3.2   The Proposed Smart Model for Medical Waste Management.

The transport vehicles must be equipped with a GPS and identification system to track the waste throughout the transport process. In addition, they must be equipped with humidity and temperature sensors to ensure the right transport conditions.

At the warehouse level: Firstly, it is necessary to choose the number and strategic location of the warehouse by taking into consideration two essential factors, the location of hospitals and treatment units.

Choosing a strategic location for a warehouse is a multi-criteria decision-making problem. This problem is of interest to several researchers, because it has an important weight for the improvement of the process. In this context, Emeç and Akkaya (2018) develop a stochastic and multi-criteria approach to solve the problem of warehouse location. In fact, they combine two methods, SAHP (Stochastic Analytical Hierarchy Process) and VIKOR. The authors Emeç and Akkaya (2018) andRaut et al. (2017) use the AHP (Analytic hierarchy process) model to rank the warehouse selection criteria and identify the best location among potential sites. According to Abo-Elnaga et al. (2017), to solve the warehouse location problem, they use an active set strategy, a penalty method, and a Trust Region approach. On the other hand, the work (Jacyna-Gołda & Izdebski, 2017) to choose the optimal location of warehouses is based on genetic algorithms.

The contribution of our work related to this point, in order to solve the multi-objective optimization problem, is to take into account the uncertainties on the transport costs from the hospital to the warehouse and on the operating costs:

- The uncertainty of travel is taken into account in order to allow for the addition of potential kilometers related to detours and indirect routes.
- Each of the units (Health or treatment) receive transportation (for loading or unloading waste) at a different rate per week, and the calculation of distance per week includes a travel uncertainty factor.
- We add to our model additional uncertainties related to wages, costs, investments, and resources.

Furthermore, we propose to equip the warehouse with smart conveyor belts capable of sorting the coloured bags using object recognition sensors. This sorting has to be according to the choice of the type of treatment for each type of waste:

- Microwave thermal system
- Encapsulation
- Chemical disinfection
- Extraction or destruction of needles
- Shredding
- Ozonation
- Landfill, Landfill pit.
  This choice is linked to several criteria such as:
- Cost
- Risks
- Sustainability of the solution
- Amount of waste
- Type of waste (liquid or solid)
- Possibility of generating dangerous reactions
- Regulations and laws
- Impact on the environment

We are in front of a multicriteria decision-making problem, which we propose to solve with the help of XAI. We have chosen this technology because it is the only one that will allow us to generate intelligent solutions that are efficient and even explainable to the authorities and controllers so that they can modify the criteria according to the results obtained and analyze the impact of each criterion in relation to the solution.

At the level of the treatment unit: It is necessary to ensure the waste treatment operations while respecting the regulations and the environment. In order to evaluate the performance of each treatment unit, we propose to implement air and soil control sensors, in order to generate real-time data, which will be processed and analyzed using XAI. XAI at this stage will allow us to identify the decisions and measures that will be taken to comply with the regulations, while explaining these measurements to the authority in order to evaluate each treatment unit and to be able to assign or not fines to them.

So, the process of our proposal is as follows: Once the medical waste has been generated, the first sorting in coloured bags is done. The bags are then transported to secure sites. In these areas, the bags are stored in a specific smart bin for each type of waste (the bin rejects unsuitable types of bags). If the bin has reached the defined maximum capacity (the capacity is defined according to the arrival time of the transport and the filling frequency) a notification will be sent to the transport unit to affect the transport.

A message about a pick-up schedule will be sent to the central warehouse to anticipate the management.

If the bin reaches two-thirds of its capacity before the arrival of the transport, it will be closed automatically.

After the arrival of the transport, the bins will be sent to the warehouse for final sorting, in order to specify the waste that can be recycled and the waste that must be disposed of (for the waste to be disposed of, the sorting must also be in relation to the appropriate type of disposal).

Then the waste will be reloaded to send it to the specific treatment unit according to the type of treatment recommended.

In the treatment units, an air and soil quality control must be installed. This control must ensure the real-time air and soil quality data, so that the authorities can take the necessary preventive and corrective measures.

This system will ensure the collaboration between the transport units, the central warehouse, and the treatment units.

3.3.3 Architecture of the Proposed Smart Solution

The proposed smart waste management architecture includes eight layers: (1) Data acquisition layer, (2) the IoT gateway, (3) Data preprocessing layer, (4) Data processing layer, (5) Application Layer, (6) Network layer, (7) Management layer, and (8) Security, privacy, and safety layer. A brief overview of the proposed architecture is provided in the following subsection, followed by a detailed description of the proposed architecture layers.

3.3.3.1 Overview

The proposed smart medical waste management architecture is illustrated in Figure 3.3, from the acquisition layer to the application and user interface. The main objective of this study is to propose a smart solution to manage medical waste, in particular hazardous waste.

Sensors are responsible for the collection of heterogeneous data within the hospitals, warehouses, and treatment units and thus constitute the lower level of the proposed architecture. In addition, these components are linked to the data prepossessing and processing layer through heterogeneous access technologies.

Upon receipt of the collected data, intelligent decisions are made by the black box AI model, located at the middle level of the architecture. This architecture is based not only on AI but rather on XAI, in order to generate in parallel with the decisions its explanations. These explanations can be used to justify the decisions taken, to control the model identifying its potential defects, to improve the model accuracy and efficiency, to discover new knowledge and also to learn model relationships. Upon receipt of autonomous decisions, event generation takes place at the higher level (Application level).

The main objective of this study was to exploit the smart city concept and XAI tools to improve medical waste management to enable real-time decision making.

3.3.3.2 Smart Devices

The goal of smart medical waste management is to optimize the process and provide efficient services. Data acquisition is therefore essential to achieve these objectives. However, data acquisition has become fastidious and difficult due to the massive amount of data created by connected devices. As a result, low-cost, low-power consumption sensors have become a promising mechanism for heterogeneous data acquisition.

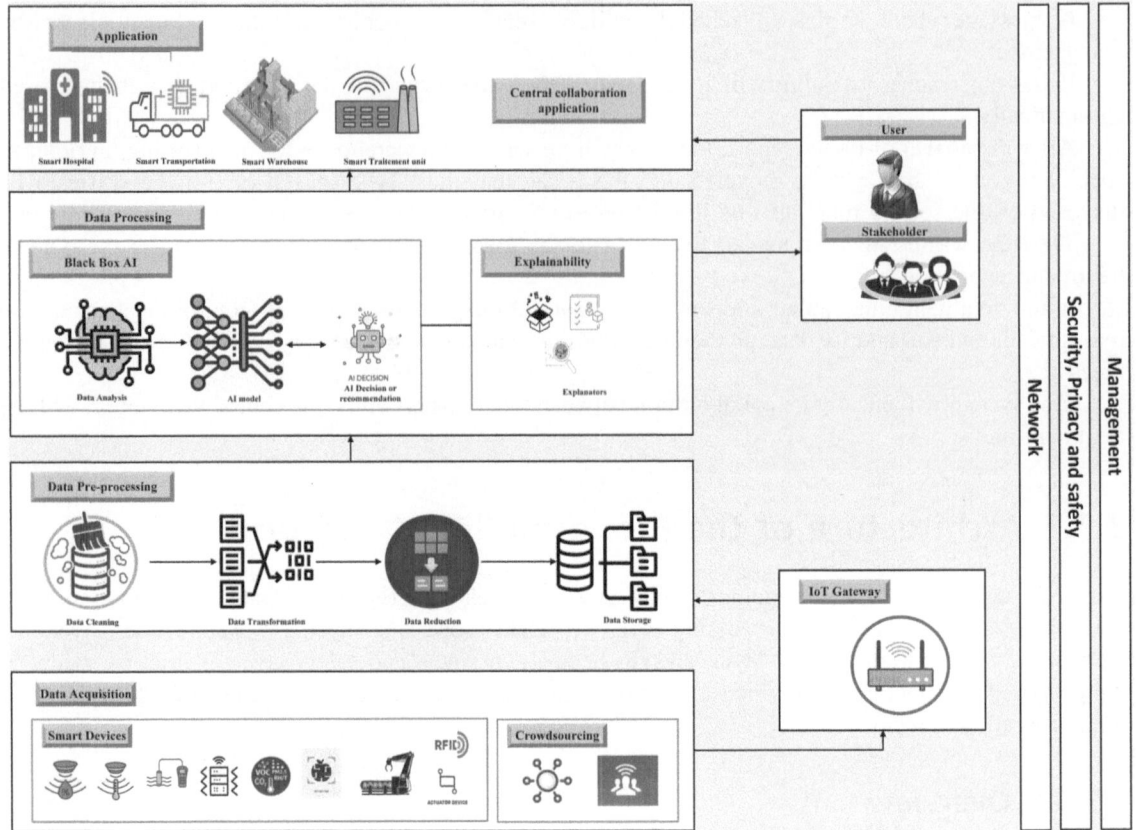

FIGURE 3.3 The Architecture of Smart Medical Waste Management Solution.

Therefore, the realization of the proposed architecture starts with the large-scale deployment of heterogeneous sensors in the hospital, warehouse, and treatment units. These sensors are responsible for collecting real-time data from the neighboring environment.

In addition, devices in this layer not only provide data, but also perform computing tasks and make decisions, if the data does not require large processing capacities and does not need to be transmitted to the Data management layer. This reduces network latency and traffic between end devices and the cloud.

For example, at a hospital, if the bin reaches its maximum defined capacity, it must be closed to avoid exceeding its load. In addition, the bin must be able to recognize the appropriate bag to store it, in order to refuse other bags.

The integration of these smart devices is inspired by the patent Tran and Tran (2020), in which they invented an IoT device that includes a camera coupled to a processor, and a wireless transceiver coupled to the processor. To facilitate secure operation, they used blockchain smart contracts with the device.

3.3.3.3 IoT Gateway

This layer acts as a data bridge between smart devices and the data preprocessing layer. This means that it is used to transfer data or information from one network to another. It also serves as a protocol adapter and communication bridge. In addition, it provides data volume filtering and security enforcement (see Figure 3.4).

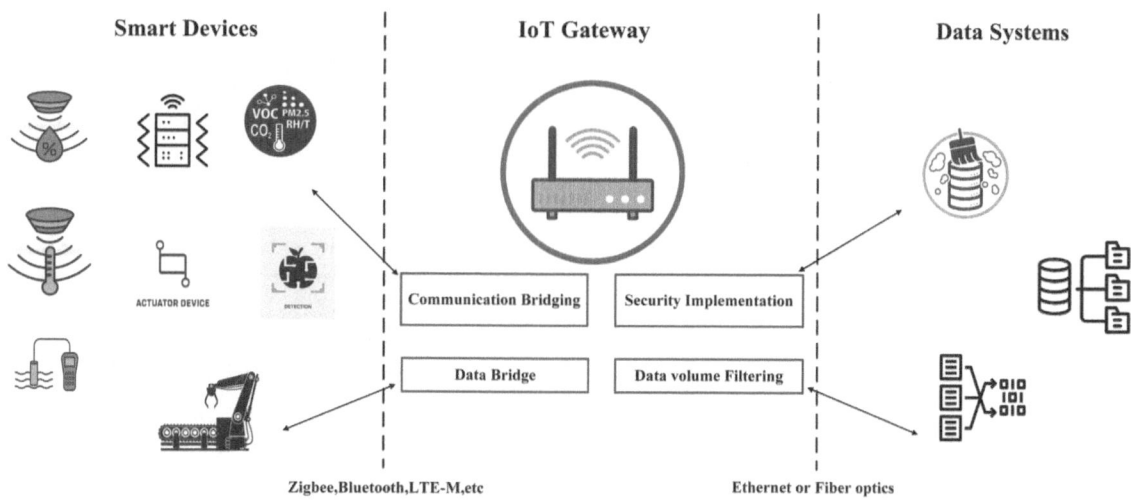

FIGURE 3.4 The IoT Gateway Layer.

IoT gateways are a key element of the IoT architecture because they establish a Device to Device or Device to Cloud communication, and also perform other tasks as described above. The IoT Gateway is therefore one of the most essential elements in our architecture.

3.3.3.4 Data Preprocessing Layer

Data preprocessing is one of the most interesting tasks in order to extract knowledge from the data. It consists of three main steps (see Figure 3.5).

The first step is data cleaning. All samples are cleaned to provide a simple, complete, and clear set. This means that it is checked if there are empty or missing values and if there is noise in the data set.

If there are missing data in the dataset, additional datasets are searched for or more observations are collected.

Noisy data refer to duplicate data or segments of data, which have no value for a particular search.

The second step is data transformation which includes methods for transforming the data into an appropriate format for the computer to learn from.

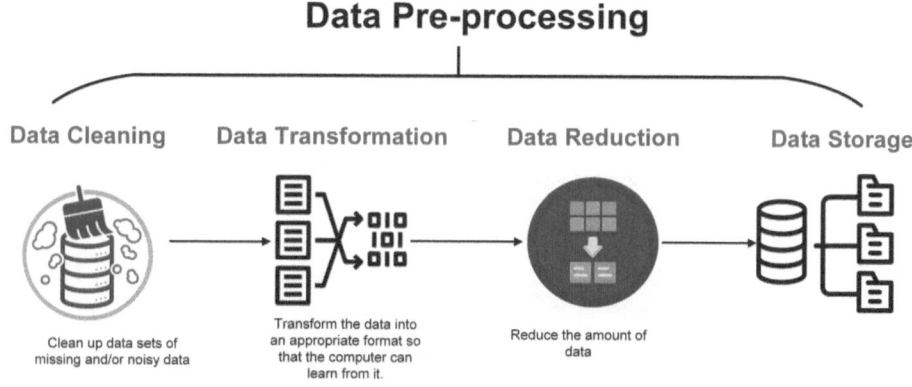

FIGURE 3.5 Data Preprocessing Layer.

There are different data transformation techniques: Data Aggregation (the data is aggregated and presented in a unified format), Normalization (scaling the data into a range to avoid building incorrect models during the training and/or execution of the data analysis), Characteristic Selection (selecting the variables in the data that are the best predictors of the variable we want to predict), Discretization (transforming data into sets of small intervals), Concept Hierarchy Generation (generating a hierarchy between attributes when not specified), and Generalization (converting low-level data characteristics into high-level data characteristics).

The last step is data reduction, which is important when working with large amounts of data because it becomes more difficult to find reliable solutions. Different techniques are applied to reduce data, such as attribute feature selection, dimension reduction (reducing the number of features used), and Numerosity reduction (replacing the original data with a smaller form of data representation).

When the data preprocessing is complete, the data is ready to be entered into any AI model.

3.3.3.5 Data Processing Layer

First of all, data related of the observed process (at the hospital, warehouse and treatment unit level) is collected using sensors and measuring devices or other data sources. Then, the detected data is filtered and preprocessed to obtain the appropriate form for the AI techniques that will be used. The pre-processed data is sent as input to the component using AI techniques (such as neural networks, machine learning, computer vision, and data mining). This component performs the data analysis and sends the results to the component that builds their semantic descriptions.

The knowledge base, ontology definitions, and possible decisions and recommendations are used with the explanation method to allow the user to see the explanations of the results and to select the desired decisions to be executed. An overview of the proposed solution based on XAI is given in Figure 3.6.

3.3.3.6 Application Layer

The application layer is the mediator between the data management layer and the end user. Indeed, it is in charge of generating the actions corresponding to the intelligent decision transmitted.

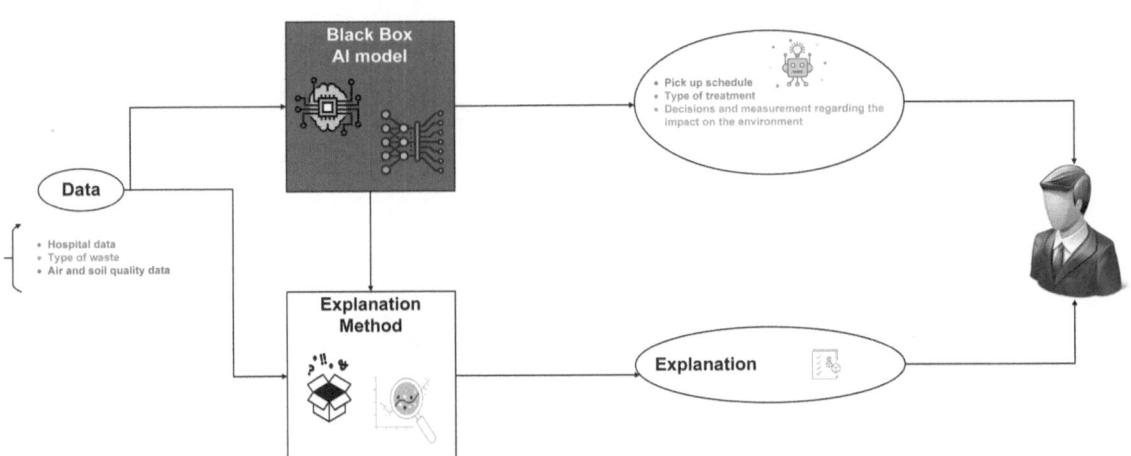

FIGURE 3.6   Explainable Artificial Intelligence.

Indeed, this layer uses the IoT data to create useful applications such as a waste management application in hospitals and warehouses, an application to track transport vehicles, an application to evaluate waste treatment operations.

In this layer, there is also a central application for collaboration between process units. In fact, this application is an interface capable of integrating all the information from the hospital, warehouse, vehicles, and waste treatment units, allowing stakeholders and users to effectively control the process by centralizing all relevant data in one place. This tool exposes the information produced by the process to decision-makers, thus facilitating decision-making.

In addition, the information collected is available via a common dashboard, enabling real-time monitoring of waste management. This application also enables the correlation of different information to be analyzed, for example, the impact of waste pickup schedules on waste management.

The information collected and the explanations resulting from the XAI are also available to the developer community, allowing the development of new applications for medical waste management.

Furthermore, this application allows the real-time modification of the status of certain equipment due to the possible evolution of external factors.

Through this application, process management is more effective and transparent because it is supported by concrete data collected. The manager and the controller obtain a global view of the state of the process, thanks to a common tool, gathering all the information, decisions, and explanations generated.

3.4 EXPECTED RESULTS

The expected contribution of the work will be to design and develop an explainable intelligent platform model for medical waste management described in the previous section, focusing on several points:

- Outsourcing the totality of waste treatment services, in order to give them to the units specialized in this field, allowing the hospital staff to focus on health services.
- Automatize the process as much as possible in order to reduce the risk of human exposure to hazardous waste.
- Ensure the right conditions for storage, transport, and treatment of medical waste.
- Track waste from production to disposal.
- Optimize the process based on intelligent decisions.
- Integrate the human factor in the success of the solution and explain the majority of the decisions to human beings so that he or she can make improvements.
- Make the majority of the units involved in this process collaborate.
- Visualize, parameterize, control, and evaluate the whole process via a common central application.
- Reduce the impact on the environment at the level of the treatment units.

3.5 CONCLUSION

In this chapter, we propose a solution that can smartly manage the medical waste disposal process.

We address several issues in this process. Indeed, we centralize the medical waste management of hospital units in a central warehouse. We completely outsource the management of this waste, in order to improve the performance of hospital staff. We integrate IoT to automate the process as much as possible.

Furthermore, our solution complies with all regulations regarding the disposal of medical waste. We also decided to establish a collaboration between the existing treatment units, the warehouse and the transport unit in order to facilitate the management of medical waste in a more efficient way.

We have also used XAI as a tool to allow humans to understand the reasoning involved in the process and to understand the impact of external factors on each of the decisions taken. This technology has allowed us to gain transparency as well as confidence in the capabilities of our solution.

For future work, we aim to model the medical waste management flow of our smart solution with multi-objective optimization. In this modelling we aim to integrate the majority of risks and probabilities related to this process, while optimizing the objective functions identified during this study (Minimizing cost, maximizing revenue, minimizing risk, and maximizing waste treatment outsourcing).

REFERENCES

Abo-Elnaga, Y., El-Sobky, B., & Al-Naser, L. (2017). An active-set trust-region algorithm for solving warehouse location problem. *Journal of Taibah University for Science, 11*(2), 353–358.

Ahlaqqach, M., Benhra, J., & Mouatassim, S. (2017, Jan). Optimisation des tournées de collecte et de desserte des déchets médicaux transitant par un entrepôt commun. *Logistique & Management, 25*(1), 25–33. 10.1080/125 07970.2017.1310601

Alshraideh, H. & Abu Qdais, H. (2017, Apr). Stochastic modeling and optimization of medical waste collection in Northern Jordan. *Journal of Material Cycles and Waste Management, 19*(2), 743–753. 10.1007/s10163-016-0474-3

Barredo Arrieta, A., Díaz-Rodríguez, N., Del Ser, J., Bennetot, A., Tabik, S., Barbado, A., Garcia, S., Gil-Lopez, S., Molina, D., Benjamins, R., Chatila, R., & Herrera, F. (2020, Jun). Explainable artificial intelligence (XAI): Concepts, taxonomies, opportunities and challenges toward responsible AI. *Information Fusion, 58*, 82–115. 10.1016/j.inffus.2019.12.012

Boudanga, Z., Benhadou, S., & Medromi, H. (2019, Jul). *Development perspective of a Moroccan smart city*. In 2019 Third World Conference on Smart Trends in Systems Security and Sustainablity (WorldS4), London, UK, IEEE, (pp. 247–254).

Brindha, S., Praveen, V., Rajkumar, S., Ramya, V., & Sangeetha, V. (2020). *Automatic medical waste segregation system by using sensors* (No. 3615). EasyChair.

Budak, A., & Ustundag, A. (2017, Jul). Reverse logistics optimisation for waste collection and disposal in health institutions: The case of Turkey. *International Journal of Logistics Research and Applications, 20*(4), 322–341. 10.1080/13675567.2016.1234595

Çetinkaya, A. Y., Kuzu, S. L., & Demir, A. (2020, Oct). Medical waste management in a mid-populated Turkish city and development of medical waste prediction model. *Environment, Development and Sustainability, 22*(7), 6233–6244. 10.1007/s10668-019-00474-6

Chen, J., Mao, J., Thiel, C., & Wang, Y. (2020, Jul). iWaste: Video-based medical waste detection and classification. In 2020 42nd Annual International Conference of the IEEE Engineering in Medicine & Biology Society (EMBC) (pp. 5794–5797). IEEE. 10.1109/EMBC44109.2020.9175645

Decree No. 2-07-253 on the classification of waste and the list of hazardous waste (July 18, 2008).

Decree No. 2-09-139 on the management of medical and pharmaceutical waste (May 21, 2009).

Decree No. 2-12-172, laying down the technical requirements for the disposal and recovery processes of waste by incineration (May 4, 2012).

Decret No. 2-07-253 du 14 rejeb 1429 (18 juillet 2008) portant classification des déchets et fixant la liste des déchets dangereux

Decret No. 2-14-85 du 28 rabii I 1436 (20 janvier 2015) reIatif a Ia gestion des dechets dangereux

Devi, G., Yasoda, K., Manian, D., & Balasubramanian, K. (2020, May). Automatic health care waste segregation and disposal system. *Xian Dianzi Keji Daxue Xuebao/Journal of Xidian University, 14*, 5281–5290. 10.37896/jxu14.5/573

Emeç, S., & Akkaya, G. (2018, Oct). Stochastic AHP and fuzzy VIKOR approach for warehouse location selection problem. *Journal of Enterprise Information Management, 31*(6), 950–962. 10.1108/JEIM-12-2016-0195

Gergin, Z., Tunçbilek, N., & Esnaf, S. (2019, Jan). Clustering approach using Artificial Bee Colony Algorithm for healthcare waste disposal facility location problem. *International Journal of Operations Research and Information Systems*, *10*(1), 56–75. 10.4018/IJORIS.2019010104

Guo, W. (2020, Jun). Explainable artificial intelligence for 6G: Improving trust between human and machine. *IEEE Communications Magazine*, *58*(6), 39–45. 10.1109/MCOM.001.2000050

He, Z., Li, Q., & Fang, J. (2016). The solutions and recommendations for logistics problems in the collection of medical waste in China. *Procedia Environmental Sciences*, *31*, 447–456. 10.1016/j.proenv.2016.02.099

Jacyna-Gołda, I., & Izdebski, M. (2017, Jan). The multi-criteria decision support in choosing the efficient location of warehouses in the logistic network. *Procedia Engineering*, *187*, 635–640, Jan. 10.1016/j.proeng.2017.04.424

Kirimtat, A., Krejcar, O., Kertesz, A., & Tasgetiren, M. F. (2020). Future trends and current state of smart city concepts: A survey. *IEEE Access*, *8*, 86448–86467. 10.1109/ACCESS.2020.2992441

Kuzlu, M., Cali, U., Sharma, V., & Güler, Ö. (2020). Gaining insight into solar photovoltaic power generation forecasting utilizing explainable artificial intelligence tools. *IEEE Access*, *8*, 187814–187823. 10.1109/ACCESS.2020.3031477

Law No. 12-03 relating to environmental impact studies (12 May 2003).

Law No. 13-03 on the Prevention of Air Pollution (June 18, 2015).

Law No. 28-00 relating to the management of waste and its elimination promulgated by the dahir n° 1-06-153 of 30 chaoual 1427 (November 22, 2006).

Law No. 30-05 on the transport of dangerous merchandise by road (June 2, 2011).

Law No. 99-12 on the National Charter for the Environment and Sustainable Development (March 20, 2014).

Mantzaras, G., & Voudrias, E. A. (2017, Nov). An optimization model for collection, haul, transfer, treatment and disposal of infectious medical waste: Application to a Greek region. *Waste Management*, *69*, 518–534. 10.1016/j.wasman.2017.08.037

Mbarki, A., Kabbachi, B., Ezaidi, A., & Benssaou, M. (2013, Aug). Medical waste management: A case study of the Souss-Massa-Drâa Region, Morocco. *Journal of Environmental Protection*, *4*(9), 6. 10.4236/jep.2013.49105

Murugesan, S., Ramalingam, S., & Kanimozhi, P. (2021). Theoretical modelling and fabrication of smart waste management system for clean environment using WSN and IOT. *Materials Today: Proceedings*, *45*, 1908–1913. 10.1016/j.matpr.2020.09.190

Nolz, P. C., Absi, N., & Feillet, D. (2014, Jan). A stochastic inventory routing problem for infectious medical waste collection. *Networks*, *63*(1), 82–95. 10.1002/net.21523

Nygaard, D. F., & David, F. (1994). *World population projections, 2020. No. 5*. International Food Policy Research Institute (IFPRI).

Osaba, E., Yang, X.-S., Fister, I., Del Ser, J., Lopez-Garcia, P., & Vazquez-Pardavila, A. J. (2019, Feb). A discrete and improved Bat Algorithm for solving a medical goods distribution problem with pharmacological waste collection. *Swarm and Evolutionary Computation*, *44*, 273–286. 10.1016/j.swevo.2018.04.001

Pardini, K., Rodrigues, J. J. P. C., Diallo, O., Das, A. K., de Albuquerque, V. H. C., & Kozlov, S. A. (2020, Apr). A smart waste management solution geared towards citizens *Sensors*, *20*(8), 2380. 10.3390/s20082380

Petrovic, N. & Tosic, M. (2020). Explainable artificial intelligence and reasoning in smart cities. https://www.researchgate.net/profile/Nenad-Petrovic/publication/339433681_Explainable_Artificial_Intelligence_and_Reasoning_in_Smart_Cities/links/5e67981592851c7ce0589d65/Explainable-Artificial-Intelligence-and-Reasoning-in-Smart-Cities.pdf

Ramaa, A., Guptha, C. K. N., & Subramanya, K. N. (2020). IoT enabled biomedical waste management system, *RV Journal of Science Technology Engineering Arts and Management*, *1*(1), 49–58, RVJ06.

Raut, RD, Narkhede, BE, Gardas, BB, & Raut, V, (2017) "Multi-criteria decision making approach: a sustainable warehouse location selection problem," *International Journal of Management Concepts and Philosophy*, *10*(3), p. 260, doi: 10.1504/IJMCP.2017.085834.

Ristvej, J, Lacinák, M, & Ondrejka, R, (2020). On smart city and safe city concepts, *Mobile Networks and Applications*, *25*(3), pp. 836–845, Jun. 10.1007/s11036-020-01524-4.

Rolewicz-Kalińska, A. (2016, Jan). Logistic constraints as a part of a sustainable medical waste management system. *Transportation Research Procedia*, *16*, 473–482. 10.1016/j.trpro.2016.11.044

Thakker, D., Mishra, B. K., Abdullatif, A., Mazumdar, S., & Simpson, S. (2020, Dec). Explainable artificial intelligence for developing smart cities solutions. *Smart Cities*, *3*(4), Art. no. 4. 10.3390/smartcities3040065

Tjoa, E., & Guan, C. (2020). A survey on explainable artificial intelligence (XAI): Toward medical XAI. *IEEE Transactions on Neural Networks and Learning Systems*, 1–21. arXiv:1907.07374. 10.1109/TNNLS.2020.3027314

Tran, B., & Tran, H. (2020, Jan 14). *Smart device* (US10532268B2).

Wang, C., Ma, Y., & Meng, F. (2020, Sep). Monitoring method of internet of things for classified recovery of medical waste. *Journal of Physics: Conference Series*, *1646*, 012099. 10.1088/1742-6596/1646/1/012099

Wang, Z., Huang, L., & He, C. X. (2019, Dec). A multi-objective and multi-period optimization model for urban healthcare waste's reverse logistics network design. *Journal of Combinatorial Optimization.* 10.1007/s10878-019-00499-7

Yu, H., Sun, X., Solvang, W. D., & Zhao, X. (2020, Jan). Reverse logistics network design for effective management of medical waste in epidemic outbreaks: Insights from the Coronavirus Disease 2019 (COVID-19) Outbreak in Wuhan (China). *International Journal of Environmental Research and Public Health, 17*(5), Art. no. 1770. 10.3390/ijerph17051770

The Impact and Usage of Smartphone among Generation Z: A Study Based on Data Mining Techniques

Mohamed Uwaiz Fathima Rushda[1] and Lakshika S. Nawarathna[2]

[1]*Postgraduate Institute of Science, University of Peradeniya, Peradeniya, Sri Lanka*
[2]*Department of Statistics and Computer Science, University of Peradeniya, Peradeniya, Sri Lanka*

Contents

DOI: 10.1201/9781003172772-4

4.1 INTRODUCTION

The Smartphone technology and the Internet become the main driving forces to the current society. This handheld technology transforms the society's culture, business, education, health, and social life. Mobile technology has drastically changed individual behaviors. Smartphones in our hands make the world in our pocket. This is impact with in our all activities and becomes like a part of our body like hand, leg, head and etc.

The smartphone technology has its positive side and negative side. Smartphones have brought a massive change in the lifestyle of people and they feel comfortable in offering users a vast platform for communication and access to a wide range of applications (John, 2013). Smartphones are popular among people for the applications they offer to users. Smartphones make communication with people quite easier. People enjoy a lot of benefits in various forms of their daily work and it turns to affect parts of the human body like brain or heart. This smartphone technology is slowly trapping people into addiction. People use smartphones at least 5 hours a day and the use of Apps increases significantly and with the result caused degradation in Physical Social interaction, Distraction, Addiction, health problems, etc. (Mount, 2012).

On the other hand, Smartphone has become a very important and communicative tool among the younger generation. Almost a quarter of young people are so dependent on their smartphones that it becomes like an addiction. Smartphone addiction has developed an unavoidable place in one's life while they wake, sleep, or go to restroom or classroom using smartphones. This bond with smartphone become unbreakable such as addictive behaviors could be linked to other problems such as stress, a depressed mood, lack of sleep, and reduced achievement in school. It seems that the lives of youth without the use of smartphones cannot run smoothly and it impacts on the physical and social lives of individuals.

This motivates us to know how the future generation people will behave following the path of the current generation. Current young generation people will move for the next role of parenting and their children will follow the path of their parents. Current youngsters' behavior will impact the next generation and it will make the new evolution to change body posture, behavior, and culture and family bonds. Mobile phones will change the way that imaging, sensing, and diagnostic measurements/tests are conducted, and fundamentally impacting the existing practices in medicine, engineering and sciences, while also creating new ones (Ozcan, 2014). This is the biggest issue and totally new evolution in traits. This lift the question thatsmartphone addicted current young people; how raise their kids.

This research is about current and future parents, how they developed the mindset which meant they allowed or disallowed their children to use smartphones. The research is based on data mining techniques. This study investigates the factors which impact the future generation where the current generation depends on smartphones and how they pre-plan to deal with their children and smartphone usage. It showed what kind of data could be collected, how the data could be preprocessed, how data mining methods could be applied on the data, and finally how they could be benefited from the discovered knowledge. A vast amount of knowledge can be discovered from the data.

The objective of this research is as follows:

- To identify the factors for how smartphones impact on the next generation
- To identify the factors regarding young parents and the decision of smartphone usage by children

The remainder of this chapter is organized as follows: In Section 4.2, related studies about using different machine learning techniques for analyzing data sets are discussed. In Section 4.3, the methodology, data set, data preprocessing, features extracted, method of balancing and scoring, and the data mining techniques used are described. In Section 4.4, the results and discussion of this study are presented. Section 4.5 presents the conclusions.

4.2 LITERATURE REVIEW

Smartphones are wonderful navigational tools and facilitate the world of information resources on one's fingertips. Smartphones have brought a massive change in the lifestyle of people and they feel comfortable in offering users a vast platform for communication and access to a wide range of applications; anything, anytime, and anywhere happens/happening in the world comes to be known within no time (Rather & RAther, 2019). The adoption of smartphones has been tremendous all over the world. Surveys show that 80% of the world population use mobile devices and 42% of mobile subscribers in the US use Smartphones (Web Design Company, 2020).

The mobile phone is a status symbol for young people. The features of the phone, the appearance, and personalized accessories all attest to the phone's status, with 60% of adolescents reporting they were keen to upgrade their mobile phones (Netsafe, 2005). While children's mobile phones are marketed to parents for security reasons this is not the reason most children want phones. It is known that children feel strong pressures to consume and various studies have pointed to rising materialism among children (Goldberg et al., 2003). The concern is that while mobile phones may be bought by parents to serve the basic functions of communications, for children phones could be servicing a broader range of materialistic and self-identity desires (Mayo, 2005). Seventy-five percent of families own some type of mobile device. In families of young children, for example, 63% own a smartphone and 40% own a tablet. This increased access to mobile devices has also influenced the amount of time the devices are used by parents and their children (Rideout, 2013). Technology is now mobile, allowing people the ability to use technology anywhere, anytime (Child in the City, 2020; Kukulska-Hulme, 2010; Ling, 2007).

Most studies did show a consistent link with smartphone addiction and mental health. For example, six of seven studies on sleep found that children and young people who exhibited problematic smartphone use had poorer sleep. This was also the case for problematic smartphone usage and experiencing higher levels of anxiety, stress, and depressive symptoms (Zysman et al., 2000).

Different patterns can be mined using different data mining techniques such as concept or class description, association analysis, classification and regression, cluster analysis, trend analysis, deviation analysis, and similarity analysis (Dunham, 2003). With the application of data mining tools in the spreadsheet of the program that analyzes data to identify patterns and relations, user profiling and development of business strategies can be started (Milovic, 2012). Most data mining software include online analytical processing, traditional statistical methods, such as cluster analysis, discriminant analysis and regression analysis, and non-traditional statistical analysis such as neural networks, decision trees, link analysis, and association analysis (Koh & Tan, 2005). This wide range of techniques is not surprising because of the fact that data mining is derived from three different disciplines: database management, statistics, and computer science, including the use of artificial intelligence and machine learning (Witten et al., 1999).

4.3 METHODOLOGY

4.3.1 Overview

Research design is an arrangement of conditions for data collection and data analysis (Logasakthi & Rajagopal, 2013). Also, a successful research design can help the researcher to specify the objectives and strategies to be used as well as the approaches for the research. Therefore, in this research by selecting the most relevant factors as attributes and by conducting a cross-sectional research design using online survey method the required data and information were collected in order to identify the factors influencing how smartphones will impact future generation by getting opinions from the current and future parents.

After collecting the data, it was preprocessed as the second step in the KDD process. After applying the transformation techniques the dataset was prepared for data mining. The data mining tool, WEKA, used to apply the data mining techniques for the data set.

4.3.2 KDD Process

Data Mining came into existence in the middle of the 1990s and appeared as a powerful tool that is suitable for fetching previously unknown patterns and useful information from huge datasets. Various studies highlighted that Data Mining techniques help the data holder to analyze and discover unsuspected relationships among their data which in turn help in making decisions (Han & Kamber, 2006). In general, Data Mining and Knowledge Discovery in Databases (KDD) are related terms and are used interchangeably but many researchers assume that both terms are different as Data Mining is one of the most important stages of the KDD process (Fayyad et al., 1996; Hand et al., 2001).

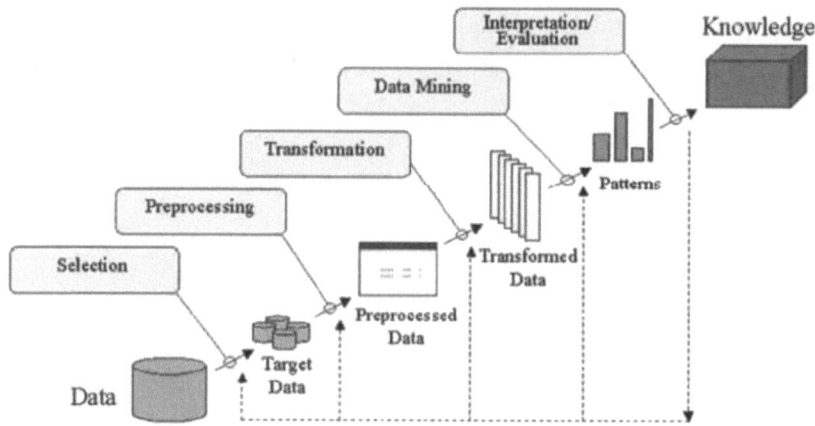

FIGURE 4.1 KDD Process.

KDD Process can be explained as follows (Figure 4.1):

- Learning the application domain: The problem must be clearly defined. Includes relevant prior knowledge and the goals of the application
- Creating a target dataset: It includes selecting a dataset or focusing on a subset of variables or data samples on which discovery is to be performed
- Data Cleaning and preprocessing: It includes removing noisy and inconsistent data.
- Data Transformation: Where data are transformed or consolidated into forms appropriate for mining by performing summary or aggregation operations.
- Data mining: It is an essential process where intelligent methods are applied in order to exact data patterns.
- Pattern Evaluation: To identify the truly interesting patterns representing knowledge based on some interestingness measures.
- Knowledge presentation: It is where visualization and knowledge representation techniques are used to present the mined knowledge to the user (Han & Kamber, 2006).

4.3.3 Data Mining Methods

4.3.3.1 Association Rule

An association rule extract interesting correlation, frequent patterns, associations or casual structures among set of items in the transaction database or other data repositories (Kotsiantis et al., 2006).

4.3.3.2 Classification

Classification is a data mining task that predicts group membership for data instances. It allows the classification of data in one of several classes (Kaur et al., 2015).

4.3.3.3 Weka Data Mining Tool

The Weka workbench contains a collection of visualization tools and algorithms for data analysis and predictive modeling, together with graphical user interfaces for easy access to this functionality. It is a freely available software. It is portable and platform-independent because it is fully implemented in the Java programming language and thus runs on almost any modern computing platform.

4.3.3.4 Approach

The project is based on data mining techniques. This research follows the KDD process. Use the WEKA data mining tool for mining the hidden patterns of the data set.

4.3.4 Implementation

4.3.4.1 Data Cleaning

Real-world data tend to be noisy, incomplete, and inconsistent so it should be corrected for the accurate performance. It is the process of detecting and correcting or removing or corrupt or inaccurate records from a record set, table, or database and refers to identifying incomplete, incorrect, inaccurate, or irrelevant parts of the data and then replacing, modifying, or deleting the dirty or coarse data. Filling the Missing Value- attribute mean replaced all missing attribute values by the global constant, such as a label like 'Unknown'. It was used to fill the missing values for categorical attributes. Smooth out noisy data and inconsistent data means, It is smoothed by combined computer and human inspection techniques to Detect suspicious values and checked each and every tuple by human manually. To avoid this missing value, noisy and inconsistent data, before the formal survey was carried out, a total of 100 respondents choose who belong to different fields was distributed questionnaires survey form and the pilot survey was carried out to minimize biased and ambiguous question in order to ensure the validity and reliability of questionnaires. Also few mistakes were identified and corrected in the questionnaire from the respondents' feedback such as grammatical errors, unclear question statements, and also typing errors.

4.3.4.2 Data Transformation

In this step, data is transformed or consolidated into forms appropriate for mining, by performing smoothing, aggregation, generalization, and normalization.

4.3.4.3 Generalization

Low-level data are replaced by high-level concepts through the use of concept hierarchy. Educational level, age, and correct age for smartphone usage were generalized. Sample Educational Status:

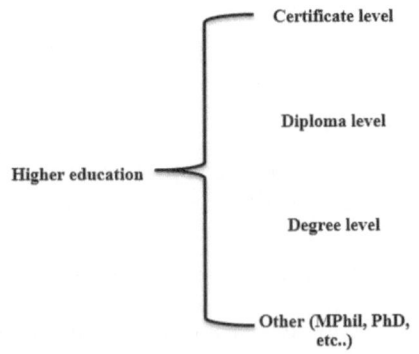

4.3.4.4 Data Reduction

These techniques can be applied to obtain a reduced representation of the data set that is much smaller in volume, yet closely maintains the integrity of the original data.

4.3.4.5 Attribute Subset Selection

It is one of the techniques that the Decision tree induction was used to select the attributes. A tree is constructed from the given data. The attributes that appeared in the tree are considered as selected attributes. Ten attributes were selected such as gender, age group, civil status, educational level, field of IT or non-IT, knowledge scale fore side effect of smartphone usage, allow or deny to use smartphones, time period taken for decision, necessity of smartphone usage, and correct age to use smartphones.

4.3.4.6 Discretization

The data set which will be used by classification algorithms may have some attributes that cannot be handled by the algorithm itself without applying some transformations. Such as, numerical scaled values are needed to be converted into nominal or discrete values in order to make some of the algorithms work correctly. This conversion step is considered in the data discretization part of classification. It is the process of dividing the continuous value into intervals. It is a process that transforms quantitative data into qualitative data. Weka tool was used to discretize the attribute age and correct age to give smartphones. An unsupervised discretization technique was used.

4.3.4.7 Association Rules

Association rule learning is a rule-based ML method for discovering interesting relations between variables in large databases. An association rule is an implication of the form X â†' Y. The standard definition of association rule is if a transaction includes X then it also has Y, where X and Y are item sets. Many algorithms are generated for association rules. Apriori uses a breadth-first search strategy to count the support of item sets and uses a candidate generation function which exploits the downward closure property of support. It uses a bottom-up approach, where frequent subsets are extended one item at a time a step known as candidate generation, and groups of candidates are tested against the data. Apriori algorithm computes the frequent item sets in several rounds; it is used to predict the rules.

4.4 RESULTS AND DISCUSSION

4.4.1 Data Collection

Data were collected from all over the world. Thousand records were collected to analyze. In that 16 attributes were taken to analyze.

4.4.2 Data Summarization

For data preprocessing to be successful, it is essential to have an overall picture of the data. Data summarization technique can be used to identify the properties of the attributes and also helped to identify the noise and outlier data. Data summarization techniques can be used to identify the typical properties of the data. Pie chart and bar chart were used to draw the graphs.

4.4.2.1 Distribution of Gender

The gender distribution of the online survey participants (Respondent) from the total 1000 respondents of the sample group, 543 of the respondents are male which represents 54% and 457 respondents are female, which represents 46%. It shows that males are more actively willing to filling the questionnaire than females.

4.4.2.2 Distribution of Age

The age group of the sample is represented by Figure 4.2. The age group that carries the highest percentage range from 25 to 35 years which is 52.6% followed by 21.4% for the age group of 36–50, and 15.8% for the age group of 18–24 years, and 7.4% for the age group of above 50 years. The least feedbacks were received from Respondents aged under 18 which is only 2.8%.

4.4.2.3 Distribution of Marital Status

According to the respondents' marital status, the majority of respondents are single which consists of 506 (51%) from the 1000 respondents. And then it is followed by respondents who are married which consists of 494 (49%) respondents.

4.4.2.4 Distribution of Education Level (Higher Education)

According to Figure 4.3, education-wise, 89.7% of the sample have obtained a degree, 3.9% have obtained a diploma, 3.4% have completed a certificate level, and another 3% have completed higher studies like postgraduate and doctorates. It is noticeable that more than three-quarters of the sample have completed their degree-level education.

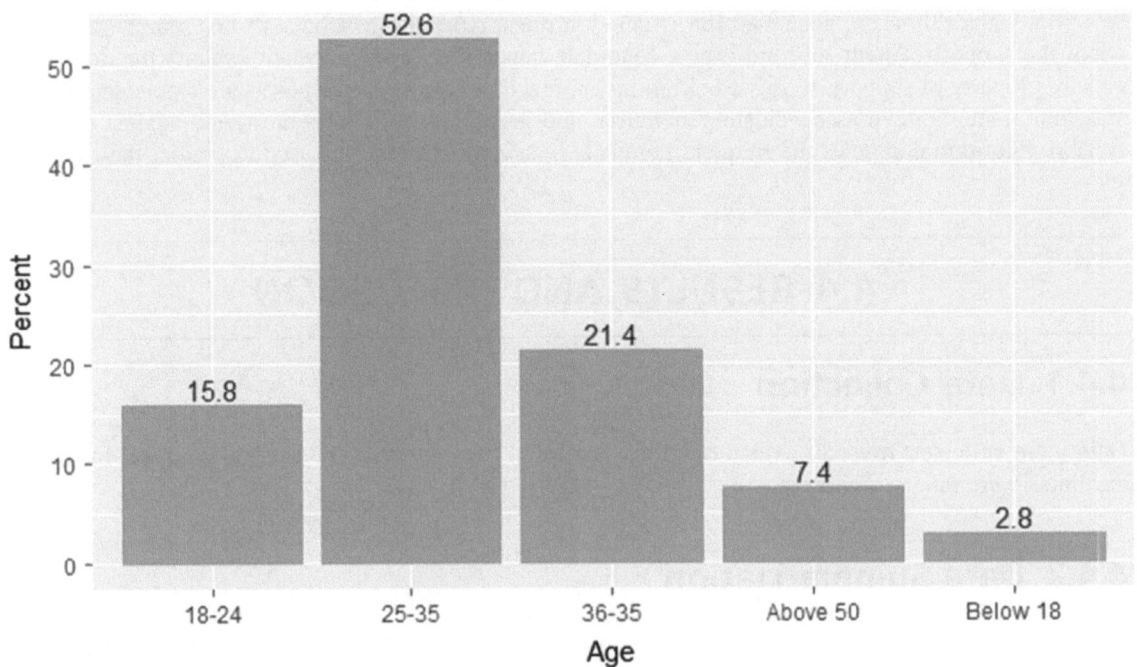

FIGURE 4.2   Age Distribution of the Respondents.

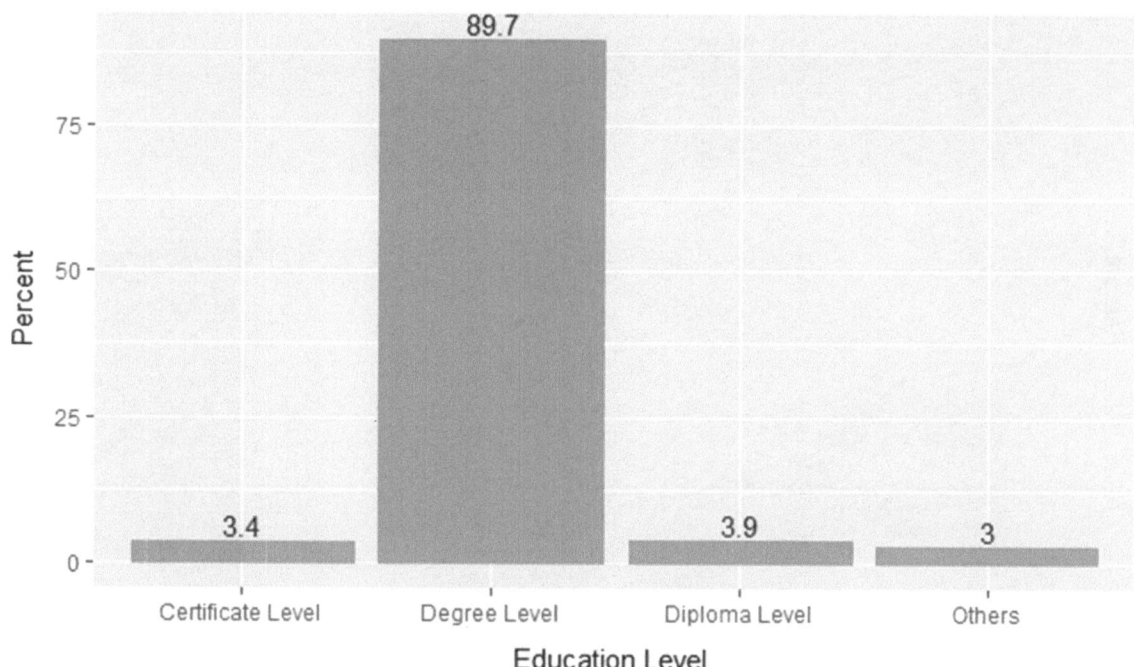

FIGURE 4.3   Higher Education Level Distribution of the Respondents.

4.4.2.5 Distribution of Field Which Related to Information Technology or Non-Information Technology

The majority of the respondents are Information Technology (IT)-relevant fields consisting of 605 respondents (60%) and 395 respondents are from non-Information Technology (Non-IT) (40%).

4.4.2.6 Distribution of Respondents' Awareness Regarding the Side Effects of Smartphone Usage

Respondents were asked about their knowledge regarding the side effect of smartphone usage. From the results (Figure 4.4), it is found that the majority of the respondents (49.8%) have a good understanding of smartphone usage, and 34% of respondents very good understanding of smartphone usage. Whereas 15.3% of respondents have a fair knowledge and 0.9% of the respondents have no knowledge about the side effects of smartphone usages.

4.4.2.7 Respondents' Opinion of Smartphone Use by Their Kids

Respondents' opinion regarding children using their smartphone: the majority of the respondents (62.9%) do not allow their children to use the smartphone while 37.1% of the respondents allow using a smartphone.

4.4.2.8 Time Period Taken for the above Decision

Respondents were asked to mark how long it takes to decide about giving or not a smartphone to their children and it is found that the majority of the respondents (45.3%) took more than 1 year and 35.1% of

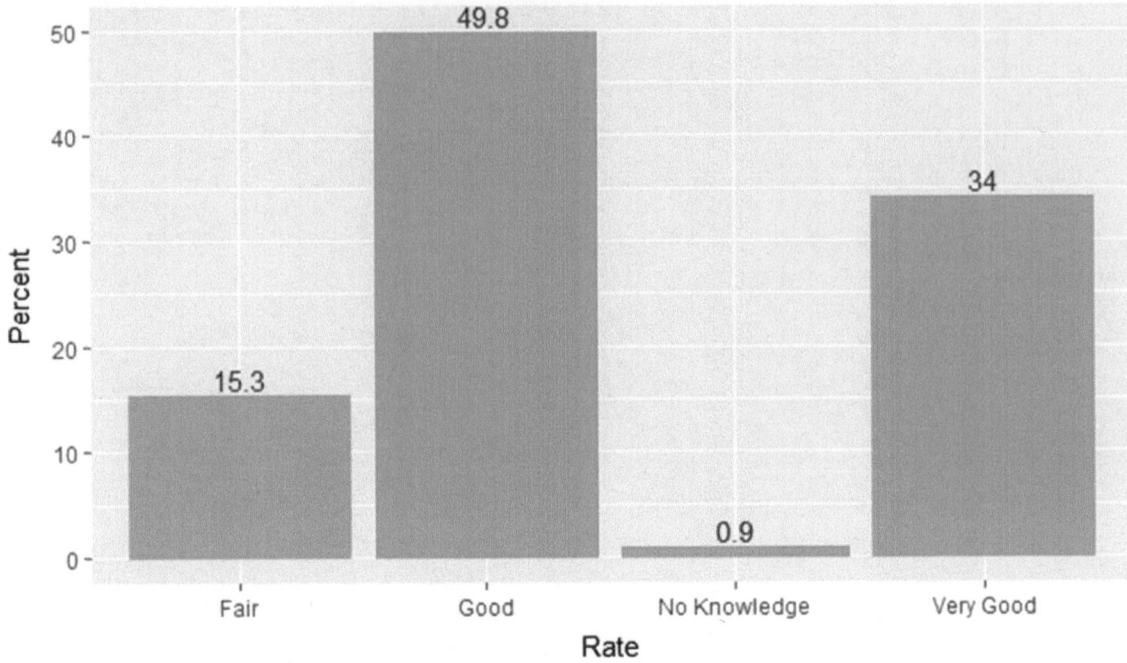

FIGURE 4.4   Awareness Rate about Side Effect of Smartphone Usage.

respondents took within 8 hours to decide. Whereas 10.4% of respondents decided before 6 months, 4% decided between 8 and 24 hours and 2.6% of the respondents decided before 1 month and 1 year (Figure 4.5).

4.4.2.9 Respondents' Opinion about Smartphone Technology Necessary to Children's Lives or Not

According to Figure 4.6, from the total of 1000 respondents, the majority of the respondents (40.6%) represent technology is important which meant they being neutral and followed by 32.9% of the respondents who chose technology is not important and 26.5% respondents choosing technology strongly important.

The respondents were asked to choose the correct age to give mobile phones to their children. Among most of the respondents who chose above 24 years old (46.2%), 45.6% respondents chose between 20 and 24 years, and 5.4% respondents chose between 15 and 19 years old. 0.8%, 0.6%, and 1.4% were represented by the sample who chose below 5, between 5 and 9, and between 10 and 14 years old, respectively (Figure 4.7).

4.4.2.10 Distribution of Respondents That Worry about the Society Underestimating When Their Child Doesn't Know How to Handle a Smartphone

A majority of respondents, 805 (80.5%) from the 1000 respondents, do not worry about the society's underestimation regarding children not handling smartphones, followed by 195 (19.5%) respondents who worry about society's underestimation regarding children not handling smartphones.

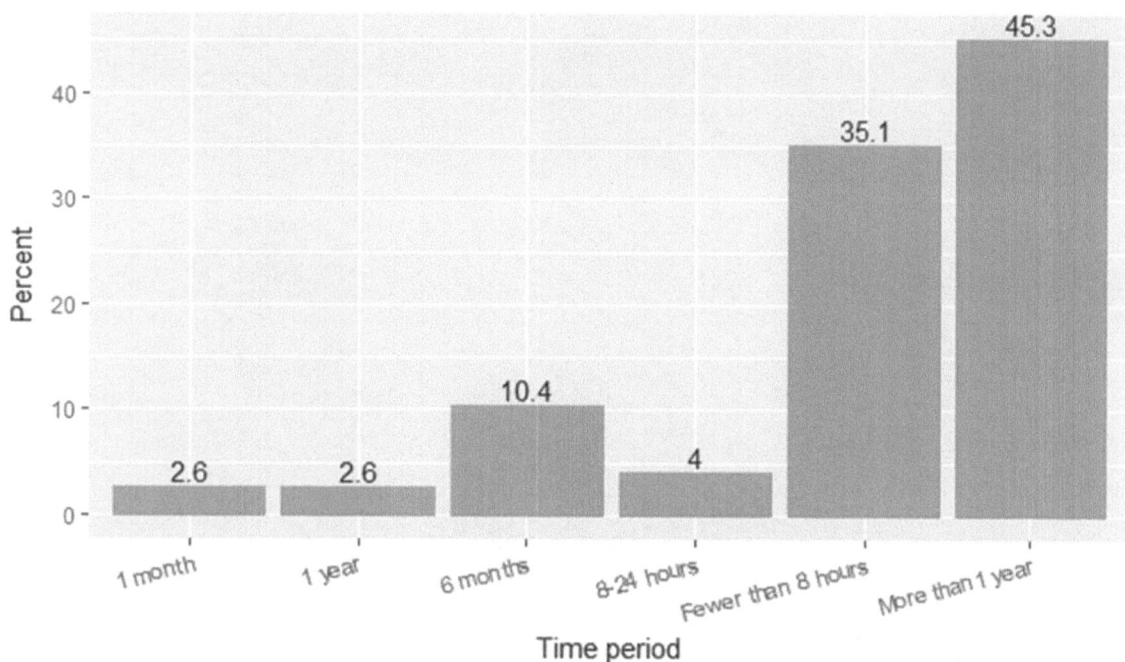

FIGURE 4.5 Time Period Taken for the Decision Made by Respondents.

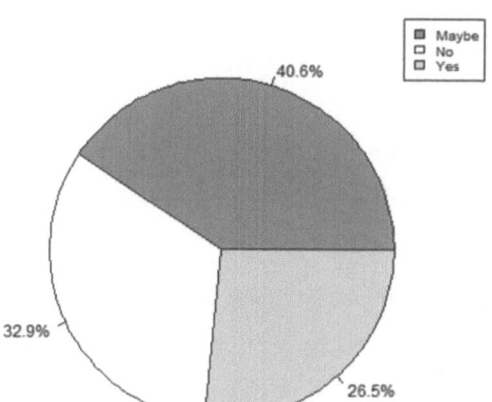

FIGURE 4.6 Opinion Regarding Smartphone Technology Important for Children or Not.

4.4.2.11 Respondents' Opinion on Monitoring Children's Mobile Activities

Figure 4.8 represents that most of the respondents preferred random monitoring which comprised of 59.20%, 592 respondents. And 294 (29.40%) respondents said they monitor mobile phone use while the remaining 114 (11.40%) respondents said they do not monitor.

4.4.2.12 Type of Mobile Activities and Monitoring Rate

To monitor mobile activity, respondents were asked to rate the children's mobile activities. A list of four types of mobile activities was given to the respondents. The list included phone call info, text message, website history, and photos taken. In the phone call category info, the highest rated the neutral (33.3%),

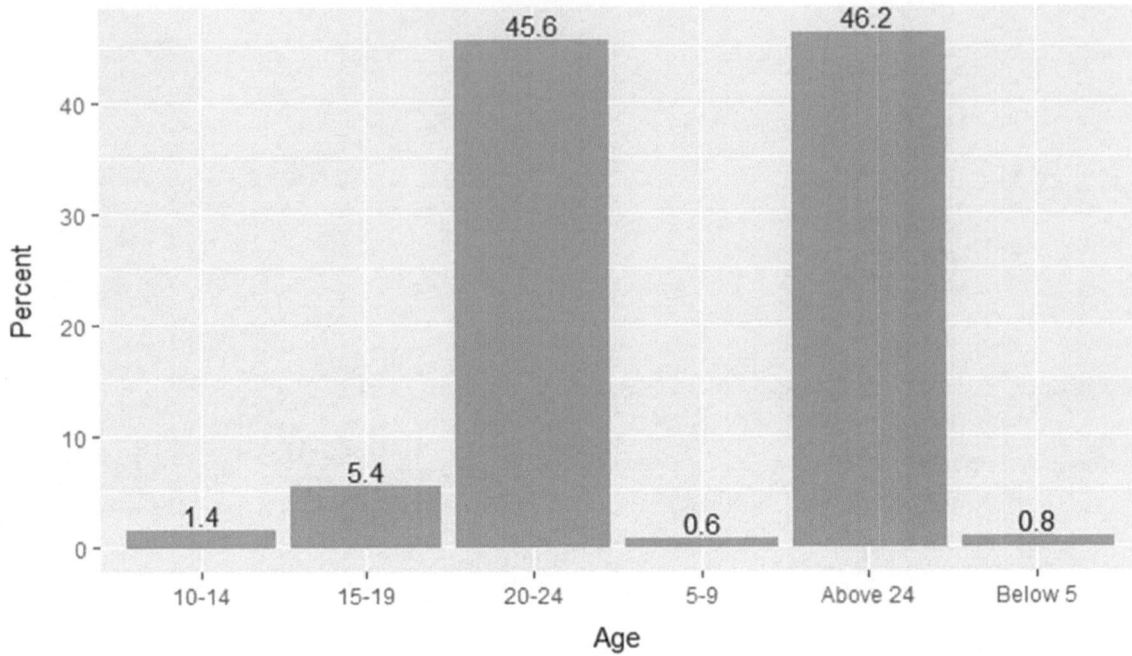

FIGURE 4.7   Correct Age to Give Mobile Phone to Children.

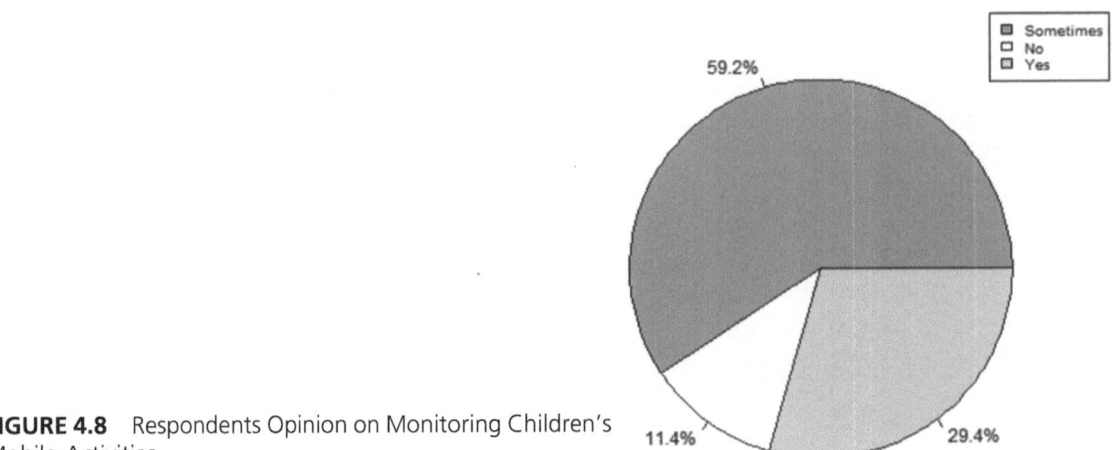

FIGURE 4.8   Respondents Opinion on Monitoring Children's Mobile Activities.

and the second-highest rate is agreed (31.1%). Strongly agree to monitor and disagree to monitor rate were taken same points which 15.9% and least rate points for strongly disagree to monitor children mobile activity (Figure 4.9).

In the text message category, most respondents (38.3%) rate neutral for monitor and 28.2% of respondents agree to monitor mobile activity. Where else 18% of respondents disagree to monitor, 9.5% strongly agree to monitor and 3.8% of the respondents strongly disagree to monitor children's mobile activities.

Most of the respondents strongly agree (31.9%) to monitor the website history of their children. Next, agree to monitor mobile activity which consists of 22.1% of respondents. And then it is

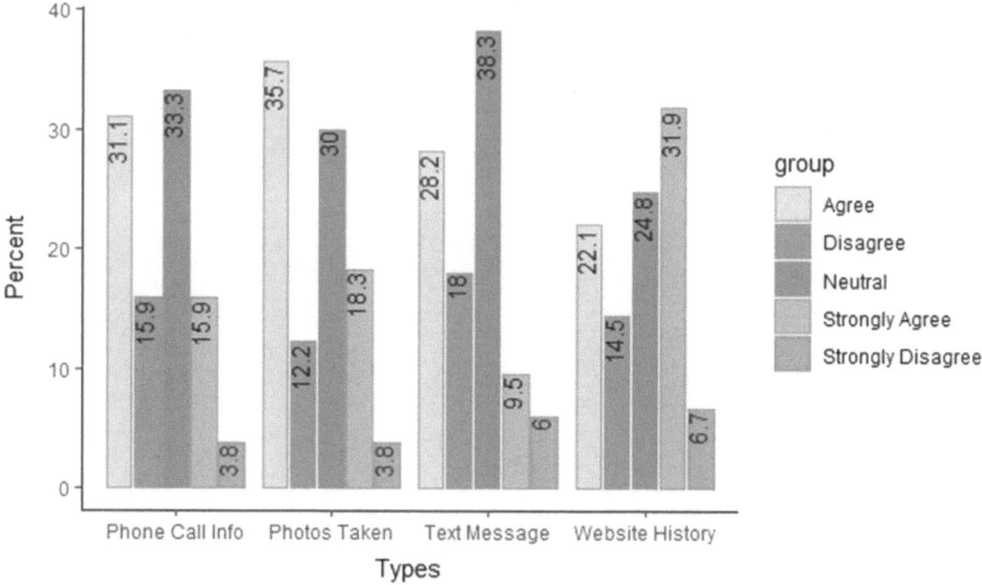

FIGURE 4.9 Mobile Activity Monitoring Rate.

followed by 14.5% disagreeing and finally, 6.7% of respondents strongly disagree to monitor children's mobile activity.

In the photos taken category, most respondents (35.7%) agreed to monitor, and 30% of respondents' rate neutral to monitor mobile activity. Whereas 18.3% of respondents strongly agree to monitor, 12.2% disagree to monitor, and 3.8% of the respondents strongly disagree to monitor children's mobile activities.

4.4.2.13 Respondents' Distribution in Preventing or Limiting Children's Smartphone Usage

According to Figure 4.10, prevention or limitation wise, 20.8% of the sample had selected time limitation, 20.3% had decided to spend more time with family, 18% of the sample had decided to bring them outside, and 15.7% had decided to set family rules. As follows 10.4%, 8.9%, and 5.9% were represented by the sample who had decided not right before sleep, Limit your child's cellphone use during certain times, for example, dinner, homework, and family and sleep time.

Figure 4.11 illustrated that most of the respondents chose to encourage to read books, which consists of 376 (37.6%) respondents. And the next majority of the respondents selected to involve children in useful and interesting hobbies which consists of 29%. And then it is followed by respondents who chose to show a better life without a smartphone which consists of 186 (18.65%) respondents and finally, 14.8% of respondents chose a child to involve in social activities.

4.4.2.14 Respondents' Opinion on Children Socializing More by Using Smartphones via Social Media

The majority of respondents said no to social media socialization via smartphone which consists of 766 (76.6%) from the 1000 respondents. And then it is followed by respondents who said yes to social media socialization via smartphones which consists of 234 (23.4%) respondents.

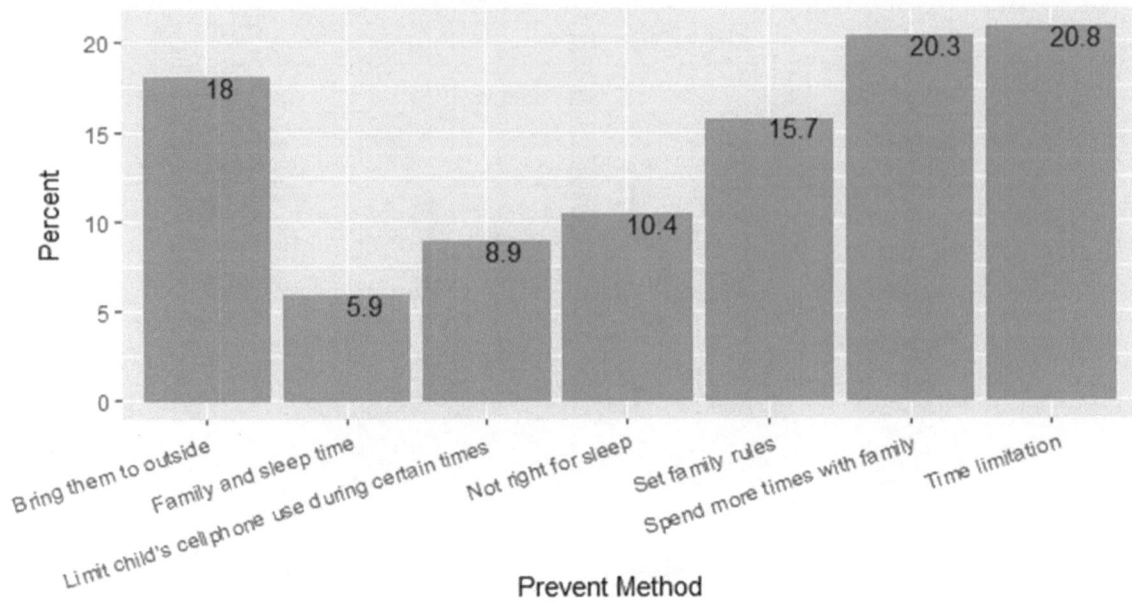

FIGURE 4.10 Respondents Distribution in Preventing or Limiting Children's Usage.

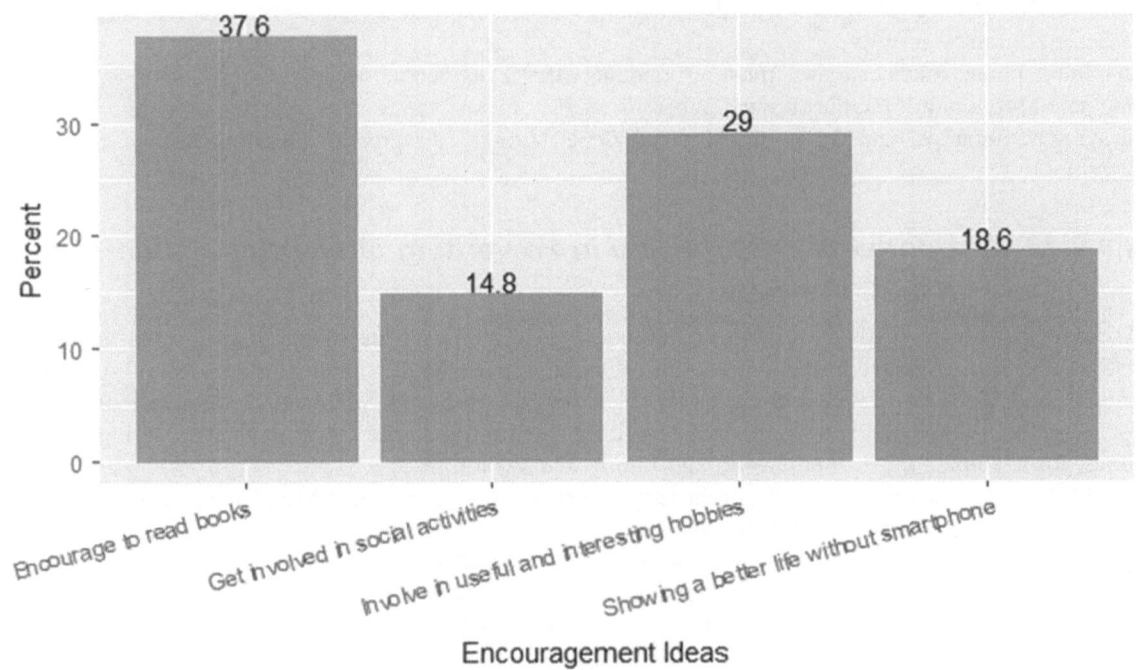

FIGURE 4.11 Respondents Opinion Regarding Encouraging Children to Live without Smartphone.

4.4.3 Evaluation

Association rule was used to test the data set and found the best rules and justification as follows:

Rule 1: Education Level = Degree level, what is the correct age to give mobile phones? (Your opinion) = Above 24, Do you allow your children to use your smartphones = No (399 respondents) confidence: (0.95)

Justification: When we considered the 399 respondents who obtained degree level education who disallowed smartphones until children are above 24 years and have 95% confidence level.

Rule 2: Civil-Status = Married, Do you allow your children to use your smartphone = No and Education Level = Degree level (355 respondents) confidence: (0.94)

Justification: From this rule, 355 respondents were married who disallowed to use smartphones and those holding a degree level education and have 94% confidence level.

Rule 3: Age Group = 25–35 and Education Level = Degree level (486 respondents) confidence: (0.92)

Justification: According to this rule, 486 respondents aged between 25 and 35 and having a degree in higher education and have 92% confidence level.

Rule 4: Are you related to IT (Information Technology) field? = Yes and Education Level = Degree level 549 confidence: (0.91)

Justification: When we considered the 539 respondents belonging to Information Technology-related field and having a degree and have 91% confidence level.

4.4.4 Classification

Decision trees are powerful classification algorithms that are becoming increasingly more popular with the growth of data mining in the field of information (Logasakthi & Rajagopal, 2013). As this is decision tree approach it is very much useful in predicting the values. Ten-fold cross-validation was used when applying classification. The J48 algorithm is used to implement decision tree approach (Figure 4.12).

4.5 CONCLUSION

The main purpose of the present study is to identify the factors of how smartphones influence generation Z by getting the opinion from the current and future parents. A total of 1000 data were obtained and 16 attributes were taken for analysis. According to association rule results, most of the respondents obtained degree level education who disallowed the use of smartphones until the child becomes 24 years old. The filter method was used to select attributes in classification. The info gain attribute evaluator was chosen as the best default attribute. Such a technique evaluates every single feature in the attribute. Finally the 10 best attributes (Gender, Age Group, Civil-Status, Education Level, Related field, Scale your knowledge about a side effect of smartphone usage, Allow/deny to use your smartphone to your Children, Taking this decision according to the knowledge you gained from your education level [Time take for your decision], Smartphone Technology is an Important Part of your Children's lives, The correct age to give mobile phone) were selected out of the 20. Through the association rules, most of the respondents are degree holders and disallowed children to use mobile until they grew up to above 20 years old. Information Technology-related field degree holders dislike providing smartphones to children and on the other hand, married degree holders aged 25–35 allow a smartphone to their children when they grow to above 20 years. Therefore, the association rule was used to test the data set and classification was used to validate it.

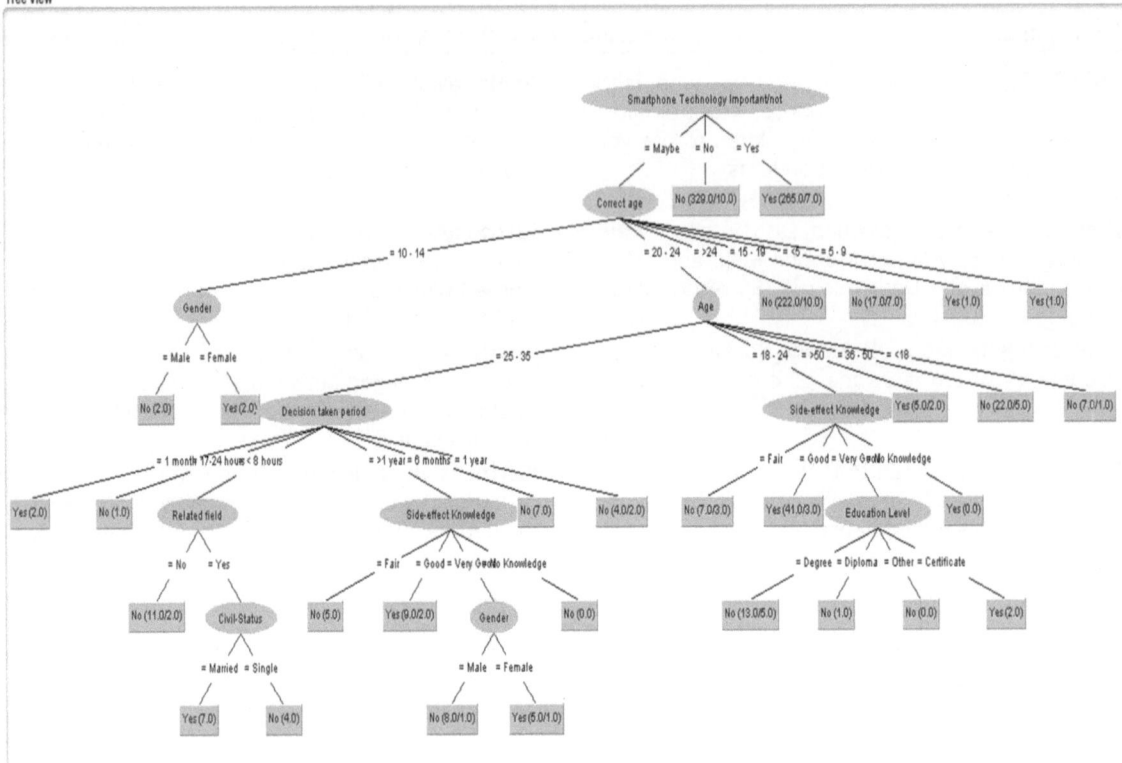

FIGURE 4.12   Decision Tree of the Classification in Weka Data Mining Tool.

REFERENCES

Child in the City. (2020). *One in four children and young people show signs of smartphone addiction* [online]. https://www.childinthecity.org/2019/12/02/one-in-four-children-and-young-people-show-signs-of-smart-phone-addiction

Dunham, M. H. (2003). *Data mining introductory and advanced topics*. Pearson Education, Inc.

Fayyad, U., Piatetsky-Shapiro, G., & Smyth, P. (1996). The KDD process for extracting useful knowledge from volumes of data. *Communications of the ACM, 39*, 27–34. 10.1145/240455.240464

Goldberg, M., Gorn, G., Peraccio, L., & Bamossy, G. (2003). Understanding materialism among youth. *Journal of Consumer Psychology, 13*(3), 278–288.

Han, J., & Kamber, M. (2006). *Data mining: Concepts and techniques*. Morgan Kaufmann publishers, 47–97.

Hand, D., Mannila, H., & Smyth, P. (2012). *Principles of data mining*. New Delhi: PHI Learning Private Limited.

John, J. (2013). *Positive impact of smartphones on social society*. https://www.trffcmedia.com/topics/positive-impacts-of-smartphones-on-social-society/

Kaur, P., Singh, M., & Josan, G. S. (2015). Classification and prediction based data mining algorithms to predict slow learners in education sector. *Procedia Computer Science, 57*, 500–508. Elsevier Masson SAS. 10.1016/j.procs.2015.07.372

Koh, H. C., & Tan, G. (2005). Data mining applications in healthcare. *Journal of Healthcare Information Management, 19*(2), 64–72.

Kotsiantis, S. B., Kanellopoulos, D., & Pintelas, P. E. (2006). Data preprocessing for supervised learning. *International Journal of Computer Science*, *1*(1), 111–116.

Kukulska-Hulme, A. (2010). Learning cultures on the move: Where are we heading? *Educational Technology and Society*, *13*(4), 4–14.

Ling, R. (2007). Children, youth and mobile communication. *Journal of Children and Media*, *1*(1), 60–67. 10.1080/1 7482790601005173

Logasakthi, K., & Rajagopal, K. (2013). A study on employee health, safety and welfare measures of chemical industry in the view of Salem region. *International Journal of Research in Business Management*, *1*(1), 1–10.

Mayo, E. (2005). *Shopping generation*. National Consumer Council, PD 43/05, London.

Milovic, B. (2012). Prediction and decision making in health care using data mining. *International Journal of Public Health Science (IJPHS)*, *1*(2), 69–78.

Mount, R. (2012). *Advantages and disadvantages of smartphone technology*. Retrieved from https://www.mobilecon2012.com/8-advantages-and-disadvantages-of-smartphonetechnology/

Netsafe. (2005, January). *The text generation: Mobile phones and New Zealand youth: A report of result from the Internet Safety Group's survey of teenage mobile phone use*. Netsafe.

Ozcan, A. (2014). Mobile phones democratize and cultivate next-generation imaging, diagnostics and measurement tools. *Lab on a Chip*, *14*, 3187–3194. 10.1039/c4lc00010b

Rather, M., & RAther, S. (2019). Impact of smartphones on young generation. *Library Philosophy and Practice*.

Rideout, V. J. (2013). *Zero to eight: Children's media use in America 2013*. Common Sense Media.

Web Design Company. (2020). *Impact of smartphones on society - Use of mobile phones - Key ideas* [online]. https://www.keyideasinfotech.com/blog/impact-of-smartphone-on-society/#:~:text=It%20is%20true%20that%20the,and%20other%20aspects%20of%20life.&text=The%20prominent%20areas%2C%20where%20impacts,cultural%20norms%20and%20individual%20behaviors [Accessed September 1, 2020].

Witten, I. H., Frank, E., Trigg, L. E., Hall, M. A., Holmes, G., & Cunningham, S. J. (1999). Weka: Practical machine learning tools and techniques with Java implementations. *Computer Science Working Papers*. https://hdl.handle.net/10289/1040 [Accessed September 1, 2020].

Zysman, G. I., Tarallo, J. A., Howard, R. E., Freidenfelds, J., Valenzuela, R. A., & Mankiewich, P. M. (2000). Technology evolution for mobile and personal communications. *Bell Labs Technical Journal*, *5*, 107–129. 10.1002/bltj.2210

Explainable Artificial Intelligence: Guardian for Cancer Care

5

Vishal Ahuja

Himachal Pradesh University, Himachal Pradesh, India

Contents

5.1 INTRODUCTION

Rapidly growing population, urbanization and industrialization are the utmost requirements of 'smart cities', having adequate opportunities for food, livelihood, healthcare, shelter, employment, security, and transportation. However, manual handling and interpretation of massive data from each sector are almost impossible and time-consuming. Deployment of artificial intelligence concept (XAI), upgrade the performance of systems, and contribute for the automation of resources. Figure 5.1 shows the dominance of artificial intelligence in various sectors. Technical innovation and ICTs create an amiable market for high-speed internet, and data cloud servers, which can be integrated and used at multiple sites together for the formulation of policies and decision making by XAI based on the previous trends. It has gripped itself to the healthcare sector so fast and become reliable for doctors and researchers. Traditional healthcare relies on the physical examination of the patient/subject every time without going to molecular diagnostics as it proves costly. Physicians use previous training and experience for formulating the treatment strategy but personal observation and responses may become slower that affect the clinical

DOI: 10.1201/9781003172772-5

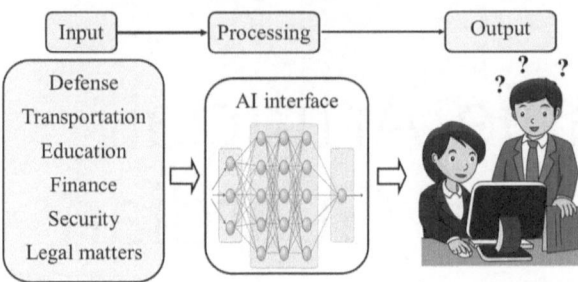

FIGURE 5.1   Explainable Artificial Intelligence.

diagnosis. XAI algorithms are quite friendly for disease diagnosis, treatment, and risk assessment (Cook et al., 2018; Golubchikov and Thornbush, 2020; Jo et al., 2021; Pesquita, 2021).

The growing population also gives birth to pollution and poses new challenges in front of healthcare practitioners with new ailments and health complications. Smoking habits and an unhealthy lifestyle become fuel for cancer, obesity, diabetes, and other metabolic disorders. Oncogenesis/cancer is the process of achieving stemness by the differentiated cells due to mutation or some other factors. The affected cells may remain localized or may relocate themselves to infect other tissues as well (Siegel et al., 2019). Identification of underlying genetic factor/s, changes associated with cancer development and progression is one of the biggest breakthroughs for effective treatment. The task becomes easier after the beginning of cancer genome sequencing and establishment of associated databases which aid in tracking the possible mutations in various human cancers (Bray et al., 2018; Chaturvedi et al., 2019; Kudo et al., 2014; Miller et al., 2018; Sayiner et al., 2019).

The second important aspect of cancer treatment is to identify the role of mutation in cancer development and progression as all the mutations do not equally lead to cancer. Only a few rare/lethal mutations are cancer drivers (Siegel et al., 2019; Torres et al., 2017), which are possible targets for drug discovery and treatment (Rao et al., 2017; Vandenbulcke et al., 2016). The manual analysis of target mutation, establishing the affected pathway and linkage network, is not possible. Methods developed to date for the purpose have been discriminated into three major categories: (i) frequency-based approaches, (ii) structure-based methods, and (iii) statistical and machine learning methods.

Frequency-based approaches identify the most common mutation in the majority of cancer patients as a causative factor (Juengpanich et al., 2020). *Structure-based methods* rely on the impact of mutations and three-dimensional structural change in mutated residues (Coudray et al., 2018; Darcy et al., 2016; Li et al., 2017; H. Lin et al., 2019). The third approach, *statistical and machine learning methods,* is the most recent and dynamic approach. It is dynamic and pragmatic enough to tackle the large and diverse data associated with oncology (Agarwal et al., 2017; Bera et al., 2019; Ehteshami Bejnordi et al., 2018; Kather et al., 2019; Liang et al., 2020; H. Lin et al., 2019; Skrede et al., 2020; Zaman et al., 2017).

5.2 EXPLAINABLE ARTIFICIAL INTELLIGENCE

5.2.1 Evolution of Explainable Artificial Intelligence

Statistical and machine learning approach is good enough for interpretation but insufficient in healthcare data collection as it represents wide subsets of data in each case namely, breast cancer case study includes morphological changes (appearances, size, shape, colour, pigmentation, etc.), physiological changes (pain, secretions), cellular and genetic changes (gene suppression, overexpression, regulating factors, etc.), and so on. The collection of data followed by its interpretation and then training of models for its future applications are challenging tasks. Technological advancements incorporate various modules and IT resources to build a new system that mimics the human brain that can deal with challenges and get updated as per new conditions. After the evolution of 'Artificial intelligence' around the 1950s, with a publication of Alan Turing's 'Computing machinery and intelligence,' a lot of advancements have been adopted in form of subsets and algorithms. Followed by the origin of 'Explainable AI' where decisions can be understood and justified with concepts and trends unlike the black box of AI (do Nascimento Souza, 2021). XAI is formulated with algorithms, with text and visual objects. Similar to the human brain, artificial intelligence also incorporates the IT resources/advanced tools for direct data acquisition, collection, interpretation, and representation as shown in Figure 5.2. Different subsets of 'XAI' are summarized below:

a. *Speech Recognition*: The speech recognition module of AI acts for the conversion of text to speech and vice versa. The speech recognition system is either speaker-dependent or -independent. It is a sort of data collection and representation system.

b. *Robotics*: Robotics is supposed to be the physical response module of AI that acts based on the program feed in the memory.

c. *Vision*: Vision is an application that captures, recognizes the object, analyzes, and communicates visually through video cameras, analogue-to-digital conversations, and digital signal processing. It is analogous to speech recognition and acts for collection and data presentation.

d. *Expert system*: Expert system is an artificially trained decision-making system that responds as per the conditions provided, and responses are suggested as per the trained modules. Figure 5.3 (a) shows the channelizing of information from one end to another through interpretation passage.

e. *Natural Language Processing*: Natural language processing (NLP) module allows the conversion of human language to machine instruction so that machine can function accordingly.

f. *Deep Learning*: It prepares algorithms, creates layers from historical data, and uses them to make intelligent decisions on its own.

g. *Machine Learning*: Machine learning, one of the subsets of artificial intelligence offers a wide variety of algorithms to handle a huge amount of data and facilitates decision-making models. Machine learning algorithms also trained themselves from the historical data to make decisions based on the given previous information/cases.

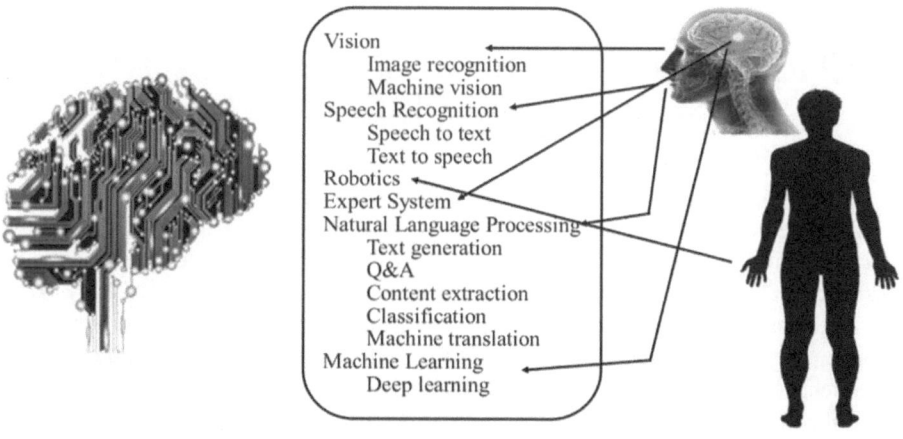

FIGURE 5.2 Subsets of Explainable Artificial Intelligence.

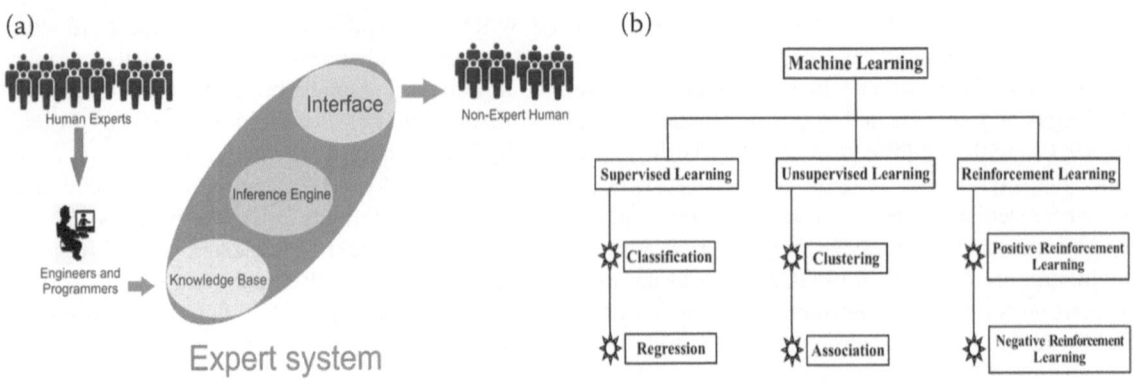

FIGURE 5.3   (a) Components of Expert System and (b) Machine Learning.

5.2.2 Artificial Intelligence: Mechanism

'AI' module gathers data through different IT resources like human data collected by eyes and ear, translates the collected data using the language processing unit, interprets and analyzes it by machine learning, deep learning, and expert system. Final step of response generation is done by robotics, vision and speech module. The expert system generates the foundation onto which deep learning and machine learning design algorithms to analyze, and trained models for unknown conditions and challenges like drug designing, target prediction, docking analysis, and drug repurposing, etc. Figure 5.3 (b) shows different types of learning methods applied by machine learning for data analysis and interpretation. The selection of learning methods depends upon the type of available data and situations (Linardatos et al., 2021).

Based on the available and past cancer cases, morphological, physiological, and genetic data, algorithms can differentiate between normal vs. oncogenic, stages of oncogenesis, benign vs. metastasis, etc., and further can be used with unknown samples without any interference and with high precision. The model can also predict the gene/s, changes associated with mutations and phenotypes under varying circumstances, and the prediction is as accurate as by a well-trained physician or cancer specialist (Zaman et al., 2017).

5.3 ARTIFICIAL INTELLIGENCE IN CANCER TREATMENT AND DRUG DISCOVERY

Initially, the computational approach is simply based on a 'Yes' or 'No' judgment, which is not feasible in each case like diseases, symptoms, mutations, and gene functions (Dietz and Pantanowitz, 2019). Advanced machine learning algorithms predict and identify cancer-associated mutations with improved accuracy (Maddox et al., 2019; Tizhoosh and Pantanowitz, 2018; Zaman et al., 2017). The basic mechanism for model training is to discriminate the abnormal and rare changes in structure, function, expression, or phenotype accurately and establishes the relation between changes and genes. Gene, RNA, and protein sequence, common mutations, associated changes, etc., can be retrieved from different databases and utilized for model construction. In the upcoming section, the contribution of AI and its subtypes have been discussed in cancer diagnosis, drug discovery, and repurposing.

5.3.1 Drug Discovery

Currently, the availability of databases has reduced the research burden up to some extent. Still, the analysis and interpretation of vast genomic data is another big challenge in itself, which is necessary for drug development. Incomplete and error-prone analysis make conventional drug development systems prone to failure. Especially when it comes to a novel drug molecule, the development is a cost-intensive and time-consuming process. Figure 5.4 showed the four important and mandatory phases that every drug molecule has to pass to warrant the formulation of an effective drug, correct dosage size, subjects' safety.

Among them, Phase 0 is the preclinical stage, and the next three phases (I–III) are clinical trials. The whole process takes approximately 5–10 years but may extend up to 15 years (Eliopoulos et al., 2008; Schuhmacher et al., 2016; Xue et al., 2018). The application of human trials data further increased the cost of investment (Khanna, 2012). During 1998–2008, more than 54% of drugs failed in clinical trials mainly due to lack of efficacy (57%), followed by safety concerns and side effects (17%), which also remain prominent during the next decades as well (Réda et al., 2020; Schuhmacher et al., 2016). As per an estimate, the average cost of drug development was $487 million for the approved drug, which increased in 2018 to $2,168 million (Réda et al., 2020). Therefore an alternative procedure is necessary to reduce the time and cost along with the increase in success rate (Eliopoulos et al., 2008; Fogel, 2018). In oncogenesis, the prediction of a target is much more tedious as even a single type of cancer may be associated with multiple factors and genetic mutations. As soon as the underlying mechanisms and molecular basis of oncology evolution and development are explored, drug discovery becomes easier for researchers. The formulation of precise medicine increased the survival rate in cancer patients. In the last few years, oncology has been recorded the highest number of approved drugs (Santos et al., 2017; Workman and Al-Lazikani, 2013). Another advantage of molecular evaluation of diseases is the identification of molecular markers, early identification of disease onset, and direct targetting the root cause of oncogenesis. With the advancement of technology, massive data has been generated from the molecular study of life in humans with respect to genomics, proteomics, metabolism, and oncogenesis, which is stored in respective databases. Table 5.1 summarizes some of the databases and associated information.

However, the challenge is still the same, the interpretation of the massive data for diagnosis, target identification, and drug development. In the 1950s, AI has been launched in various sectors which analyze approximately any size of data by various algorithms like ML and DL and generate results as per

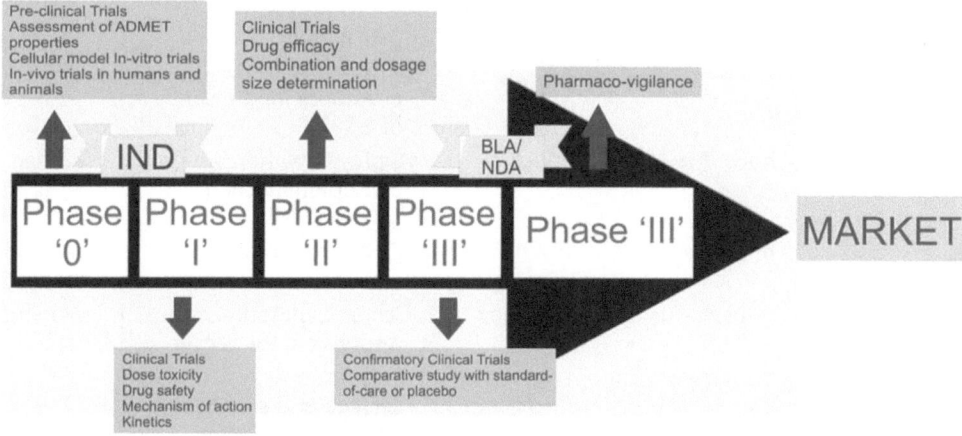

FIGURE 5.4 Phases in Drug Development (IND: Investigational New Drug; BLA: Biologics License Application; NDA: New Drug Application).

TABLE 5.1   Databases for sequence retrieval and drug discovery

DATABASE	WEB ADDRESS/URL	DESCRIPTION	REFERENCES
OpenTargets	https://www.opentargets.org/	Genetically validated targets for drugs	Carvalho-Silva et al. (2019)
Comparative Toxicogenomics Database (CTD)	https://ctdbase.org/	Relation between environment and human health, chemical–gene, chemical–disease gene-disease, chemical-protein	Davis et al. (2019)
COSMIC	https://cancer.sanger.ac.uk/cosmic	Somatic mutations leading cancer	Forbes et al. (2011)
DisGeNET	http://www.disgenet.org/	Human gene-disease relationship	Piñero et al. (2017)
STRING database	https://string-db.org/	Protein-protein interactions (physical and functional)	Szklarczyk et al. (2019)
Kyoto Encyclopedia of Genes and Genomes (KEGG Pathway)	https://www.genome.jp/kegg/pathway.html	Database for genome, proteome, biological pathways, disease, and impact of chemicals and drugs	Kanehisa et al. (2017)
BioModels	https://www.ebi.ac.uk/biomodels/	Mathematical models of biological and biomedical systems	Chelliah et al. (2015)
Connectivity Map (CMap)	https://portals.broadinstitute.org/cmap/	Deviation of cellular system in human disease	Lamb (2006)
LINCS	https://clue.io/lincs The Library of Integrated Network-Based Cellular Signatures	Change in gene expression, cellular physiology after the exposure of the cell to a variety of perturbing agents	Subramanian et al. (2017)
Therapeutic Target Database (TTD)	http://bidd.nus.edu.sg/group/cjttd/	Nucleic acid targets, targeted disease, disease-associated pathway, therapeutic protein	Li et al. (2018)
RepoDB	http://apps.chiragjpgroup.org/repoDB/	Visualization and data querying regarding the clinical trials	Brown and Patel (2017)
ClinicalTrials.gov	https://clinicaltrials.gov	Clinical trial setup, procedure, and results of US	Réda et al. (2020)
Drug Bank	https://www.drugbank.ca/	Approved/withdrawn drugs, and potential chemical compounds	Wishart et al. (2018)
ChEMBL	https://www.ebi.ac.uk/chembl/	ADMET properties	Gaulton et al. (2017)
The Cancer Genome Atlas Program (TCGA)	https://portal.gdc.cancer.gov	Genomic mutations responsible for cancer	Workman et al. (2019)
Genomics of Drug Sensitivity in Cancer	https://www.cancerrxgene.org	Drug responses on cancer cell line and sensitivity of genomic biomarkers	
BindingDb	https://www.bindingdb.org	Interaction between a protein with drug-like molecules, ligands, and their binding affinities	

(Continued)

TABLE 5.1 *(Continued)*   Databases for sequence retrieval and drug discovery

DATABASE	WEB ADDRESS/URL	DESCRIPTION	REFERENCES
canSAR	https://cansarblack.icr.ac.uk	Complete database from various streams like biology, chemistry, pharmacology, structural biology, cellular networks, and clinical annotations for drug-discovery	
Chemical Probes Portal	http://www.chemicalprobes.org	Identify and report potential chemical probes for biological research and drug discovery	
Probes & Drugs	https://www.probes-drugs.org	Libraries of probes, drugs, and inhibitor	

requirement. In the beginning, ML was restricted to data analysis, which was referred to as 'Passive Machine Learning'. Still with time, it also upgraded and now it can also be used for predictions based on historical data with a high accuracy rate. This updated version of ML is referred to as 'Active Machine Learning'. Figure 5.5 depicts the layers on 'AI', worked on algorithms, forming artificial neural networks which analyze genomic and clinical data and extracting the hidden information or pattern carrying key critical importance.

The extracted information is presented in a simple and understandable form for its efficient utilization. For the prioritization of bio-active molecules, 'Quantitative Structure-Activity Relationships' (QSAR) was introduced in the 1960s with a supervised learning approach (Sellwood et al., 2018). Later it was also realized that the majority of drugs failed in clinical trials due to unaddressed 'Absorption, Distribution, Metabolism, Excretion, and Toxicity'. In the 1990s, another application/algorithm, 'ADMET modelling', was introduced to determine the physicochemical properties of drug molecules (van de Waterbeemd and Gifford, 2003; Ye et al., 2019). With time, new advanced models evolved and exploited to harness the big data for drug discovery.

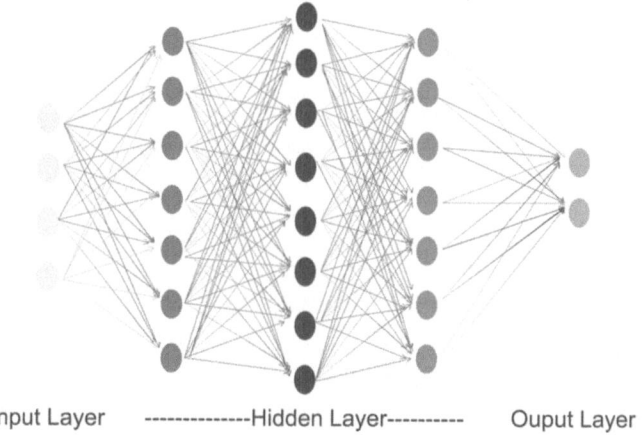

Input Layer   --------------Hidden Layer----------   Ouput Layer

FIGURE 5.5   A Diagrammatic Illustration of the ANN Structure and Interaction between the Node.

5.3.2 Prediction of Oncogenesis

Drug before approval analyzed for its safety and short-term as well as long-term side effects. However, there is still a possibility to have differential side effects in different physiological conditions. For example, special attention is required while prescribing medicine to pregnant subjects due to teratogenic effects and probable abnormalities in the fetus. Challa and colleagues have received more than 9000 drug molecules from the drug bank and screened for teratogenic side effects. Drug-repository offers detail about the physical, chemical, and bio-active characteristics of each entry. Unsupervised and supervised models were developed considering three important characteristics of a drug molecule: chemical fingerprint, meta-structural features, and side effects as shown in Figure 5.6. Accordingly, pattern and cluster relationship were established to identify the fingerprint features of a drug that determine its teratogenicity effect. Structural and functional patterning revealed chemical functionalities with teratogenic and protective effects (Challa et al., 2020).

On the genetic/molecular level, the progressive accumulation of mutations is the major source of cancer initiation. But still, all the mutations do not equally contribute to cancer, as only a few of them are rare and disease drivers. The human genome contains around 518 protein kinase genes that are collectively known as kinome. Multiple classifier systems were used to segregate the mutations in positive and negative classes. Positive and negative mutations were retrieved from the 'Catalogue of Somatic Mutations in Cancer' (COSMIC) database and SNP@Domain database. Also, protein characteristics concerning point mutations were determined from various sources, including KinBas, Uniprot, EMBOSS, etc. Eleven machine learning methods, including J48 (Tree), Random Forest, NB Tree, Functional Tree, Decision Table, DTNB, LWL (J48+KNN), Bayes Net, Naive Bayes, SVM, and Neural Network, were applied for the development of classifier system to identify EGFR mutations. For the first time, the connection between EGFR mutations T725M and L861R was established with cancers (U et al., 2014).

In cancer, normal differentiated cells lose their phenotypic differentiation and regain stem-cell-like features due to mutations. 'One-class logistic regression' (OCLR) ML algorithm was prepared from transcriptomic and epigenetic features of non-transformed pluripotent stem cells and their differentiated progeny. The OCLR model can identify underlying undiscovered biological mechanisms of dedifferentiated oncogenic states. For the efficient functioning of the OCLR algorithm, it is necessary to extract the unique morphological and physiological features of stem cells. The algorithm was trained on embryonic stem cell, induced pluripotent stem cell, and their differentiated progenitors (ecto-, meso-, and

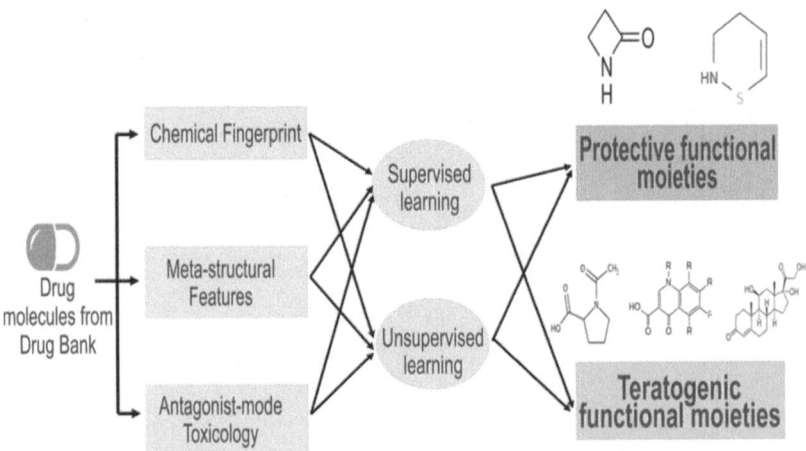

FIGURE 5.6   Screening of Pre-Approved Drugs for Teratogenic Effect.

endoderm) to build stemness indices which aided in differentiating between normal, differentiated, and stem cells.

The prepared stemness indices were precisely trained with the epigenetic and expression-based pattern, tumour, and immuno-microenvironment characteristics to identify dedifferentiation stemness and identification of drug targets. Cancer stemness was found to be correlated with immune checkpoint expression and escape immune cells. It was also revealed that the dedifferentiated oncogenic phenotype is associated with metastatic tumours. Developed indices and algorithms improved the identification of novel targets, therapies, and respective targets to prevent tumour progression (Malta et al., 2018).

Hepatocellular carcinoma (HCC) is the most common subtype of liver cancer and the fourth leading cause of cancer-related mortality which causes more than one million casualties worldwide annually (Bray et al., 2018; Malta et al., 2018; Siegel et al., 2019). An experienced pathologist can identify HCC by careful visual inspection besides tissue-based examinations. However, it is still a lengthy and cost-intensive process to identify the mutations associated with HCC and relate them with visual histological changes. Histological images and mutations linked to HCC were retrieved from GDC-portal (https:// portal.gdc.cancer.gov/) and compared with normal hepatic histological features. Normal tissues were collected from Sir Run-Run Shaw Hospital. Based on the analysis, the 'tiles dataset' was prepared for liver cancer. Each image was split into tiles of 256×256 pixels, and the collected data set was segregated into training, testing, internal validation, and external validation sets. Matthews correlation coefficient for the prepared model was close to the diagnosis by a 5-year experienced pathologist. Moreover, the model was up to 96% accurate in identifying the malignancy of the tumour and up to 89.6% accurate in determining the tumour differentiation stage. The model was also trained to identify the mutation in the most common gene, including CTNNB1, FMN2, TP53, and ZFX4 from histopathology images with external AUCs from 0.71 to 0.89 (Chen et al., 2020). The findings from previous literature suggested the possible application of artificial intelligence, neural networks, and machine learning in determining the possible mutations, drug targets, and tissue differentiation. However, for the efficient functioning of any model, identification of unique changes is necessary as to which model is going to be trained.

5.3.3 Drug Repurposing

Investment cost and time are the two major factors that remain crucial for the screening of novel drug molecules. As discussed earlier, the cost associated with a single drug molecule to enter into the market is about 700–800 $ and takes around 5–7 years. Besides, a low success rate is also a major hurdle for researchers and investors. Drug repurposing offers little advantage over working for a new drug molecule. It refers to finding new targets for a drug molecule that has already been approved for its safety and efficacy. As a disease may affect multiple pathways, a bioactive molecule may work on symptoms or root cause. Drug repurposing seems a realistic and attractive way for drug development. It can bring new drugs to light with reduced risks, cost, and time and also with a high success rate. Earlier, drug discovery and repurposing were majorly accidental and serendipitous due to a lack of knowledge and proper resources. Slowly, the accidental process is replaced with an experimental approach and then a computation approach involving in-silico analysis and artificial intelligence.

Recently, deep learning algorithms were used for novel drug repurposing with the help of transcriptomic data and chemical structures. Biochemical data of seventy-five drugs of six classes, vasodilator agents, anti-dyskinesia agents, anticonvulsants, hypolipidemic agents, anti-asthmatic agent, and anti-neoplastic agents data were retrieved from the database. Drug data includes gene expression, pathway activation, and affected pathways. All drugs were assigned scores based on the chemical structural similarities and pathway activation. Drugs having the closest relation to cancer and associated pathway were selected. Among 75 retrieved drugs, Pimozide was identified as a potential candidate for drug purposing. Usually, Pimozide is an anti-dyskinesia agent for the treatment of Tourette's Disorder, which was also identified as a strong anti-neoplastic potential against non-small cell lung cancer. The proposed results were also validated successfully against A549 cell lines (Li et al., 2020). In cancer, symptoms and cure procedures change with the type of cancer and affected genes. This gave rise to a new challenge to develop personalized diagnostics and drugs for each tumour. Nilotinib, Venetoclax, and Abemaciclib were screened for their efficacy in Colon Adenocarcinoma (COAD). In humans, a group of nine low molecular weight proteins 'FABP1-9' facilitates the transfer of long-chain bioactive fatty acids and is responsible for the development of cancer cells. Among these fatty acids, FABP6 is overexpressed in COAD. It is known that FABP6 belongs to a group of low molecular-weight proteins related to the transport of long-chain bioactive fatty acids in cells. In humans, there are nine different subgroups (FABP1-9). This group of proteins play a role in the development of different types of cancer cells (Jing et al., 2000; Nieman et al., 2011; Shen et al., 2016; Zhang et al., 2019). From the previous researches, three signatures with significant prognostic value have been identified and well documented in Colon Adenocarcinoma (COAD) (D. Sun et al., 2018; Wen et al., 2018; Xu et al., 2017). RNASeq2 data and differential expression for the COAD cohort were retrieved from the TCGA repository and edgeR package. For better understanding, samples were segregated into normal vs patients and then as per different stages of cancer. The ML approach identifies a new meta-signature for COAD by random forest (RF) and generalized linear model (glmnet) algorithm. Newly identified targets were docked with 81 anti-cancer drugs, which have already been approved for the treatment of other cancer cases. Out of these, four effective interactions, GLTP-Nilotinib, PTPRN-Venetoclax, VEGFA-Venetoclax, and FABP6-Abemaciclib, were found effective. In-silico analysis suggested Abemaciclib, CDK4/6 protein inhibitor as the most potent among all selected anti-cancer drugs. The ML approach identified new molecular markers and potential drug specificity for colonic tumour tissue (Liñares-Blanco et al., 2020). Some more drugs have been enlisted in Table 5.2 that have been identified for anti-cancer potential.

5.3.4 Modulation and Set Up Therapy Plans

Non-specific action of chemo and radiotherapeutics further induces side effects and complications. Therefore the identification of target and site-specific treatment is necessary to prevent damage to aligned tissues. AI algorithms support radiologists to track and plan to target oncogenic tissue with radiation regimens (Fiorino et al., 2020; Lou et al., 2019; Meyer et al., 2018). DL algorithms integrated with radionics to determine the radiotherapy responses in bladder cancer (Cha et al., 2017). Such ideas were carried forward to develop automated software. The software work on algorithms that were trained on

TABLE 5.2 Some of the repurposed drugs for cancer treatment

DRUG	KNOWN TARGET	REPURPOSED TARGET	REFERENCE
Metformin	Type-2 diabetes	Breast cancer	Zhang et al. (2020)
Indomethacin	Non-steroidal anti-inflammatory drug	Cox-1/2-dependent angiogenesis inhibitor	Zhang et al. (2020)
Irinotecan	Colon cancer	Neuroblastoma	Kim et al. (2019)
Podophyllotoxin	genital warts and molluscum contagiosum	Leukaemia	Kim et al. (2019)
Hydroxychloroquine (HCQ)	Antimalarial drug	Pancreatic cancer	Abarientos et al. (2011), Al-Rawi et al. (2018)
Rapamycin	Immunosuppressant and anti-restenosis agent	Myeloid leukaemia	Altman et al. (2011), Brown et al. (2003)
Prazosin	Hypertension	pheochromocytoma.	Wallace and Gill (1978)
Thalidomide	Sedative; morning sickness in pregnancy	Refractory multiple myeloma	Singhal et al. (1999)

previous history/cases and based on the detection it aids in designing the subject-specific therapy within hours (Babier et al., 2018). In succession, precancerous lesions of women's cervix (Hu et al., 2019) and breast cancer lesions (Bahl et al., 2018) were analyzed with a diagnostic ML algorithm which also guides during the treatment. The AI and automated approach were validated successfully for delineation of nasopharyngeal carcinoma with up to 79% accuracy by using a three-dimensional convolutional neural network (3D CNN) (L. Lin et al., 2019).

Overtreatment is another side effect associated with radio and chemotherapy. Determination of treatment efficacy and dosage size, frequent analysis and modification of therapeutic plan are the major challenges (Chen et al., 2018; Levine et al., 2019; Smaïl-Tabbone et al., 2019; Zhu et al., 2012). A similar approach was used to evaluate the role of immunotherapy in cancer treatment (Abbasi, 2019; Jabbari and Rezaei, 2019; Tan et al., 2020; Trebeschi et al., 2019).

A drug-response model was designed with ML to determine the tolerance of breast cancer cells for chemotherapy. For that, the designed model was trained on the gene responses recorded against the different dosages of various drugs (Dorman et al., 2016). Epstein-Barr Virus-DNA-based models were the conventional approach to the formulation of a treatment strategy for nasopharyngeal carcinomas which still suffers from efficiency and efficacy. Researchers of the National University of Singapore developed the computational platform 'CURATE.AI' with integrated DL technologies for treatment plan optimization. The platform was successfully evaluated for treatment impact and determination of optimal dose of zen-3694 and enzalutamide (Pantuck et al., 2018). However, DL models designed for therapeutics have been proved superior over conventional ones for the treatment of nasopharyngeal carcinomas by reducing overtreatment and guiding the therapy (Peng et al., 2019; Tang et al., 2019). It was found that solid tumours are sensitive to 'Programmed Cell Death (PD-1) inhibitors'. Considering the fact, Sun and colleagues developed an AI platform to detect the inhibitory effect of PD-1 on solid tumours (R. Sun et al., 2018). Homologous recombination (HR) deficiency is one of the major factors causing breast cancer, which can be treated with poly ADP-ribose polymerase (PARP) inhibitors. DL algorithms detected HR affected breast cancer cells with more than 74% accuracy (Gulhan et al., 2019). The ML method based on human leukocyte antigen (HLA) mass spectrometry database was developed to identify cancer-associated neoantigens and improve immunotherapy (Bulik-Sullivan et al., 2019). The aim of the database is rapid detection and identification of antigens and aid in the prognosis of disease and cancer progression.

5.4 FUTURE PROSPECTS

'Explainable artificial intelligence' becomes a boon for society, changing the face of technology completely. In healthcare especially cancer, its role is really commendable. Two major challenges for cancer care are diagnostics and formulation/modulation of the effective therapy plan. Ultra imaging records and detects even minor morphological as well as histological changes and establishes the link with growth stages of cancers. Similarly, therapeutic databases have been used to predict side effects and then modulate the therapeutics plans accordingly. Currently, some literature also showed its role in other ailments as well like poisoning (Chary et al., 2021), arterial fibrillation (Jo et al., 2021) and even for COVID-19 (Sharma et al., 2021) besides drug repositioning and screening. Expansion of database with more cancer variants, availability of technology, are two major challenges in front of healthcare professionals. Real-time analysis of different cancers, stages, tracking of morphological, histological, molecular, as well as physiological changes, etc., are required to build a complete database that will be available free of cost for research and treatment. Along with that data collection from each end of the globe will further validate the data. The second biggest challenge is the availability of high-end equipment, networking, and connectivity to the respective databases that restrict their application either to the research lab or highly equipped labs while healthcare centres from rural areas, low and middle economy countries are devoid of such technology. The future lies in the arms of such areas where such technologies is necessary but for that developers and healthcare department has a long way to cover. It is necessary to make a one-time investment in developing the infrastructure so that the underprivileged people may also get benefitted.

5.5 CONCLUSION

In the last few years, AI has upgraded to a much efficient and logical version of XAI which can justify the logic of the decision to users. The implications of XAI to define the advancements of cities to smart cities and most importantly healthcare, an essential and inseparable part of smart cities. An unhealthy lifestyle predisposed the human body to several health issues and metabolic disorders like diabetes and cancer. The major obstacle in cancer therapy is early diagnosis and efficient therapy with minimum side effects. Integration of XAI to healthcare revolutionized cancer care and drug development by reducing the time and cost taken in novel drug discovery and optimizing the drug performance. ML classifiers and pattern recognition systems improved the detection, imaging, and diagnosis by identifying the morphological and histological changes in comparison to healthy tissues. Also, convolutional neural networks and variational autoencoders reduced the possible side effects of radio/chemotherapy by precise application of chemicals and radiation to the target sites. Limited ability and knowledge of humans restricted the key treatment opportunities which are very well complemented by the AI modules for human cancer research and future treatment. It is believed that it will bring revolutionary changes in medical technology and healthcare in the upcoming years.

REFERENCES

Abarientos, C., Sperber, K., Shapiro, D. L., Aronow, W. S., Chao, C. P., & Ash, J. Y. (2011). Hydroxychloroquine in systemic lupus erythematosus and rheumatoid arthritis and its safety in pregnancy. *Expert Opinion on Drug Safety*, *10*, 705–714. 10.1517/14740338.2011.566555

Abbasi, J. (2019). "Electronic nose" predicts immunotherapy response. *Journal of the American Medical Association*, *322*, 1756–1756. 10.1001/jama.2019.18225

Agarwal, R., Narayan, J., Bhattacharyya, A., Saraswat, M., & Tomar, A. K. (2017). Gene expression profiling, pathway analysis and subtype classification reveal molecular heterogeneity in hepatocellular carcinoma and suggest subtype specific therapeutic targets. *Cancer Genetics*, *216–217*, 37–51. 10.1016/j.cancergen.2017.06.002

Al-Rawi, H., Meggitt, S. J., Williams, F. M., & Wahie, S. (2018). Steady-state pharmacokinetics of hydroxychloroquine in patients with cutaneous lupus erythematosus. *Lupus*, *27*, 847–852. 10.1177/0961203317727601

Altman, J. K., Sassano, A., Kaur, S., Glaser, H., Kroczynska, B., Redig, A. J., Russo, S., Barr, S., & Platanias, L. C. (2011). Dual mTORC2/mTORC1 targeting results in potent suppressive effects on acute myeloid leukemia (AML) progenitors. *Clinical Cancer Research*, *17*, 4378–4388. 10.1158/1078-0432.CCR-10-2285

Babier, A., Boutilier, J. J., McNiven, A. L., & Chan, T. C. Y. (2018). Knowledge-based automated planning for oropharyngeal cancer. *Medical Physics*, *45*, 2875–2883. 10.1002/mp.12930

Bahl, M., Barzilay, R., Yedidia, A. B., Locascio, N. J., Yu, L., Lehman, C. D. (2018). High-risk breast lesions: A machine learning model to predict pathologic upgrade and reduce unnecessary surgical excision. *Radiology*, *286*, 810–818. 10.1148/radiol.2017170549

Bera, K., Schalper, K. A., Rimm, D. L., Velcheti, V., & Madabhushi, A. (2019). Artificial intelligence in digital pathology—New tools for diagnosis and precision oncology. *Nature Reviews Clinical Oncology, 16*, 703–715. 10.1038/s41571-019-0252-y

Bray, F., Ferlay, J., Soerjomataram, I., Siegel, R. L., Torre, L. A., & Jemal, A. (2018). Global cancer statistics 2018: GLOBOCAN estimates of incidence and mortality worldwide for 36 cancers in 185 countries. *CA: A Cancer Journal for Clinicians, 68*, 394–424. 10.3322/caac.21492

Brown, A. S., & Patel, C. J. (2017). A standard database for drug repositioning. *Scientific Data, 4*, 170029. 10.1038/sdata.2017.29

Brown, V. I., Fang, J., Alcorn, K., Barr, R., Kim, J. M., Wasserman, R., & Grupp, S. A. (2003). Rapamycin is active against B-precursor leukemia in vitro and in vivo, an effect that is modulated by IL-7-mediated signaling. *Proceedings of the National Academy of Sciences, 100*, 15113–15118. 10.1073/pnas.2436348100

Bulik-Sullivan, B., Busby, J., Palmer, C. D., Davis, M. J., Murphy, T., Clark, A., Busby, M., Duke, F., Yang, A., Young, L., Ojo, N. C., Caldwell, K., Abhyankar, J., Boucher, T., Hart, M. G., Makarov, V., De Montpreville, V. T., Mercier, O., Chan, T. A....Yelensky, R. (2019). Deep learning using tumor HLA peptide mass spectrometry datasets improves neoantigen identification. *Nature Biotechnology, 37*, 55–63. 10.1038/nbt.4313

Carvalho-Silva, D., Pierleoni, A., Pignatelli, M., Ong, C., Fumis, L., Karamanis, N., Carmona, M., Faulconbridge, A., Hercules, A., McAuley, E., Miranda, A., Peat, G., Spitzer, M., Barrett, J., Hulcoop, D. G., Papa, E., Koscielny, G., & Dunham, I. (2019). Open Targets Platform: New developments and updates two years on. *Nucleic Acids Research, 47*, D1056–D1065. 10.1093/nar/gky1133

Cha, K. H., Hadjiiski, L., Chan, H.-P., Weizer, A. Z., Alva, A., Cohan, R. H., Caoili, E. M., Paramagul, C., & Samala, R. K. (2017). Bladder cancer treatment response assessment in CT using radiomics with deep-learning. *Scientific Reports, 7*, 8738. 10.1038/s41598-017-09315-w

Challa, A. P., Beam, A. L., Shen, M., Peryea, T., Lavieri, R. R., Lippmann, E. S., & Aronoff, D. M. (2020). Machine learning on drug-specific data to predict small molecule teratogenicity. *Reproductive Toxicology, 95*, 148–158. 10.1016/j.reprotox.2020.05.004

Chary, M., Boyer, E. W., & Burns, M. M. (2021). Diagnosis of acute poisoning using explainable artificial intelligence. arXiv:2102.01116 [cs].

Chaturvedi, V. K., Singh, A., Dubey, S. K., Hetta, H. F., John, J., & Singh, M. P. (2019). Molecular mechanistic insight of hepatitis B virus mediated hepatocellular carcinoma. *Microbial Pathogenesis, 128*, 184–194. 10.1016/j.micpath.2019.01.004

Chelliah, V., Juty, N., Ajmera, I., Ali, R., Dumousseau, M., Glont, M., Hucka, M., Jalowicki, G., Keating, S., Knight-Schrijver, V., Lloret-Villas, A., Natarajan, K. N., Pettit, J.-B., Rodriguez, N., Schubert, M., Wimalaratne, S. M., Zhao, Y., Hermjakob, H., Le Novère, N., & Laibe, C. (2015). BioModels: Ten-year anniversary. *Nucleic Acids Research 43*, D542–D548. 10.1093/nar/gku1181

Chen, G., Tsoi, A., Xu, H., & Zheng, W. J. (2018). Predict effective drug combination by deep belief network and ontology fingerprints. *Journal of Biomedical Informatics, 85*, 149–154. 10.1016/j.jbi.2018.07.024

Chen, M., Zhang, B., Topatana, W., Cao, J., Zhu, H., Juengpanich, S., Mao, Q., Yu, H., & Cai, X. (2020). Classification and mutation prediction based on histopathology H&E images in liver cancer using deep learning. *NPJ Precision Oncology, 4*, 14. 10.1038/s41698-020-0120-3

Cook, D. J., Duncan, G., Sprint, G., & Fritz, R. L. (2018). Using smart city technology to make healthcare smarter. *Proceedings of the IEEE, 106*, 708–722. 10.1109/JPROC.2017.2787688

Coudray, N., Ocampo, P. S., Sakellaropoulos, T., Narula, N., Snuderl, M., Fenyö, D., Moreira, A. L., Razavian, N., & Tsirigos, A. (2018). Classification and mutation prediction from non–small cell lung cancer histopathology images using deep learning. *Nature Medicine, 24*, 1559–1567. 10.1038/s41591-018-0177-5

Darcy, A. M., Louie, A. K., & Roberts, L. W. (2016). Machine learning and the profession of medicine. JAMA, *315*, 551. 10.1001/jama.2015.18421

Davis, A. P., Grondin, C. J., Johnson, R. J., Sciaky, D., McMorran, R., Wiegers, J., Wiegers, T. C., & Mattingly, C. J. (2019). The Comparative Toxicogenomics Database: update 2019. *Nucleic Acids Research, 47*, D948–D954. 10.1093/nar/gky868

Dietz, R. L., & Pantanowitz, L. (2019). The future of anatomic pathology: deus ex machina? *Journal of Medical Artificial Intelligence, 2*, 4–4. 10.21037/jmai.2019.02.03

do Nascimento Souza, J. (2021). Explainable artificial intelligence for human-friendly explanations to predictive analytics on big data. Masters Thesis. University of Manitoba, Canada.

Dorman, S. N., Baranova, K., Knoll, J. H. M., Urquhart, B. L., Mariani, G., Carcangiu, M. L., & Rogan, P. K. (2016). Genomic signatures for paclitaxel and gemcitabine resistance in breast cancer derived by machine learning. *Molecular Oncology, 10*, 85–100. 10.1016/j.molonc.2015.07.006

Ehteshami Bejnordi, B., Mullooly, M., Pfeiffer, R. M., Fan, S., Vacek, P. M., Weaver, D. L., Herschorn, S., Brinton, L. A., van Ginneken, B., Karssemeijer, N., Beck, A. H., Gierach, G. L., van der Laak, J. A. W. M., &

Sherman, M. E. (2018). Using deep convolutional neural networks to identify and classify tumor-associated stroma in diagnostic breast biopsies. *Modern Pathology*, *31*, 1502–1512. 10.1038/s41379-018-0073-z

Eliopoulos, H., Giranda, V., Carr, R., Tiehen, R., Leahy, T., & Gordon, G. (2008). Phase 0 trials: An industry perspective. *Clinical Cancer Research*, *14*, 3683–3688. 10.1158/1078-0432.CCR-07-4586

Fiorino, C., Guckenberger, M., Schwarz, M., Heide, U. A., & van der Heijmen, B. (2020). Technology-driven research for radiotherapy innovation. *Molecular Oncology*, *14*, 1500–1513. 10.1002/1878-0261.12659

Fogel, D. B. (2018). Factors associated with clinical trials that fail and opportunities for improving the likelihood of success: A review. *Contemporary Clinical Trials Communications*, *11*, 156–164. 10.1016/j.conctc.201 8.08.001

Forbes, S. A., Bindal, N., Bamford, S., Cole, C., Kok, C. Y., Beare, D., Jia, M., Shepherd, R., Leung, K., Menzies, A., Teague, J. W., Campbell, P. J., Stratton, M. R., & Futreal, P. A. (2011). COSMIC: Mining complete cancer genomes in the Catalogue of Somatic Mutations in Cancer. *Nucleic Acids Research*, *39*, D945–D950. 10.1093/ nar/gkq929

Gaulton, A., Hersey, A., Nowotka, M., Bento, A. P., Chambers, J., Mendez, D., Mutowo, P., Atkinson, F., Bellis, L. J. , Cibrián-Uhalte, E., Davies, M., Dedman, N., Karlsson, A., Magariños, M. P., Overington, J. P., Papadatos, G., Smit, I., & Leach, A. R. (2017). The ChEMBL database in 2017. *Nucleic Acids Research*, *45*, D945–D954. 10.1093/nar/gkw1074

Golubchikov, O., & Thornbush, M. (2020). Artificial Intelligence and Robotics in Smart city strategies and planned smart development. *Smart Cities*, *3*, 1133–1144. 10.3390/smartcities3040056

Gulhan, D. C., Lee, J. J. K., Melloni, G. E. M., Cortés-Ciriano, I., & Park, P. J. (2019). Detecting the mutational signature of homologous recombination deficiency in clinical samples. *Nature Genetics*, *51*, 912–919. 10.103 8/s41588-019-0390-2

Hu, L., Bell, D., Antani, S., Xue, Z., Yu, K., Horning, M. P., Gachuhi, N., Wilson, B., Jaiswal, M. S., Befano, B., Long, L. R., Herrero, R., Einstein, M. H., Burk, R. D., Demarco, M., Gage, J. C., Rodriguez, A. C., Wentzensen, N., &Schiffman, M. (2019). An observational study of deep learning and automated evaluation of cervical images for cancer screening. *JNCI: Journal of the National Cancer Institute*, *111*, 923–932. 10.1093/jnci/djy225

Jabbari, P., & Rezaei, N. (2019). Artificial intelligence and immunotherapy. *Expert Review of Clinical Immunology*, *15*, 689–691. 10.1080/1744666X.2019.1623670

Jing, C., Beesley, C., Foster, C. S., Rudland, P. S., Fujii, H., Ono, T., Chen, H., Smith, P. H., & Ke, Y. (2000). Identification of the messenger RNA for human cutaneous fatty acid-binding protein as a metastasis inducer. *Cancer Research*, *60*, 2390–2398.

Jo, Y.-Y., Cho, Y., Lee, S. Y., Kwon, J., Kim, K.-H., Jeon, K.-H., Cho, S., Park, J., & Oh, B.-H. (2021). Explainable artificial intelligence to detect atrial fibrillation using electrocardiogram. *International Journal of Cardiology*, *328*, 104–110. 10.1016/j.ijcard.2020.11.053

Juengpanich, S., Topatana, W., Lu, C., Staiculescu, D., Li, S., Cao, J., Lin, J., Hu, J., Chen, M., Chen, J., & Cai, X. (2020). Role of cellular, molecular and tumor microenvironment in hepatocellular carcinoma: Possible targets and future directions in the regorafenib era. *International Journal of Cancer*, *147*, 1778–1792. 10.1002/ ijc.32970

Kanehisa, M., Furumichi, M., Tanabe, M., Sato, Y., & Morishima, K. (2017). KEGG: New perspectives on genomes, pathways, diseases and drugs. *Nucleic Acids Research*, *45*, D353–D361. 10.1093/nar/gkw1092

Kather, J. N., Pearson, A. T., Halama, N., Jäger, D., Krause, J., Loosen, S. H., Marx, A., Boor, P., Tacke, F., Neumann, U. P., Grabsch, H. I., Yoshikawa, T., Brenner, H., Chang-Claude, J., Hoffmeister, M., Trautwein, C., & Luedde, T. (2019). Deep learning can predict microsatellite instability directly from histology in gastrointestinal cancer. *Nature Medicine*, *25*, 1054–1056. 10.1038/s41591-019-0462-y

Khanna, I. (2012). Drug discovery in pharmaceutical industry: Productivity challenges and trends. *Drug Discovery Today*, *17*, 1088–1102. 10.1016/j.drudis.2012.05.007

Kim, E, Choi, A, Nam, H, 2019. Drug repositioning of herbal compounds via a machine-learning approach. BMC Bioinformatics 20, 247. 10.1186/s12859-019-2811-8

Kudo, M., Han, G., Finn, R. S., Poon, R. T. P., Blanc, J.-F., Yan, L., Yang, J., Lu, L., Tak, W.-Y., Yu, X., Lee, J.-H., Lin, S.-M., Wu, C., Tanwandee, T., Shao, G., Walters, I. B., Dela Cruz, C., Poulart, V., & Wang, J.-H. (2014). Brivanib as adjuvant therapy to transarterial chemoembolization in patients with hepatocellular carcinoma: A randomized phase III trial. *Hepatology*, *60*, 1697–1707. 10.1002/hep.27290

Lamb, J. (2006). The connectivity map: Using gene-expression signatures to connect small molecules, genes, and disease. *Science*, *313*, 1929–1935. 10.1126/science.1132939

Levine, M. N., Alexander, G., Sathiyapalan, A., Agrawal, A., & Pond, G. (2019). Learning health system for breast cancer: Pilot project experience. *JCO Clinical Cancer Informatics*, *3*, 1–11. 10.1200/CCI.19.00032

Li, B., Dai, C., Wang, L., Deng, H., Li, Y., Guan, Z., & Ni, H. (2020). A novel drug repurposing approach for non-small cell lung cancer using deep learning. *Plos One*, *15*, e0233112. 10.1371/journal.pone.0233112

Li, S., Jiang, H., & Pang, W. (2017). Joint multiple fully connected convolutional neural network with extreme learning machine for hepatocellular carcinoma nuclei grading. *Computers in Biology and Medicine*, *84*, 156–167. 10.1016/j.compbiomed.2017.03.017

Li, Y. H., Yu, C. Y., Li, X. X., Zhang, P., Tang, J., Yang, Q., Fu, T., Zhang, X., Cui, X., Tu, G., Zhang, Y., Li, S., Yang, F., Sun, Q., Qin, C., Zeng, X., Chen, Z., Chen, Y. Z., & Zhu, F. (2018). Therapeutic target database update 2018: Enriched resource for facilitating bench-to-clinic research of targeted therapeutics. *Nucleic Acids Research*, *46*, D1121–D1127. 10.1093/nar/gkx1076

Liang, G., Fan, W., Luo, H., & Zhu, X. (2020). The emerging roles of artificial intelligence in cancer drug development and precision therapy. *Biomedicine & Pharmacotherapy*, *128*, 110255. 10.1016/j.biopha.2020.110255

Lin, H., Wei, C., Wang, G., Chen, H., Lin, L., Ni, M., Chen, J., & Zhuo, S. (2019). Automated classification of hepatocellular carcinoma differentiation using multiphoton microscopy and deep learning. *Journal of Biophotonics*, *12*. e201800435. 10.1002/jbio.201800435

Lin, L., Dou, Q., Jin, Y.-M., Zhou, G.-Q., Tang, Y.-Q., Chen, W.-L., Su, B.-A., Liu, F., Tao, C.-J., Jiang, N., Li, J.-Y., Tang, L.-L., Xie, C.-M., Huang, S.-M., Ma, J., Heng, P.-A., Wee, J. T. S., Chua, M. L. K., Chen, H., & Sun, Y. (2019). Deep learning for automated contouring of primary tumor volumes by MRI for nasopharyngeal carcinoma. *Radiology*, *291*, 677–686. 10.1148/radiol.2019182012

Linardatos, P., Papastefanopoulos, V., & Kotsiantis, S. (2021). Explainable AI: A review of machine learning interpretability methods. *Entropy*, *23*, 18. 10.3390/e23010018

Liñares-Blanco, J., Munteanu, C. R., Pazos, A., & Fernandez-Lozano, C. (2020). Molecular docking and machine learning analysis of Abemaciclib in colon cancer. *BMC Molecular and Cell Biology*, *21*, 52. 10.1186/s12860-020-00295-w

Lou, B., Doken, S., Zhuang, T., Wingerter, D., Gidwani, M., Mistry, N., Ladic, L., Kamen, A., & Abazeed, M. E. (2019). An image-based deep learning framework for individualising radiotherapy dose: A retrospective analysis of outcome prediction. *The Lancet Digital Health*, *1*, e136–e147. 10.1016/S2589-7500(19)30058-5

Maddox, T. M., Rumsfeld, J. S., & Payne, P. R. O. (2019). Questions for artificial intelligence in health care. *JAMA*, *321*, 31. 10.1001/jama.2018.18932

Malta, T. M., Sokolov, A., Gentles, A. J., Burzykowski, T., Poisson, L., Weinstein, J. N., Kamińska, B., Huelsken, J., Omberg, L., Gevaert, O., Colaprico, A., Czerwińska, P., Mazurek, S., Mishra, L., Heyn, H., Krasnitz, A., Godwin, A. K., & Lazar, A. J. (2018). Machine learning identifies stemness features associated with oncogenic dedifferentiation. *Cell*, *173*, 338–354.e15. 10.1016/j.cell.2018.03.034

Meyer, P., Noblet, V., Mazzara, C., & Lallement, A. (2018). Survey on deep learning for radiotherapy. *Computers in Biology and Medicine*, *98*, 126–146. 10.1016/j.compbiomed.2018.05.018

Miller, K. D., Goding Sauer, A., Ortiz, A. P., Fedewa, S. A., Pinheiro, P. S., Tortolero-Luna, G., Martinez-Tyson, D., Jemal, A., & Siegel, R. L. (2018). Cancer statistics for Hispanics/Latinos, 2018. *CA: A Cancer Journal for Clinicians*, *68*, 425–445. 10.3322/caac.21494

Nieman, K. M., Kenny, H. A., Penicka, C. V., Ladanyi, A., Buell-Gutbrod, R., Zillhardt, M. R., Romero, I. L., Carey, M. S., Mills, G. B., Hotamisligil, G. S., Yamada, S. D., Peter, M. E., Gwin, K., & Lengyel, E. (2011). Adipocytes promote ovarian cancer metastasis and provide energy for rapid tumor growth. *Nature Medicine*, *17*, 1498–1503. 10.1038/nm.2492

Pantuck, A. J., Lee, D.-K., Kee, T., Wang, P., Lakhotia, S., Silverman, M. H., Mathis, C., Drakaki, A., Belldegrun, A. S., Ho, C.-M., & Ho, D. (2018). Modulating BET bromodomain inhibitor ZEN-3694 and enzalutamide combination dosing in a metastatic prostate cancer patient using CURATE.AI, an artificial intelligence platform. *Advances in Therapy*, *1*, 1800104. 10.1002/adtp.201800104

Peng, H., Dong, D., Fang, M.-J., Li, L., Tang, L.-L., Chen, L., Li, W.-F., Mao, Y.-P., Fan, W., Liu, L.-Z., Tian, L., Lin, A.-H., Sun, Y., Tian, J., & Ma, J. (2019). Prognostic value of deep learning PET/CT-based radiomics: Potential role for future individual induction chemotherapy in advanced nasopharyngeal carcinoma. *Clinical Cancer Research*, *25*, 4271–4279. 10.1158/1078-0432.CCR-18-3065

Pesquita, C. (2021). Towards *semantic integration for explainable artificial intelligence in the biomedical domain*. Proceedings of the 14th International Joint Conference on Biomedical Engineering Systems and Technologies. Presented at the 14th International Conference on Health Informatics, SCITEPRESS - Science and Technology Publications, Online Streaming, — Vienna Austria 11-13 Feb 2021, pp. 747–753. 10.5220/0010389707470753

Piñero, J., Bravo, A., Queralt-Rosinach, N., Gutiérrez-Sacristán, A., Deu-Pons, J., Centeno, E., García-García, J.,

Sanz, F., & Furlong, L. I. (2017). DisGeNET: A comprehensive platform integrating information on human disease-associated genes and variants. *Nucleic Acids Research*, *45*, D833–D839. 10.1093/nar/gkw943

Rao, C. V., Asch, A. S., & Yamada, H. Y. (2017). Frequently mutated genes/pathways and genomic instability as prevention targets in liver cancer. *Carcinogenesis*, *38*, 2–11. 10.1093/carcin/bgw118

Réda, C., Kaufmann, E., & Delahaye-Duriez, A. (2020). Machine learning applications in drug development. *Computational and Structural Biotechnology Journal*, *18*, 241–252. 10.1016/j.csbj.2019.12.006

Santos, R., Ursu, O., Gaulton, A., Bento, A. P., Donadi, R. S., Bologa, C. G., Karlsson, A., Al-Lazikani, B., Hersey, A., Oprea, T. I., & Overington, J. P. (2017). A comprehensive map of molecular drug targets. *Nature Reviews Drug Discovery 16*, 19–34. 10.1038/nrd.2016.230

Sayiner, M., Golabi, P., & Younossi, Z. M. (2019). Disease burden of hepatocellular carcinoma: A global perspective. *Digestive Diseases and Sciences*, *64*, 910–917. 10.1007/s10620-019-05537-2

Schuhmacher, A., Gassmann, O., & Hinder, M. (2016). Changing R&D models in research-based pharmaceutical companies. *Journal of Translational Medicine*, *14*, 105. 10.1186/s12967-016-0838-4

Sellwood, M. A., Ahmed, M., Segler, M. H., & Brown, N. (2018). Artificial intelligence in drug discovery. *Future Medicinal Chemistry*, *10*, 2025–2028. 10.4155/fmc-2018-0212

Sharma, V., Piyush, Chhatwal, S., & Singh, B. (2021). An explainable artificial intelligence based prospective framework for COVID-19 risk prediction. *medRxiv* 2021.03.02.21252269. 10.1101/2021.03.02.21252269

Shen, X., Yue, M., Meng, F., Zhu, J., Zhu, X., & Jiang, Y. (2016). Microarray analysis of differentially-expressed genes and linker genes associated with the molecular mechanism of colorectal cancer. *Oncology Letters*, *12*, 3250–3258. 10.3892/ol.2016.5122

Siegel, R. L., Miller, K. D., & Jemal, A. (2019). Cancer statistics, 2019. *CA A Cancer Journal for Clinicians*, *69*, 7–34. 10.3322/caac.21551

Singhal, S., Mehta, J., Desikan, R., Ayers, D., Roberson, P., Eddlemon, P., Munshi, N., Anaissie, E., Wilson, C., Dhodapkar, M., Zeddis, J., & Barlogie, B. (1999). Antitumor activity of thalidomide in refractory multiple myeloma. *New England Journal of Medicine*, *341*, 1565–1571. 10.1056/NEJM199911183412102

Skrede, O.-J., De Raedt, S., Kleppe, A., Hveem, T. S., Liestøl, K., Maddison, J., Askautrud, H. A., Pradhan, M., Nesheim, J. A., Albregtsen, F., Farstad, I. N., Domingo, E., Church, D. N., Nesbakken, A., Shepherd, N. A., Tomlinson, I., Kerr, R., Novelli, M., Kerr, D. J., & Danielsen, H. E. (2020). Deep learning for prediction of colorectal cancer outcome: A discovery and validation study. *The Lancet*, *395*, 350–360. 10.1016/S0140-6736(19)32998-8

Smaïl-Tabbone, M., Rance, B., Section Editors for the IMIA Yearbook Section on Bioinformatics and Translational Informatics. (2019). Contributions from the 2018 Literature on Bioinformatics and Translational Informatics. *Yearbook of Medical Informatics*, *28,* 190–193. 10.1055/s-0039-1677945

Subramanian, A., Narayan, R., Corsello, S. M., Peck, D. D., Natoli, T. E., Lu, X., Gould, J., Davis, J. F., Tubelli, A. A., Asiedu, J. K., Lahr, D. L., Hirschman, J. E., Liu, Z., Donahue, M., Julian, B., Khan, M., Wadden, D., Smith, I. C., Lam, D....Golub, T. R. (2017). A next generation connectivity map: L1000 Platform and the first 1,000,000 profiles. *Cell*, *171*, 1437–1452.e17. 10.1016/j.cell.2017.10.049

Sun, D., Chen, J., Liu, L., Zhao, G., Dong, P., Wu, B., Wang, J., & Dong, L. (2018). Establishment of a 12-gene expression signature to predict colon cancer prognosis. *PeerJ*, *6*, e4942. 10.7717/peerj.4942

Sun, R., Limkin, E. J., Vakalopoulou, M., Dercle, L., Champiat, S., Han, S. R., Verlingue, L., Brandao, D., Lancia, A., Ammari, S., Hollebecque, A., Scoazec, J.-Y., Marabelle, A., Massard, C., Soria, J.-C., Robert, C., Paragios, N., Deutsch, E., & Ferté, C. (2018). A radiomics approach to assess tumour-infiltrating CD8 cells and response to anti-PD-1 or anti-PD-L1 immunotherapy: An imaging biomarker, retrospective multicohort study. *The Lancet Oncology*, *19*, 1180–1191. 10.1016/S1470-2045(18)30413-3

Szklarczyk, D., Gable, A. L., Lyon, D., Junge, A., Wyder, S., Huerta-Cepas, J., Simonovic, M., Doncheva, N. T., Morris, J. H., Bork, P., Jensen, L. J., & von Mering, C. (2019). STRING v11: Protein–protein association networks with increased coverage, supporting functional discovery in genome-wide experimental datasets. *Nucleic Acids Research*, *47*, D607–D613. 10.1093/nar/gky1131

Tan, S., Li, D., & Zhu, X. (2020). Cancer immunotherapy: Pros, cons and beyond. *Biomedicine & Pharmacotherapy*, *124*, 109821. 10.1016/j.biopha.2020.109821

Tang, X., Huang, Y., Lei, J., Luo, H., & Zhu, X. (2019). The single-cell sequencing: New developments and medical applications. *Cell & Bioscience*, *9*, 53. 10.1186/s13578-019-0314-y

Tizhoosh, H., & Pantanowitz, L. (2018). Artificial intelligence and digital pathology: Challenges and opportunities. *Journal of Pathology Informatics*, *9*, 38. 10.4103/jpi.jpi_53_18

Torres, H. A., Shigle, T. L., Hammoudi, N., Link, J. T., Samaniego, F., Kaseb, A., &Mallet, V. (2017). The oncologic burden of hepatitis C virus infection: A clinical perspective: Hepatitis C and Cancer. *CA: A Cancer Journal for Clinicians*, *67*, 411–431. 10.3322/caac.21403

Trebeschi, S., Drago, S. G., Birkbak, N. J., Kurilova, I., Călin, A. M., Delli Pizzi, A., Lalezari, F., Lambregts, D. M.

J., Rohaan, M. W., Parmar, C., Rozeman, E. A., Hartemink, K. J., Swanton, C., Haanen, J. B. A. G., Blank, C. U., Smit, E. F., Beets-Tan, R. G. H., & Aerts, H. J. W. L. (2019). Predicting response to cancer immunotherapy using noninvasive radiomic biomarkers. *Annals of Oncology*, *30*, 998–1004. 10.1093/annonc/mdz108

ManChon, U., Talevich, E., Katiyar, S., Rasheed, K., & Kannan, N. (2014). Prediction and prioritization of rare oncogenic mutations in the cancer kinome using novel features and multiple classifiers. *PLoS Computational Biology*, *10*, e1003545. 10.1371/journal.pcbi.1003545

Vandenbulcke, H., Moreno, C., Colle, I., Knebel, J.-F., Francque, S., Sersté, T., George, C., de Galocsy, C., Laleman, W., Delwaide, J., Orlent, H., Lasser, L., Trépo, E., Van Vlierberghe, H., Michielsen, P., van Gossum, M., de Vos, M., Marot, A., Doerig, C....Deltenre, P. (2016). Alcohol intake increases the risk of HCC in hepatitis C virus-related compensated cirrhosis: A prospective study. *Journal of Hepatology*, *65*, 543–551. 10.1016/j.jhep.2016.04.031

van de Waterbeemd, H., & Gifford, E. (2003). ADMET in silico modelling: Towards prediction paradise? *Nature Reviews Drug Discovery 2*, 192–204. 10.1038/nrd1032

Wallace, J. M., & Gill, D. P. (1978). Prazosin in the diagnosis and treatment of pheochromocytoma. *JAMA*, *240*, 2752–2753.

Wen, J.-X., Li, X.-Q., & Chang, Y. (2018). Signature gene identification of cancer occurrence and pattern recognition. *Journal of Computational Biology*, *25*, 907–916. 10.1089/cmb.2017.0261

Wishart, D. S., Feunang, Y. D., Guo, A. C., Lo, E. J., Marcu, A., Grant, J. R., Sajed, T., Johnson, D., Li, C., Sayeeda, Z., Assempour, N., Iynkkaran, I., Liu, Y., Maciejewski, A., Gale, N., Wilson, A., Chin, L., Cummings, R....Wilson, M. (2018). DrugBank 5.0: A major update to the DrugBank database for 2018. *Nucleic Acids Research*, *46*, D1074–D1082. 10.1093/nar/gkx1037

Workman, P., & Al-Lazikani, B. (2013). Drugging cancer genomes. *Nature Reviews Drug Discovery*, *12*, 889–890. 10.1038/nrd4184

Workman, P., Antolin, A. A., & Al-Lazikani, B. (2019). Transforming cancer drug discovery with Big Data and AI. *Expert Opinion on Drug Discovery*, *14*, 1089–1095. 10.1080/17460441.2019.1637414

Xu, G., Zhang, M., Zhu, H., & Xu, J. (2017). A 15-gene signature for prediction of colon cancer recurrence and prognosis based on SVM. *Gene*, *604*, 33–40. 10.1016/j.gene.2016.12.016

Xue, H., Li, J., Xie, H., & Wang, Y. (2018). Review of drug repositioning approaches and resources. *International Journal of Biological Science*, *14*, 1232–1244. 10.7150/ijbs.24612

Ye, Z., Yang, Y., Li, X., Cao, D., & Ouyang, D. (2019). An integrated transfer learning and multitask learning approach for pharmacokinetic parameter prediction. *Molecular Pharmaceutics*, *16*, 533–541. 10.1021/acs.molpharmaceut.8b00816

Zaman, G. J. R., de Roos, J. A. D. M., Libouban, M. A. A., Prinsen, M. B. W., de Man, J., Buijsman, R. C., & Uitdehaag, J. C. M. (2017). TTK Inhibitors as a targeted therapy for *CTNNB1* (β -catenin) mutant cancers. *Molecular Cancer Therapeutics*, *16*, 2609–2617. 10.1158/1535-7163.MCT-17-0342

Zhang, Y., Zhao, X., Deng, L., Li, X., Wang, G., Li, Y., & Chen, M. (2019). High expression of FABP4 and FABP6 in patients with colorectal cancer. *World Journal of Surgical Oncology*, *17*, 171. 10.1186/s12957-019-1714-5

Zhang, Z., Zhou, L., Xie, N., Nice, E. C., Zhang, T., Cui, Y., & Huang, C. (2020). Overcoming cancer therapeutic bottleneck by drug repurposing. *Signal Transduction and Targeted Therapy*, *5*, 1–25. 10.1038/s41392-020-00213-8

Zhu, X., Lin, M. C. M., Fan, W., Tian, L., Wang, J., Ng, S. S., Wang, M., Kung, H., & Li, D. (2012). An intronic polymorphism in GRP78 improves chemotherapeutic prediction in non-small cell lung cancer. *Chest*, *141*, 1466–1472. 10.1378/chest.11-0469

ANN-Based Brain Tumor Classification: Performance Analysis Using K-Means and FCM Clustering With Various Training Functions

6

A. Biswas and M.S. Islam

Chittagong University of Engineering and Technology, Chittagong, Bangladesh

Contents

DOI: 10.1201/9781003172772-6

CHAPTER HIGHLIGHTS

- Two algorithms are approached based on k-means and FCM to classify brain tumor applying ANN to find better performance.
- Neural Network was trained with nine training functions separately to evaluate the finest result for both proposed approaches and 'Levenberg–Marquardt' function provides the best result.
- Obtained classification accuracy is 98.0% for first algorithm and 99.8% for second algorithm.

6.1 INTRODUCTION

Brain tumor is a critical disease for human beings which can cause death. A survey from American Society of Clinical Oncology (ASCO) expresses that yearly 18020 adults are dying from brain tumor. Modern medical science finds out 120 types of brain tumors. The functionality of brain becomes hampered when unexpected tissue growth starts (Alkassar & Abdullah, 2019). Cancerous-type tumors are more hazardous. Glioma and pituitary are common tumor types. Glioma tumor arises from the place of glial cells and pituitary tumor takes place in the pituitary gland. Whatever the brain tumor type, it is initial to detect the tumor and initiate the treatment at an early stage. For brain tumor treatment different research is going on using technology. Magnetic Resonance Imaging is an effective way as it is easy for locating the tumor area from background (Alkassar & Abdullah, 2019; Shanmuga Priya & Valarmathi, 2018). For automatic brain tumor detection, a proper clustering algorithm is essential. It is extremely complicated to classify the brain tumor types from MRI manually without expert radiologist (Kaur & Oberoi, 2018). With the advancement of technology, a revolutionary change in our society can be re-marked. Hence, the 'smart city' concept arises where this concept expressed that technology leads the way of solving real-life complications. Smart city concept also includes the health care system of society in the process of modernizing the medical sector. A combination of technology and medical science is capable of ensuring upgradation of citizens' lives. It is more efficient, less time-consuming, and pre-eminence manner. Artificial Intelligence changes the conventional process vastly and brings pragmatic changes in health care system. In that case artificial neural network (ANN) can perform more accurately than manual process. Before ANN training, images are segmented with the help of renowned image cluster process to separate the tumor locality from image background. After features extraction and reduction, data are taken for training. Network architecture as well as training functions are also important for improvement of classification accuracy. The proposed process is faster and more accurate than manual process. But it is challenging to turn out higher accuracy by training the network using sizeable patient data. In this theme, this research work is done. Authors used two distinguished clustering methods for image segmentation and the main classification was done by artificial neural network.

Many authors work in this brain tumor detection field and they showed several approaches. Artificial Neural Network, Deep Neural Network, CNN, SVM, and ML (Hemanth et al., 2019; Siar & Teshnehlab, 2019) are the most common and renowned approaches for brain tumor classification. Authors Sanjeev Kumar and Chetna Dabas proposed a brain tumor detection process using 25 images (Kumar et al. 2017). They claimed it as hybrid approach where their image preprocessing steps consisted of image format conversion, feature extraction, reduction of extracted features and classification with SVM. For feature extraction, authors utilized DWT and for features reduction, they used Genetic Algorithm. For classification, authors used SVM and highest accuracy was found to be 90% (Kumar et al. 2017). Author V. Amsaveni approached for the same detection process using ANN (Amsaveni & Albert Singh, 2013). Histogram and median filter were used for the preprocessing step and feature extraction was done by Gabor filer. Classification accuracy was found to be 89.9% for 500 iterations by using 40 brain tumor images (Amsaveni & Albert Singh, 2013). Combination of templated-based K-means and FCM (TKFCM) was used by authors Md Shahariar Alam and Md Mahbubur Rahman for brain tumor detection (Alam et al., 2019). Before image clustering, images were preprocessed by enhancement and grey level conversion. Their identification technique provided 97.5% accuracy from a total 40-image dataset. Authors aimed for future to reduce the computational time and better feature analysis for improving accuracy. Another approach came from authors Y. Zhang and L. Wu for brain tumor detection and they used K-means algorithm, DWT, and PCA for preprocessing steps (Zhang & Wu, 2012). But for classification, they utilized SVM and the accuracy was obtained 99.38% accuracy using 160 MRI images. Authors Milica M. Badza and Marko C. Barjaktarovic developed an innovative CNN architecture for brain tumor classification (Badza & Barjaktarovic, 2020Badza 2020). Images were preprocessed by resizing and rotation and a total of 22 layers were used for CNN architecture. Four approaches were done to test the network performance and 96.56% accuracy was found. Another machine learning approach was done by authors G. Hemanth, M. Janardhan and L. Sujihelen (Hemanth et al., 2019) for tumor detection. Authors used data resizing, noise reduction, smoothing filter, and also used the average filter. Segmentation was done on pixel basis. Tumor recognition approach was done by convolutional layers and pooling layers. This proposed method concluded 91% final accuracy for tumor classification (Hemanth et al., 2019). Another brain tumor classification research work was presented by authors Pushpa B. R. and Flemin Louies (Pushpa & Louies, 2019Pushpa 2019). Preprocessing steps consisted of median filter to denoise the image, Gaussian filter for sharpening edge, and morphological filter for clearly visualize the region of tumor. Image segmentation took place using masking and features were extracted using GLSM. Classification was done by SVM and the proposed method found 99% accuracy (Pushpa & Louies, 2019Pushpa 2019). Authors J. Vijay and J. Subhashini presented their detection process of brain tumor applying the clustering process named k-means (Vijay & Subhashini, 2013). Image enhancement technique and morphological filter operation were used for preprocessing. Their CAD detection method provided 95% accuracy to detect abnormal cells. Back propagation neural network (BPNN) was implemented by Mustafa R. Ismael and Ikhlas Abdel-Qader (Ismael & Abdel-Qader, 2018) for classification. Before classification, the region of interest (ROI) was the segmentation process. 2D DWT and Gabor filter combinedly utilized for feature extraction. Authors explained that 91.9% was the accuracy of the proposed method (Ismael & Abdel-Qader, 2018). Parveen and Amritpal Singh proposed a method using FCM and SVM (Parveen, 2015Parveen 2015). Image enhancement and skull stripping were applied in 120 patient's data. Feature extraction process was done using GLRLM and finally, liner support vector provided the highest accuracy about 91.66%. A combination of TKFCM and FCM was used by authors (Ahmmed & Hossain, 2016Ahmmed 2016). Their proposed method provided 97.1% accuracy using ANN classifier where the database consisted of 30 MRI images.

Again, some authors also researched to compare clustering algorithms to detect brain tumor more accurately (Jose et al., 2014; Rakesh & Ravi, 2012). Authors used both k-means and FCM to analyze the performance and to observe which clustering algorithm provides the vital role. ANN is an efficient brain tumor classifier and it was utilized by many authors. Accuracy of brain tumor classification also depends on ANN architecture. It varies because it is important to know which training algorithm is applied for specific architectures. Several pieces of research were done previously to observe which training algorithm contributes the most. Most of the authors (Batra, 2014, Kurukilla & Gunavathi, 2014; Mishra et al., 2015; Saji & Balachandran, 2015) found that 'Levenberg–Marquardt' provided the best accuracy than other training algorithms.

The novelty of this proposed method is to compare both k-means and FCM clustering algorithm for the same dataset and the utilization of nine training algorithms separately to explore the best accuracy. Various evaluations were done to observe the variation of experimental results. The concern of this work is to evaluate the best value that will be beneficial for brain tumor treatment. The rest of this chapter discusses on the proposed methodology with two algorithms in Section 6.2, result and other comparisons in Section 6.3. Finally Section 6.4 includes the conclusion part.

6.2 PROPOSED METHODOLOGY

Brain tumor classification is one of the challenging tasks and author-proposed classification techniques which include two different clustering algorithms and ANN. K-means and FCM are the renowned clustering processes which are generally used in medical image processing. These two clustering processes are used in this experiment in separate algorithm. In the proposed techniques, input images were preprocessed at the beginning and then images were used as segmented by the clustering algorithm. After the feature extraction and reduction, ANN was used for tumor type classification. Figure 6.1 indicates the proposed Algorithm 1 and Figure 6.2 indicates the proposed Algorithm 2. Algorithm 1 is based on k-means clustering and Algorithm 2 is based on FCM clustering. Working procedure is expressed through the given flowcharts in Figures 6.1 and 6.2.

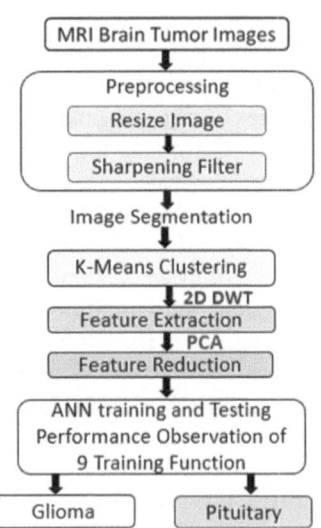

FIGURE 6.1 Workflow of Proposed Methodology Using K-Means.

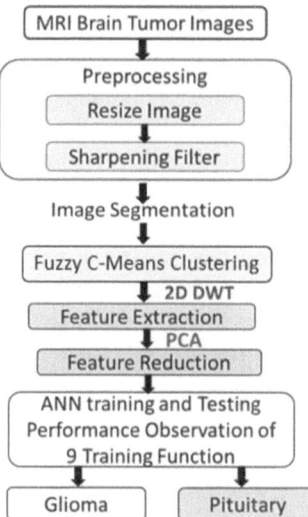

FIGURE 6.2   Workflow of Proposed Methodology Using FCM Clustering.

The workflow in Figures 6.1 and 6.2 is separated into four steps. They include preprocessing, image segmentation (clustering), extraction and reduction of features, and classification step (ANN). Firstly, images were taken from the database to preprocess those images. The preprocessing step consisted of resizing and sharpening filter. Secondly, images were segmented through both k-means and fuzzy C means clustering separately. Thirdly, 2D DWT was utilized to extract features and PCA was used to compress the feature quantity effectively. Finally, the artificial neural network was used to classify tumor types. In this case, performance of different nine training functions was observed to evaluate the best result. Elaboration of the workflow is discussed below.

6.2.1 Preprocessing

Image preprocessing has great significance as it improves the image quality for further processing and helps to increase the classification accuracy (Hemanth et al., 2019). The proposed preprocessing step consisted of two sub-steps: resizing and sharpening filtering.

6.2.1.1 Resizing

All the MRI images are not of the same size but to work with all data it is essential to resize them (Badza & Barjaktarovic, 2020Badza 2020). Different images were resized to the same 200 × 200 pixels. This sub-step modifies the pixel information.

6.2.1.2 Sharpening Filter

All images of the database did not consist of better quality so the sharpening filter was used to improve the contrast quality (Alam et al., 2019). Sharpening filter smoothes the image. Sharpening filter strengthens the object edge (Abdalla & Esmail, 2018Abdalla 2018) and enhances the image contrast slightly.

6.2.2 Image Segmentation

Proposed method is approached for clustering to segment the preprocessed images. The clustering process creates a group of points to separate the object from the boundary.

6.2.2.1 K-Means Clustering

This is the most frequently used clustering algorithm and K-means is unsupervised algorithm (Alam et al., 2019;Dhanachandra et al., 2015). This is an uncomplicated algorithm to separate the data into a definite number of cluster groups (Alam et al., 2019). The main reason for naming it K-means clustering is each cluster group contains a K center which is selected desultorily. Measuring distance is a foremost matter and 'Euclidean distance' was calculated for this experiment. For brain tumor detection, K-means clustering is often used by authors (Alam et al., 2019;Hazra et al., 2017) as it is effective to point out the tumor edge (Dhanachandra et al., 2015). This algorithm can be classified into two phases. Firstly, k centroid is intended and then at the second phase, clustering groups are constructed from centroid to nearest data point (Dhanachandra et al., 2015). This iterative algorithm diminishes the total distance from every object to the centroid of cluster. Each step of this K-means algorithm is described below.

K-Means Algorithm:

Step 1. Declare the K number cluster and center.
Step 2. Calculate the distance between center and image's all pixel which is known as Euclidean distance 'd',

$$d = \|p(x, y) - c_k\| \tag{6.1}$$

where, image resolution is x × y, input image pixel = p (x, y) and cluster center = c_k.

Step 3. Allot each pixel to the closest center based on d.
Step 4. Calculate the current position again after allotting all pixels,

$$c_k = \frac{1}{k} \sum_{y \epsilon c_k} \sum_{x \epsilon c_k} p(x, y) \tag{6.2}$$

Step 5. Repeat the above steps until it fulfills the error estimate.
Step 6. Reshape the image's cluster pixels.

6.2.2.2 Fuzzy C-Means Clustering

Bezdek is the first person who brought up unsupervised FCM algorithm in the middle of 1981. This algorithm is the modernized version of other precedent clustering methods (Preetha & Suresh, 2014) and within one pixel, FCM permits more than one cluster (Shasidhar et al., 2011). Clusters are created in the FCM process from the distance of data points to cluster centers (Brundha & Nagendra, 2015). FCM works based on optimizing Objective Function (Dhanachandra et al., 2015).

$$J(U, V) = \sum_{i=1}^{n} \sum_{j=1}^{c} (\mu_{ij})^m \|x_i - v_j\|^2 \tag{6.3}$$

where, J (U, V) = Objective Function, n = data point, m = fuzziness index, μ_{ij} = membership, j^{th} cluster center = v_j, No. of cluster center = c, Euclidean distance of ith data and j^{th} cluster center = $\|x_i - v_j\|$, k = iteration step, d_{ij} = Euclidean distance.

Algorithm of Fuzzy C-Means:

Step 1. Selection of 'c' cluster center arbitrary.
Step 2. Computation of the function of fuzzy membership,

$$\mu_{ij} = 1 / \sum_{k=1}^{c} (d_{ij} - d_{ik})^{\left(\frac{2}{m} - 1\right)} \tag{6.4}$$

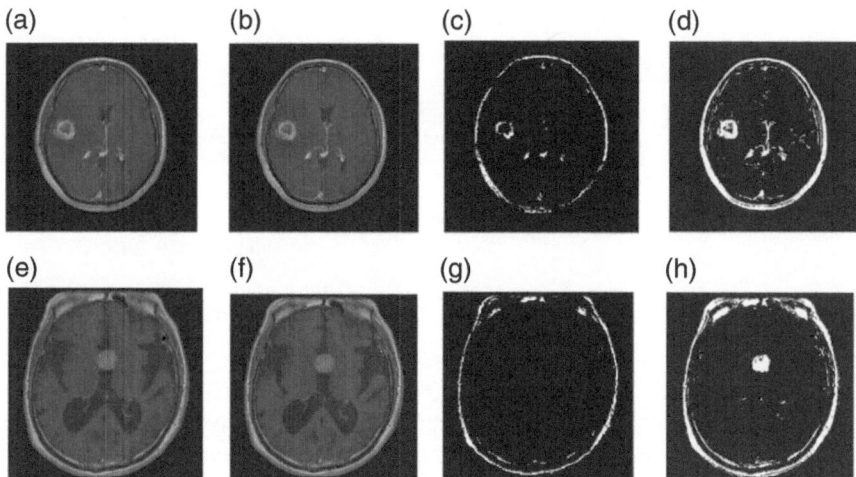

FIGURE 6.3   (a) Glioma MRI; (b) Sharpening Filtered Image; (c) K-Means Clustered Image; (d) FCM Clustered Image; (e) Pituitary MRI; (f) Sharpening Filtered Image; (g) K-Means Clustered Image; (h) FCM Clustered Image.

Step 3. Determining the fuzzy clusters,

$$v_j = (\sum_{i=1}^{n} (\mu_{ij})^m) x_i)/(\sum_{i=1}^{n} (\mu_{ij})^m), \ v_j = 1, 2, \ ...,c \tag{6.5}$$

Step 4. Reiterating Steps 2 and 3 to gain minimal Objective Function value.

The MRI image to segmentation process is shown in Figure 6.3. Preprocessing step and clustering view of the experiment are given here. After preprocessing, segmentation was done based on clustering. But in both cases, glioma and pituitary, fuzzy C-means clustering was able to locate the tumor region more perfectly than k-means. Pituitary tumor is smaller than others and it is difficult to detect perfectly. So, a better clustering algorithm can provide higher classification accuracy.

6.2.3 Feature Extraction and Reduction

After performing the image segmentation, feature extraction from image was done. Fourier Transform or (FT) is the conventional implementation for signal analyticity. For analysis purposes, FT fragments the time domain signal and converts it into a frequency domain. But the major limitation of FT is it misses out on time-related information (Zhang & Wu, 2012). For that reason, the implementation of Wavelet transform is beneficial for brain tumor feature extraction. Because of multi-resolution diagnostic functionality, WT permits several levels of resolution analysis (Zhang & Wu, 2012). Discrete wavelet transform (DWT) of WT uses both scale and position. 2D DWT is beneficial for image processing and this 2D DWT was utilized for experimental feature extraction. DWT splits the signal into sub-bands. The third level of 2D DWT was used for feature extraction.

Figure 6.4 represents the DWT levels and sub-band view. DWT splits up into four sub-bands and they are LL, LH, HL, and HH. Here, the sub-band LL is known as component's approximation and other sub-bands (LH, HL, and HH) are known as detailed components. Thirteen features in particular are extracted from segmented images and they are: Contrast, Entropy, Energy, Mean, Kurtosis, Correlation, RMS, Smoothness, Variance, Homogeneity, IDM, Standard Deviation, and Skewness.

Notwithstanding, DWT shows some complications such as the necessity of sizeable memory and expensive calculation process (Zhang & Wu, 2012). This conveys the theme of vector dimension

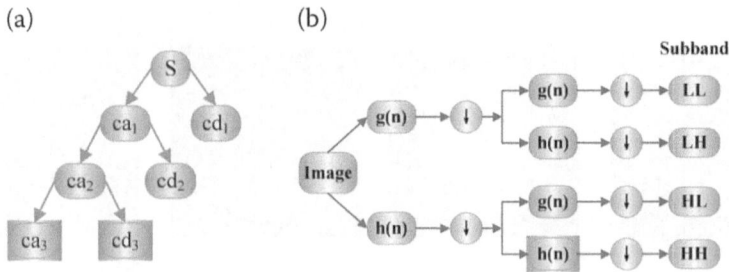

FIGURE 6.4   DWT Representation: (a) Tree of 3 Level DWT; (b) Diagrammatic Representation of 2D DWT (Zhang & Wu, 2012).

reduction and for this reason, the principal component analysis (PCA) comes in. PCA can diminish data dimension productively and can approximate the real information. This experiment works with a sizeable dataset and it becomes complex when feature extraction takes place from each image. PCA is utilized for minimizing large data dimensions but it contains considerable information in the new dataset.

6.2.4 Classification

For classification, the authors proposed Artificial Neural Network which is an efficient apparatus for image recognition. Implementation of ANN in biomedical tumor MRI classification can provide more accurate results than manual processes. Artificial neural network is the part of Artificial Intelligence and ANN works based on the learning process of human brain (Kukreja et al., 2016). Artificial intelligence arises as a black-box concept which indicates that AI is a complex computation machine process for human perception. The AI concept bequeaths lacking a transparency degree. Different researches are going on with regard to this manner to improve this transparency degree for human perception. This concept is known as 'Explainable Artificial Intelligence (XAI)'. This concept provides the basic reasons for AI functionality. XAI is getting popularity for its clear interpretation of AI framework. XAI is increasing the application of AI in various sectors and serving for more innovation. It is essential to know how AI actually works through machines and performs decision making. ANN is a machine-based work but it is understandable for humans. Artificial neural network has the capability of learning new objects and it has the adaptation potentiality to handle new conditions (Kukreja et al., 2016). The major element of this ANN modeling is 'neuron' and this model is built up with the connection of neurons (Shanmuganathan, 2016). ANN connection topology can be fully connected or partially connected (Shanmuganathan, 2016). Learning process of ANN is classified into three categories: supervised, un-supervised, and reinforcement learning (Shanmuganathan, 2016). Classification process of ANN consists of two phases. One is training phase and the other one is recall phase (Shanmuganathan, 2016). In the training phase, artificial neural network learns from a set of data and then at the recall phase, a similar type of training data is inputted to test the performance. Propagation algorithm helps to reduce the error and it helps to set the weight value of neurons (Ahmmed et al., 2017). In this way, ANN reduces the error of classification and gradually becomes capable of recognizing objects more perfectly. The proposed network architecture is shown in Figure 6.5.

Figure 6.5 is expressing that network architecture consisted of 13 features of images as input layers, 10 hidden neurons, and 1 output layer. Total network had three layers (input, hidden, and output). After network setting, network was trained and tested using a features value of 489 MRI images.

Proposed ANN architecture was constructed with multilayer feed forward network. Basic architecture layers are three and layers are: input, hidden, and output. Proposed hidden layer consisted of 10 neurons. Input data is firstly splitted up for training, validation, and test. Seventy percent of the sample was used for training, 15% sample was taken for validation, and rest of the 15% data was utilized for

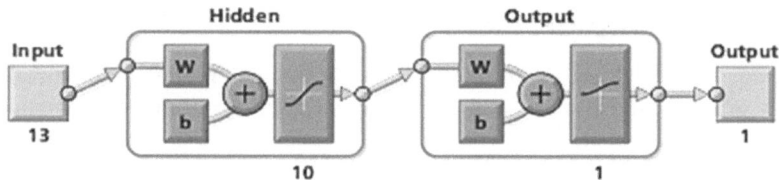

FIGURE 6.5   ANN Network Architecture of Proposed Method.

final testing. Proposed neural network is trained with features value for tumor classification. Training algorithm has a great impact on classification accuracy. 'Scaled Conjugate Gradient' is by default MATLAB® implemented training function of ANN model. But proposed method aims to use other multilayer training algorithms to evaluate the performance variation. A total of nine training algorithms were applied in this experiment and they are: Levenberg–Marquardt, Resilient Backpropagation, One Step Secant, Scaled Conjugate Gradient, BFGS Quasi-Newton, Polak–Ribiére Conjugate Gradient, Variable Learning Rate Backpropagation, Fletcher–Powell Conjugate Gradient, and Conjugate Gradient with Powell–Beale Restarts. After implementation of all these training algorithms, maximum precision was found for 'Levenberg–Marquardt' training algorithm. Theoritically, Levenberg–Marquardt shows the rapid convergence and fast accuracy provider (Kurukilla & Gunavathi, 2014). The main advantage of Levenberg–Marquardt is the presentation of more flawless ANN construction than others. Description of used nine training functions are given below.

6.2.4.1 Levenberg–Marquardt

This training function can be defined as 'trainlm'. The main function of this algorithm is the combination of Gauss–Newton and Steepest Descent. For optimizing updated weight as well as bias, it works. This function is fastest than others. The following Hessian Matrix is utilized by this algorithm (Kurukilla & Gunavathi, 2014):

$$X_{k+1} = X_k - (J^T J + \mu I)^{-1} J^T e \tag{6.6}$$

where, J = Jacobian Matrix, $J^T e$ = *Gradient*, *I = identity matrix*, e= network error vector.

6.2.4.2 BFGS Quasi-Newton

This training function can be defined as 'trainbfg'. Method of BFGS Quasi-Newton is utilized in this function to update bias, weight. It is faster than Conjugate gradient (Kurukilla & Gunavathi, 2014). It occupies more storage and consumes more complexity (Saji & Balachandran, 2015). Basic Newton method's formula:

$$W_{k+1} = W_k - A_k^{-1} g_k \tag{6.7}$$

where, A_k = Hessian Matrix, g_k = Gaussian function.

6.2.4.3 Resilient Backpropagation

The 'trainrb' function works based on Resilient Backpropagation algorithm and updates the weights (Kurukilla & Gunavathi, 2014). Partial derivatives have a magnitude which can create an effect and for eliminating it, Resilient Backpropagation is used. Derivative sign indicates the weight update's direction.

6.2.4.4 Scaled Conjugate Gradient

This training function can be defined as 'trainscg' and algorithm behind this function helps to reduce the time of line search. Scaled conjugate gradient algorithm is expensive from a computational perspective.

6.2.4.5 Conjugate Gradient With Powell–Beale Restarts

This training function can be defined as 'traincgb'. Search direction reset time is being determined by a test of this algorithm. The workflow of this algorithm calculates search directions: negative gradient, past search direction, and final search direction.

6.2.4.6 Fletcher–Powell Conjugate Gradient

This training function can be defined as 'traincgf'. The search of the highest slope decent direction is looked at in the first iteration (Kurukilla & Gunavathi, 2014).

$$P_0 = -g_o \tag{6.8}$$

where, P_0 = early search gradient, g_o = initial gradient
Line search helps to find out the optimal distance.

$$X_{k+1} = X_k - \alpha_k p_k \tag{6.9}$$

where, X_{k+1} = following wight vector, X_k = present weight vector, α_k = learning rate, p_k = present search direction.
The equation for variable adjustment:

$$X = X + a \times dX \tag{6.10}$$

where, a = search direction optimizer.

6.2.4.7 Polak–Ribiére Conjugate Gradient

This training function can be defined as 'traincgp'. In this algorithm, the present gradient is divided by past gradient with square. The equation is given below:

$$\beta_k = \frac{\Delta g_k^T g_k}{g_k g_{k-1}^T} \tag{6.11}$$

6.2.4.8 One Step Secant

This training function can be defined as 'trainoss'. This algorithm comes to remove previous limitations and it is the combination of the previous two algorithms: Quasi-Newton and Conjugate Gradient. Hessian Matrix doesn't require it. Calculation of search direction doesn't need matrix inverse (Kurukilla & Gunavathi, 2014).

6.2.4.9 Variable Learning Rate Backpropagation

This training function can be defined as 'traingda'.

6.2.5 Algorithm of Two Proposed Approaches

The workflow of proposed methodology from Figures 6.1 and Figures 6.2 can be represented as algorithms. Description of each step is already given and the algorithms are presented below.

6.2.5.1 Algorithm 1

Step 1. Load the brain tumor images from the database

Step 2. Preprocess the images

      a. Resize the image
      b. Use the sharpening filter

Step 3. Implement the K-means algorithm

Step 4. Apply 2D-DWT to extract features

Step 5. Extract 13 features

Step 6. Use PCA for feature reduction

Step 7. Features values were taken for ANN training, validation, and test

Step 8. Repeat Step 7 for nine different training functions and collect data

6.2.5.2 Algorithm 2

Step 1. Load the brain tumor images from the database

Step 2. Preprocess the images

      a. Resize the image
      b. Use the sharpening filter

Step 3. Implement the fuzzy C-means algorithm

Step 4. Apply 2D-DWT to extract features

Step 5. Extract 13 features

Step 6. Use PCA for feature reduction

Step 7. Features values were taken for ANN training, validation, and test

Step 8. Repeat Step 7 for nine different training functions and collect data

6.3 EXPERIMENTAL RESULTS AND ANALYSIS

6.3.1 Dataset

For this proposed method, MRI brain tumor data was accumulated from China's Nanfang Hospital, Tianjing Medical University (Figshare Database). Axial plane and coronal plane images were used. A total of 489 patient's brain tumor MRI images were used for the experiment. An unbiased dataset was prepared for neutral accuracy where Glioma data were 246 and Pituitary data were 243.

6.3.2 Classification Accuracy of the Two Proposed Methods

After training and validation, the final test result was obtained. Experimental results were explored by both K-means and FCM clustering. Observation of different training algorithm performances for classification of glioma and pituitary brain tumor is shown in Table 6.1.

Table 6.1 is the experimental evaluation. Table 6.1 shows that the highest accuracy was obtained for fuzzy C-means clustering when the applied training algorithm was Levenberg–Marquardt and that is 99.8%. This best result came within 18 iterations. But it is clear that for both clustering techniques, only Levenberg–Marquardt algorithm provided the higher accuracy. When Glioma and Pituitary brain tumors were classified by Levenberg–Marquardt training algorithm using K-means clustering, classification accuracy was found to be 98.0% within 16 iterations. Again, when the MRI was clustered by FCM and the ANN was trained with Levenberg–Marquardt algorithm, then the obtained accuracy was 99.8% within 18 iterations.

In Figure 6.6, blue bars indicate FCM classification accuracy and orange bars indicate K-means classification accuracy. A graph is constructed using data from Table 6.1. Here, X axis represents various training algorithms and Y axis represents accuracy for two clusterings. The graph shows that of all training algorithms, Levenberg–Marquardt provided the best result for both clusters. But for all training algorithms, FCM accuracy is much better than K-means cluster data. From this graph, it can be

TABLE 6.1 Obtained brain tumor classification accuracy with nine training functions for both K-means and FCM clustering

MULTILAYER TRAINING FUNCTION	FIRST APPROACH (WITH K-MEANS) ACCURACY	NO. OF ITERATIONS	SECOND APPROACH (WITH FCM) ACCURACY	NO. OF ITERATIONS
Levenberg–Marquardt	98.0%	16	99.8%	18
BFGS Quasi-Newton	96.3%	24	98.4	21
Resilient Backpropagation	96.3%	25	98.6%	29
Scaled Conjugate Gradient	95.1%	15	98.4%	15
Conjugate Gradient with Powell–Beale Restarts	96.5%	23	98.4%	15
Fletcher–Powell Conjugate Gradient	96.3%	22	98.8%	26
Polak–Ribiére Conjugate Gradient	96.1%	20	98.6%	15
One Step Secant	95.5%	19	98.8%	19
Variable Learning Rate Backpropagation	95.5%	158	99.0%	187

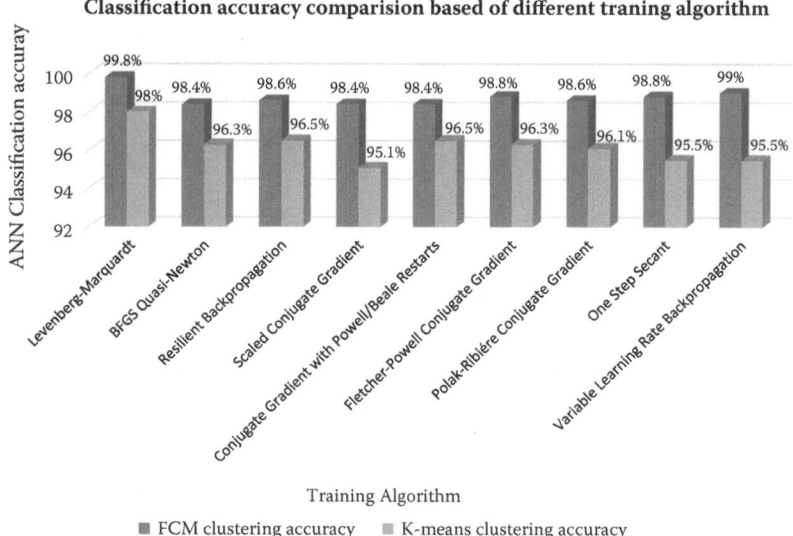

FIGURE 6.6   Graphical Representation of Classification Accuracy of the First Approach (K-Means) and Second Approach (FCM) Based on Different Nine Different Training Algorithms.

concluded that FCM provided better clustering for proposed dataset and Levenberg–Marquardt performed better for both cases.

Figure 6.7 is the confusion matrix of the first approach (Algorithm 1) with Levenberg–Marquardt training function. By performance analysis of the first approach, it was observed that Levenberg–Marquardt function provided a higher accuracy than other training functions. The obtained highest accuracy was about 98% within 16 iterations from the first approach.

6.3.3 Analysis of the Best-Obtained Accuracy

Confusion matrix of Figure 6.8(a) shows the best result with 99.8% accuracy within 18 iterations. This result was obtained when FCM clustering was used for segmentation and 'Levenberg–Marquardt' training function was applied for ANN classification. As the final accuracy provided only 0.2% error which is negligible, the result proved that the proposed method can classify glioma and pituitary brain tumor flawlessly. Performance curve is displayed in Figure 6.8(b) and this was found after training the network with data. High-performance target was achieved within 18 iterations. The finest validation performance within 12 iterations is 0.00063146. The tested data meets along with the path of validation at its best values. Proposed network architecture took 12 iterations to reach minimal error.

The process of how the gradient value decreased and met settled within 18 iterations as shown in Figure 6.8(a). At 18 iterations, the gradient value met 0.0018799. It also shows that the optimization parameter of Levenberg–Marquardt (MU) was increased and diminished up to the value of 0.001 within 18 iterations. In the end, network checked 6 validation at 18 epochs. Figure 6.9(b) indicates the histogram curve of the best result. Here, X-axis indicates the errors and Y-axis indicates the number of samples for errors. Histogram curve presents training, test, validation, and zero error. Here, at point -0.00255, the threadlike white line in the middle indicates zero error.

$$\text{Accuracy} = (TN + TP)/(TP + TN + FN + FP) \times 100 \qquad (6.12)$$

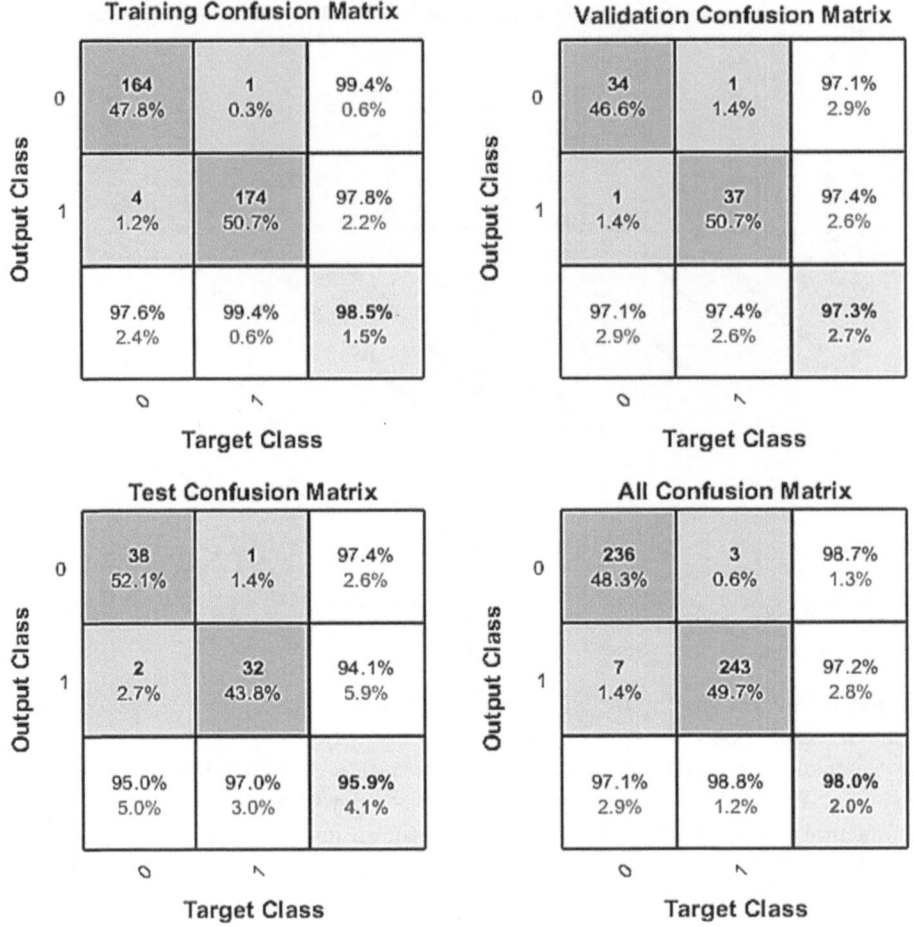

FIGURE 6.7 Confusion Matrix with 98.0% Accuracy for First Approach Used Training Function Was Levenberg–Marquardt.

$$\text{Sensitivity/Recall} = TP/(TP + FN) \times 100 \tag{6.13}$$

$$\text{Specificity} = TN/(TN + FP) \times 100 \tag{6.14}$$

$$\text{Precision} = TP/(TP + FP) \times 100 \tag{6.15}$$

$$F - \text{measure} = (2 \times Precision \times Recall)/(Precision + Recall) \tag{6.16}$$

where, TP = True Positive, FP = False Positive, TN = True Negative, FN = False Negative. Elevated Accuracy is 99.8%, Sensitivity is 100%, Specificity is 99.59%, Precision is 99.59%, and F-measure is 99.79%, which are obtained from the best result confusion matrix of Figure 6.8(a) and Equations (6.12), (6.13), (6.14), (6.15), (6.16). Network Performance was found to be 0.0033 from target data and output data.

Figure 6.10 shows the accuracy, sensitivity, specificity, precision, and F-measure comparison between the two proposed algorithms. These are the most common parameters to evaluate the performance of any method. Here, for all the parameters, Algorithm 2 shows better results than Algorithm 1.

(a)

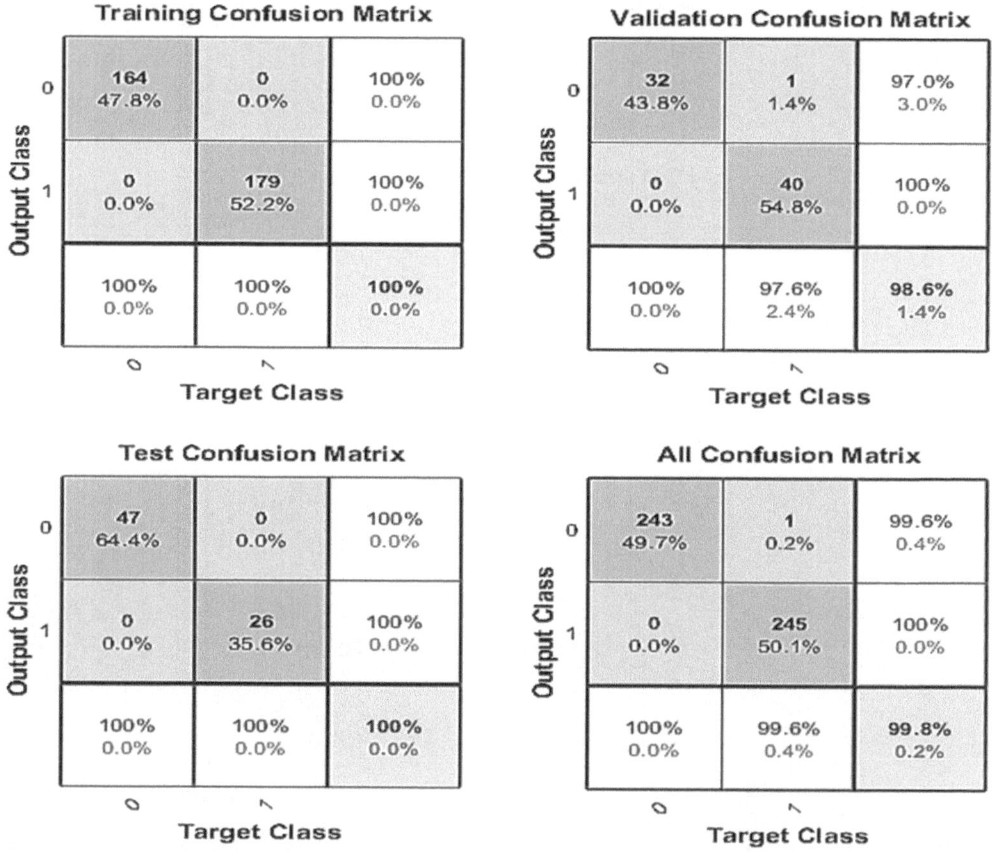

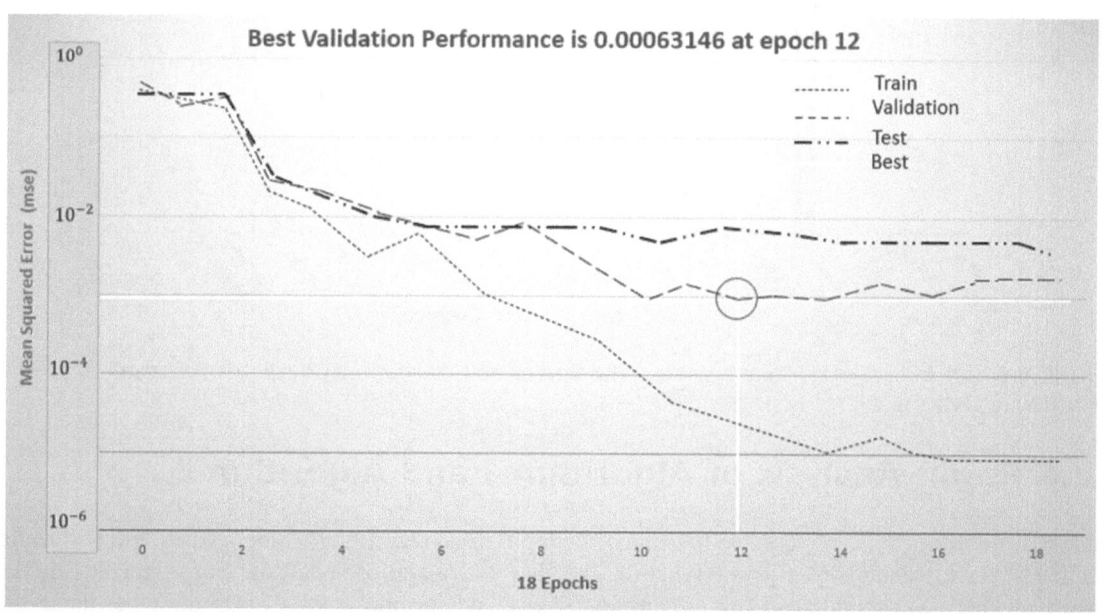

FIGURE 6.8 (a) Confusion Matrix of Obtained Best 99.8% Accuracy for Second Approach (algorithm 2) and (b) Validation Performance Graph of Best Result Used Taring Function Was Levenberg–Marquardt.

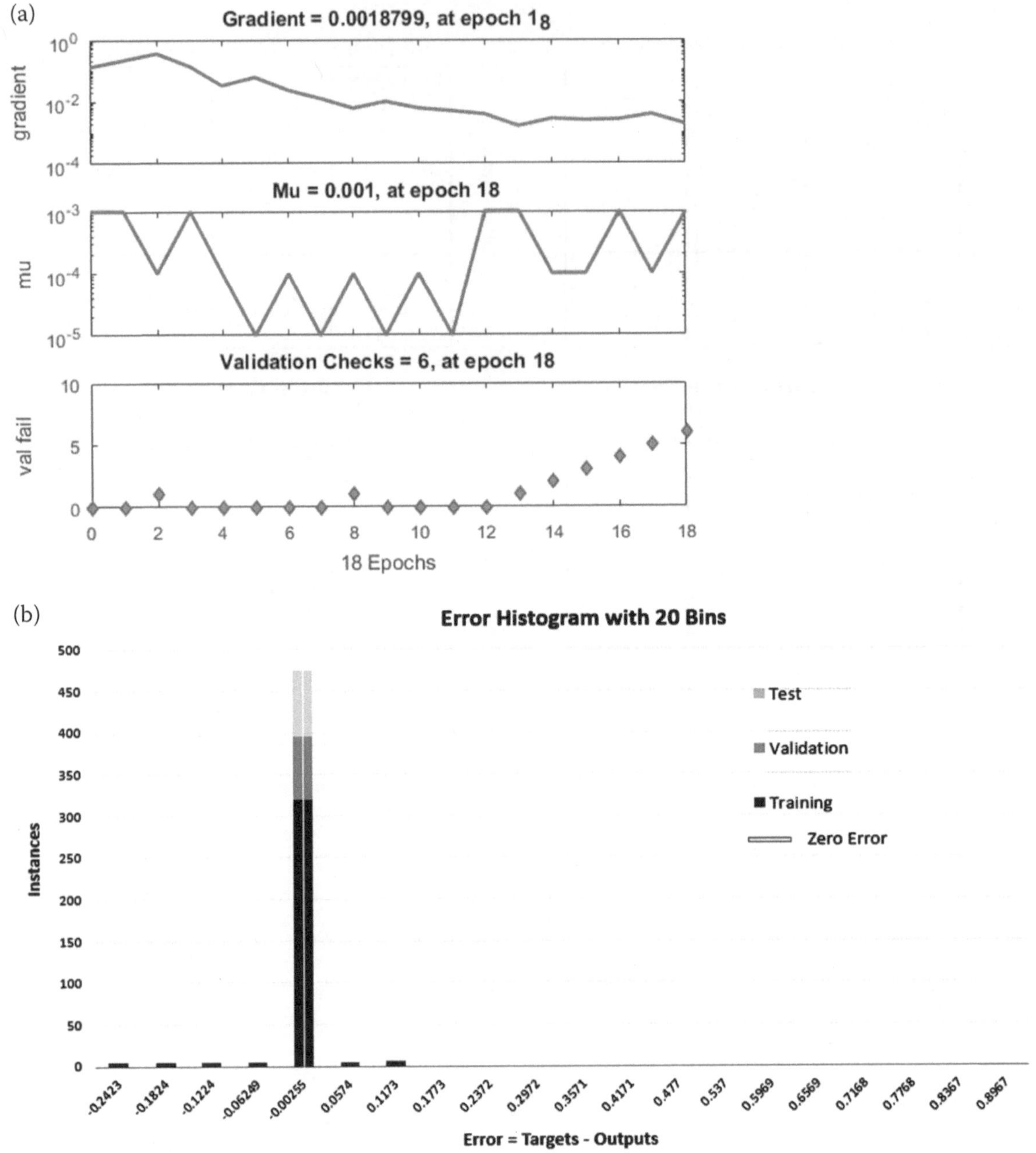

(a)

FIGURE 6.9 (a) Minimum Gradient Value, Mu Curve, and Validation Check for Proposed Network. (b) Proposed Network's Error Histogram.

6.3.4 Result Analysis of Algorithm 1 and Algorithm 2

Table 6.2 is constructed using obtained results of both proposed algorithms. In both cases, the best result was found for Levenberg–Marquardt function and that was taken for analysis. For Algorithm 1, the highest accuracy was found at 98.0%, sensitivity 97.2%, and specificity 98.7%. On the other hand, the highest accuracy was 99.8%, sensitivity 100%, and specificity 99.59%. This data shows that Algorithm 2 provides the best accuracy, sensitivity, and specificity in classification.

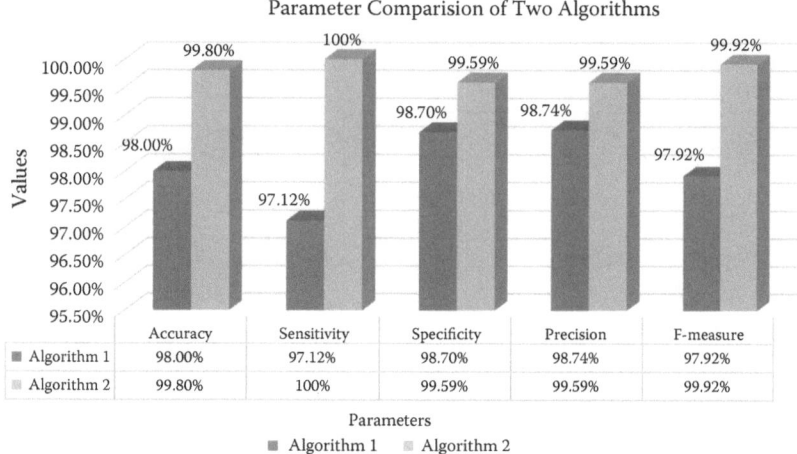

FIGURE 6.10 Accuracy, Sensitivity, Specificity, Precision and F-Measure Comparison of Two Proposed Algorithms.

TABLE 6.2 Comparative analysis of Algorithm 1 and Algorithm 2

COMPARISON	ALGORITHM 1	ALGORITHM 2
Accuracy	98.0%	99.8%
Sensitivity	97.12%	100%
Specificity	98.7%	99.59%
Precision	98.74%	99.59%
F-measure	97.92%	99.92%

6.3.5 Comparison of Proposed Work with Relevant Works

Table 6.3 represents the comparison between previous automatic brain tumor detection techniques and proposed technique. Different authors approached various ways to detect brain tumors with artificial intelligence. Authors aimed to get higher accuracy so that tumors can be detected flawlessly. This proposed method explored the best result of 99.8% accuracy within 18 iterations. It can be claimed that proposed tumor classification technique can classify more accurately than other previous techniques.

Graphical analysis of Figure 6.11 is done from Table 6.3. In the case of automatic brain tumor detection, the theme of existing works is increasing the accuracy level so that the tumor can be classified perfectly. Authors are also aware of this theme and proposed two algorithms for a better outcome. The best result of this proposed work is 99.8% which was found from the second approach and Figure 6.11 is showing that proposed work showed the better result.

TABLE 6.3   Comparison of previous analogous work and proposed method

CLASSIFICATION TECHNIQUE	CLASSIFICATION ACCURACY	REFERENCES
DWT + PCA + SVM	90%	Sanjeev Kumar et al. (2017)
Image Enhancement + Median Filter + Gabor filer + BPNN	89.9%	Amsaveni and Albert Singh (2013)
Image Enhancement + TKFCM + ANN	97.1%	Alam et al. (2019)
K-Means + DWT + PCA + Kernel SVM	99.38%	Zhang and Wu (2012)
Image Resize and Rotation + CNN	96.56%	Badza and Barjaktarovic (2020)
Median filter + Gaussian filter + Morphological filter + Masking based segmentation + GLSM + SVM	99%	Pushpa and Louies (2019)
Image Enhancement + Morphological Filter + K-means + ANN	95%	Vijay and Subhashini (2013)
Region of interest (ROI) + DWT + Gabor Filter + BPNN	91.9%	Ismael and Abdel-Qader (2018)
Image Enhancement + Skull Stripping + GLRLM + SVM	91.66%	Parveen (2015)
Proposed Method (Best result obtained for: Sharpening Filter + FCM + DWT + PCA + ANN)	99.8%	Proposed Algorithm 2

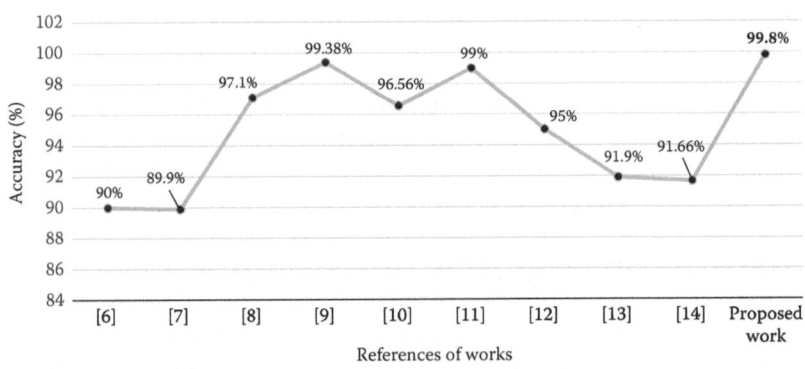

FIGURE 6.11   Graphical Representation of a Comparison of Existing Classification Techniques and Proposed Work.

6.4 CONCLUSION

The major theme of this work was to classify brain tumor types with higher accuracy to contribute to medical treatment and for that, authors proposed two algorithms with different clustering algorithms. Classification of brain tumor with higher accuracy by using the sizeable amount of patient data is a tough task. Artificial intelligence also works with the same aim to develop medical treatment. In that manner, proposed method used 489 patient's data to classify brain tumor types and examine the result for two proposed approaches with various training functions, nine to be specific. For the first approach, pre-processing steps included resizing and sharpening filter. Then k-means algorithm was implemented for the purpose of segmentation. Feature extraction was achieved by the utilization of 2D DWT and feature reduction was accomplished by applying PCA. Then proposed classification employed artificial neural network and the nine different training algorithms were applied separately to find the best result. 'Levenberg–Marquardt' provided the highest 98.0% accuracy for the first approach but that was not much satisfactory. For the second approach, after preprocessing, fuzzy C means algorithm was approached. Similar to the first approach, feature extraction and reduction were done. Then features data was used for ANN training in the same manner. Of the nine training algorithms, 'Levenberg–Marquardt' provided the best result of 99.8% accuracy, 100% sensitivity, and 99.59% specificity. In the second approach, accuracy increased sufficiently than the first approach. Finally, a flawless outcome was found and it was better than previous several existing classification accuracy. This improved accuracy is helpful for MRI brain tumor classification in the medical sector. Besides all superiority, some limitations have been found. Authors are interested to calculate tumor area numerically and this will be better if the best feature selection is possible before training.

ACKNOWLEDGEMENTS

The authors are thankful for the support which is cordially provided by Chittagong University of Engineering and Technology. Other facilities are provided by the department of Electronics and Telecommunication Engineering.

REFERENCES

Abdalla, H. E. M., & Esmail, M. Y. (2018). *Brain tumor detection by using artificial neural network. International Conference on Computer, Control, Electrical, and Electronics Engineering (ICCCEEE), IEEE*, pp. 1–6. 10.1109/ICCCEEE.2018.8515763

Ahmmed, R., & Hossain, M. F. (2016). *Tumor detection in brain MRI image using template based K-means and fuzzy C-means clustering algorithm. International Conference on Computer Communication and Informatics (ICCCI-2016), India, IEEE*. 10.1109/ICCCI.2016.7479972

Ahmmed, R., Swakshar, A. S., Hossain, M. F., & Rafiq, M. A. (2017). *Classification of tumors and it stages in brain MRI using support vector machine and artificial neural network. International Conference on Electrical, Computer and Communication Engineering (ECCE), IEEE*, pp. 229–234. 10.1109/ECACE.2017.7912909

Alam, M. S., Rahman, M. M., Hossain, M. A., Islam, M. K., Ahmed, K. M., Ahmed, K. T., Singh, B. C., & Miah, M. S. (2019). Image using template-based K means and improved fuzzy C means clustering algorithm. *Multidisciplinary Digital Publishing Institute (MDPI)*, 3(2), 27. 10.3390/bdcc3020027

Alkassar, S., & Abdullah, M. A. M. (2019). *Automatic brain tumor segmentation using fully convolution network and transfer learning. 2nd International Conference on Electrical, Communication, Computer, Power and Control Engineering ICECCPCE19, Mosul, Iraq, IEEE*, 13–14 February 2019. 10.1109/ICECCPCE46549.2019.203771

Amsaveni, V., & Albert Singh, N. (2013). *Detection of brain tumor using neural network. Fourth International Conference on Computing, Communications and Networking Technologies (ICCCNT), Tiruchengode, India, IEEE*, 4–6 July, 2013. 10.1109/ICCCNT.2013.6726524

Badza, M. M., & Barjaktarovic, M. C. (2020). Classification of brain tumors from MRI images using a convolutional neural network. *Multidisciplinary Digital Publishing Institute (MDPI)*, 10(6), 2–13. 10.3390/app10061999

Batra, D. (2014). Comparison between Levenberg-Marquardt and scaled conjugate gradient training algorithms for image compression using MLP. *International Journal of Image Processing (IJIP)*, 8(6), 412–422.

Brain Tumor Dataset. *Figshare Database*. https://figshare.com/articles/brain_tumor_dataset/1512427 [Accessed in 2017].

Brundha, B., & Nagendra, M. (2015). MR image segmentation of brain to detect brain tumor and its area calculation using K-means clustering and fuzzy c-means algorithm. *International Journal for Technological Research in Engineering*, 2(9), 1781–1784.

Dhanachandra, N., Manglem, K., & Chanu, Y. J. (2015). *Image segmentation using K-means clustering algorithm and subtractive clustering algorithm. Eleventh International Multi-Conference on Information Processing-2015 (IMCIP-2015)*, *54*, 764–771. Elsevier. 10.1016/j.procs.2015.06.090

Hazra, A., Dey, A., Gupta, S. K., & Ansari, M. A. (2017). *Brain tumor detection based on segmentation using MATLAB. International Conference on Energy, Communication, Data Analytics and Soft Computing (ICECDS-2017), Chennai, India, IEEE*, 1–2 August 2017. 10.1109/ICECDS.2017.8390202

Hemanth, G., Janardhan, M., & Sujihelen, L. (2019). *Design and implementing brain tumor detection using machine learning approach. Proceedings of the Third International Conference on Trends in Electronics and Informatics (ICOEI 2019), IEEE Xplore.* 10.1109/ICOEI.2019.8862553

Ismael, M. R., & Abdel-Qader, I. (2018). *Brain tumor classification via statistical features and back-propagation neural network. IEEE International Conference on Electro/Information Technology (EIT).* 10.1109/EIT.2018.8500308

Jose, A., Ravi, S., & Sambath, M. (2014). Brain tumor segmentation using K-means clustering and fuzzy C-means algorithms and its area calculation. *International Journal of Innovative Research in Computer and Communication Engineering*, *2*(3), 60–64. 10.13140/RG.2.1.3961.5841

Kaur, G., & Oberoi, A. (2018). Development of an efficient clustering technique for brain tumor detection for MR images. *International Journal of Computer Sciences and Engineering (JCSE)*, *6*(9), 404–409. 10.26438/ijcse/v6i9.404409

Kukreja, H., Bharath, N., Siddesh, C. S., & Kuldeep, S. (2016). An introduction to artificial neural network. *International Journal of Advance Research And Innovative Ideas In Education*, *1*(5), 27–30.

Kumar, S., Dabas, C., & Godara, S. (2017). Classification of brain MRI tumor images: A hybrid approach. *Information Technology and Quantitative Management (ITQM2017)*, *122*, 510–517. 10.1016/j.procs.2017.11.400

Kurukilla, J., & Gunavathi, K. (2014). Lung cancer classification using neural network for CT images. *Computer Methods and Programs in Biomedicine*, *113*(1), 202–209.

Mishra, S., Prusty, R., & Hota, P. K. (2015). *Analysis of Levenberg-Marquardt and scaled conjugate gradient training algorithms for artificial neural network based LS and MMSE estimated channel equalizers. International Conference on Man and Machine Interfacing (MAMI), Bhubaneswar, India, IEEE*, 17–19 December 2015. 10.1109/MAMI.2015.7456617

Parveen, A. (2015). *Detection of brain tumor in MRI Images, using combination of FCM and SVM. 2nd International Conference on Signal Processing and Integrated Networks (SPIN) IEEE.* 10.1109/SPIN.2015.7095308

Preetha, R., & Suresh, G. R. (2014). *Performance analysis of FCM in automated detection of brain tumor. World Congress on Computing and Technologies, Trichirappalli, India, IEEE*, 27 February–1 March 2014. 10.1109/WCCCT.2014.26

Pushpa, B. R., & Louies, F. (2019). Detection and classification of brain tumor using machine learning approaches. *International Journal of Research in Pharmaceutical Sciences*, *10*(3), 2153–2162. 10.26452/ijrps.v10i3.1442

Rakesh, M., & Ravi, T. (2012). Image segmentation and detection of tumor objects in MR brain images using fuzzy C-means (FCM) algorithm. *International Journal of Engineering Research and Applications (IJERA)*, *2*(3), 2088–2094.

Saji, S. A., & Balachandran, K. (2015). *Performance analysis of training algorithms of multilayer perceptrons in diabetes prediction. International Conference on Advances in Computer Engineering and Applications (ICACEA), Ghaziabad, India, IEEE*, 19–20 March 2015. 10.1109/ICACEA.2015.7164695

Shanmuganathan, S. (2016). Artificial neural network modelling: An introduction. In S. Shanmuganathan & S. Samarasinghe (Eds.), *Artificial neural network modelling. Studies in computational intelligence*, 628. Springer. 10.1007/978-3-319-28495-8_1

Shanmuga Priya, S., & Valarmathi, A. (2018). Efficient fuzzy c-means based multilevel image segmentation for brain tumor detection in MR images. *Springer*, *22*, 81–93. 10.1007/s10617-017-9200-1

Shasidhar, M., Sudheer Raja, V., & Vijay Kumar, B. (2011). *MRI brain image segmentation using modified fuzz C-means clustering algorithm. 2011 International Conference on Communication Systems and Network Technologies, IEEE*, pp. 473–478. 10.1109/CSNT.2011.102

Siar, M., & Teshnehlab, M. (2019). *Brain tumor detection using deep neural network and machine learning algorithm. 9th International Conference on Computer and Knowledge Engineering (ICCKE 2019), Mashhad, Iran, IEEE*, 24–25 October, 2019. 10.1109/ICCKE48569.2019.8964846

Vijay, J., & Subhashini, J. (2013). *An efficient brain tumor detection methodology using K-means clustering algorithm. International Conference on Communication and Signal Processing, India, IEEE*, 3–5 Apr. 2013. 10.1109/iccsp.2013.6577136.

Zhang, Y., & Wu, L. (2012). An MR brain images classifier via principal component analysis and kernel support vector machine. *Progress In Electromagnetics Research*, *130*, 369–388. 10.2528/PIER12061410

Recognition of the Most Common Trisomies through Automated Identification of Abnormal Metaphase Chromosome Cells

7

Reem Bashmail[1], Muna Al-Kharraz[1],
Lamiaa A. Elrefaei[1,2], Wadee Alhalabi[1,3], and Mai Fadel[1]

[1] *Computer Science Department, Faculty of Computing and Information Technology, King Abdulaziz University, Jeddah, Saudi Arabia*
[2] *Electrical Engineering Department, Faculty of Engineering at Shoubra, Benha University, Cairo, Egypt*
[3] *Computer Science Department, College of Engineering, Effat University, Jeddah, Saudi Arabia*

Contents

DOI: 10.1201/9781003172772-7

7.1 INTRODUCTION

7.1.1 Background

In 1956, Levan and Tjio discovered that the number of human chromosomes is 46 (Wang et al., 2005) and in 1960, the Denver group established as a classification standard. Denver group classified chromosomes into seven groups (A–G) according to their length and centromere position (Markou et al., 2012) (Table 7.1). Since then, karyotyping of human chromosomes has become an important clinical procedure in the screening and diagnosis of cancers and genetic disorders (Piper et al., 1980).

The cells used for chromosome imaging and analysis are taken mostly from blood samples, amniotic fluid, and bone marrow. These samples are cultured overnight in a mitotic arresting agent. Then, to increase cell volume and to spread the chromosomes apart, the cells are processed with hypotonic solutions (Lejeune et al., 1960). The processed cells are dropped onto a standard glass microscope slide where they are allowed to dry; then, the cell nucleus is photographed under a light microscope. These steps represent the G-banding or Giemsa banding technique (Lejeune et al., 1960). Because it is difficult to identify and group chromosomes based on simple staining (as the uniform color of the structures the differentiation process inefficient), it is used in cytogenetics on metaphase chromosomes to produce a visible karyotype. Therefore, techniques like G banding were developed that made 'bands' appear on the chromosomes. These bands were the same in appearance on the homologous chromosomes, and thus, the identification became easier and more accurate. The details of the G banding analysis process are illustrated in El-Khateeb (2013).

After the application of G banding, the sample is ready for reading and karyotyping, which is the process of pairing and ordering all the chromosomes of an organism (classified into 24 types in normal cases), thus providing a genome-wide snapshot of an individual's chromosomes. Figure 7.1 demonstrates

TABLE 7.1 Denver group classification

Chromosome class	1–3	4–5	6–12, X	13–15	16–18	19–20	21–22, Y
Denver Group	Group A	Group B	Group C	Group D	Group E	Group F	Group G

(a)   (b)

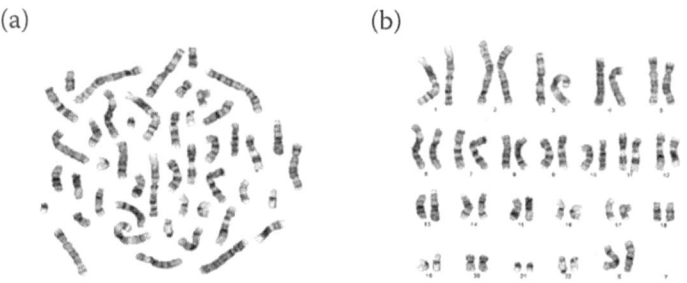

FIGURE 7.1   (a) Female Metaphase Spread Chromosome Image and (b) Its Karyotype (Bashmail et al., 2018).

a female normal metaphase spread and the corresponding karyotype of chromosomes (Bashmail et al., 2018).

Manually performing these steps takes a lot of time and effort and can reduce the efficiency and accuracy of diagnosis. Therefore, for the last 30 years, the development of automated karyotyping systems has attracted a lot of attention (Lejeune et al., 1960) to make this process easier with the help of a smart health system idea. The smart health system uses all the data coming from different sources to help make better decisions and improve healthcare (Rayan et al., 2019).

Although many efforts were done to develop automated karyotyping systems, they still need substantial human intervention. That is what motivated us to work on developing an automatic identification system to recognize the most common trisomies by determining if there is an extra copy of (13, 18, 21, X, and Y) classes. A trisomy is an abnormality type where, instead of two, there are three copies of a particular chromosome. The most common chromosomes trisomy types in humans that survive to birth are trisomy 13 (Patau syndrome), trisomy 18 (Edwards syndrome), and trisomy 21 (Down syndrome). Among them, trisomy 18 and trisomy 21 are the most common. With trisomy 13, in rare cases, a baby can survive, giving rise to Patau syndrome. Autosomal trisomy can be associated with intellectual disability, birth defects, and shortened life expectancy (Shoblak & Baraka, 2015). Trisomy of the sex chromosomes, compared to the trisomy of the autosomal chromosomes, ordinarily has less intense consequences. People may show few or no symptoms and have an ordinary life. It includes XXX syndrome (Triple X syndrome), XXY syndrome (Klinefelter syndrome), and XYY syndrome (Jacobs syndrome).

7.1.2 Contributions

Focusing on the issues discussed above, this work develops an automatic identification system to recognize the most common trisomies from metaphase chromosome images. We summarize the contributions of this study as follows:

1. Investigating three different combinations of six extracted features that can define the chromosomes: length, relative length, centromere index, area, relative area, and density profile.
2. Train and test a special type of neural network known as an Autoencoder classifier on a Diagnostic Genomic Medicine Unit (DGMU) laboratory dataset and benchmarking it against a classic neural network to classify chromosomes into six classes (13, 18, 21, X, Y, and other).
3. Recognize the most common trisomies by determining if there is an extra copy of (13, 18, 21, X, and Y) classes.

7.1.3 Chapter Organization

The rest of this chapter is structured as follows: related works are summarized and discussed in Section 7.2. The research design and methodology are explained in detail in Section 7.3. Then, the results are discussed and analyzed including experimental evaluation in Section 7.4. Finally, Section 7.5 presents conclusions and future work.

7.2 RELATED WORKS

Chromosomal classification, the first and most fundamental step of the chromosome abnormality recognition, is an important pattern recognition problem. Hence, the need to develop an automated chromosomal classification system emerged. In developing such a system, the most important challenges faced are the computation of the chromosome's features and the selection of feature classifiers.

To these means, the authors in Tso & Graham (1983) extracted and measured chromosome areas and the corresponding centromeric indices using a television-based image analyzer from 90 human cells. They adopted a maximum likelihood procedure with the transportation algorithm to obtain optimal maximum-likelihood and classifying chromosomes to ten groups ((1), (2), (3), (4–5), (6–12, X), (13–15), (16), (17–18), (19–20), (20–21, Y)). They obtained a 3.3% misclassification rate.

In Piper and Granum (1989), the segmenting of G-banded chromosomes occurs by using a threshold to the initial cell image. They acquired three datasets: Copenhagen dataset that contains 180 metaphase images, Edinburgh dataset that contains 125 metaphase images, and Philadelphia dataset that contains 130 metaphase images. From these datasets, they extracted size, area, density, length, centromere index, global band pattern, and global shape features. They classified chromosomes to 24 classes using maximum likelihood. They achieved an error rate of 8.4%, 19.6%, and 27% for Copenhagen, Edinburgh, and Philadelphia datasets, respectively.

The research work in Keller et al. (1995) separated the foreground of metaphase image from the background by locally refined thresholding. Then, chromosomes segmentation occurred by Connected Component Analysis and Thinning Objects to One Pixel Wide 'Skeletons'. The datasets contain 23400 of 16 and 18 chromosomes and they extracted centromeric index, relative length, and banding pattern features. The fuzzy logic classifier was used to differentiate between 16 and 18 chromosomes and achieved 100% classification accuracy for chromosome 16 and 87% accuracy for chromosome 18.

The authors in Popescu et al. (1999) performed a preprocessing step that involved thresholding with a combination of Kittler-Illingworth and Otsu methods, binary noise cleaning and connected component labeling on a dataset containing 219 metaphase images collected from the University of Missouri Ellis Fischel Cancer Center. Splitting touching chromosomes were carried out by pale path and cross section sequence graph (CSSG). This process was followed by feature extraction and legal segment identification. The extracted features were relative length, area, and 18 features based on weighted density distribution (WDD). To associate between segments and chromosome classes, they used two methods: benchmark for non-overlapped chromosomes and a mathematical programming for overlapped chromosomes to classify chromosomes to 24 classes and achieved 87% accuracy.

In Rungruangbaiyok and Phukpattaranont (2010), noises were removed by the average filter and the contrast was improved by histogram equalization. After enhancement, the segmentation of chromosomes is done by thresholding technique and Otsu's algorithm to segment out 60 metaphase images. Also, dilation and erosion were used to enhance the images. Area, perimeter, band's area, band profile, and singular value decomposition features were extracted and fed into a probabilistic neural network (PNN) classifier. The classification has occurred in two steps. The first step classified chromosome images into six groups; then in the second step, the six groups were classified into twenty-four classes. They obtained 61.30% classification accuracy for males and 68.18% for females.

In Roshtkhari and Setarehdan (2008), the dataset was collected from Cytogenetic Laboratory of Cancer Institute, Imam Hospital, Tehran, Iran. The images preprocessed by using intensity normalizing then, the features were extracted from the density profile of the chromosome by the Discrete Wavelet Transform (DWT). Linear Discriminant Analysis (LDA) has been applied for feature reduction. Also, they extracted the relative length and centromeric index. A three-layer feed-forward perceptron neural network was trained by means of the backpropagation procedure in order to classify three classes (16, 17, and 18) in the group E of Denver groups and by this way they achieved 99.3% average correct classification rate.

The authors in Sathyan et al. (2016) enhanced the images by using noise removal, thresholding, and median filter after that, segmentation is carried out to segment out 1628 chromosome images based on labeled region and bounding box. A Two-feedforward Artificial Neural Network (ANN) classifier was utilized to classify chromosomes into Denver groups based on chromosome global features (medial axis length, area, contour length, and average grey value of each chromosome) and chromosome centromeric features (contour length p/q ratio and medial axis length ratio of p/q). From the Denver group, each chromosome was recognized based on its textural features (homogeneity, correlation, contrast, energy, and entropy) using Gray Level Co-occurrence Matrix (GLCM). Their proposed method obtained 75% classification accuracy.

In the work by Groen et al. (1989), a local band description method (Laplace band descriptor) and global band description (WDD) techniques were evaluated by using several classification methods like non-parametric Bayes rule, Nearest Neighbor, and linear Fisher discriminant. They utilized two datasets: Copenhagen dataset that contained 7000 chromosomes and Leyden dataset that contained 920 chromosomes. The used features were length, centromere index, and 8 weighted density distribution (WDD) functions. They classified chromosomes into 4, 5, 7, 9–12, X classes. They achieved the smallest error rate (2.1%) on Copenhagen dataset when using global band description, while for Leyden dataset they achieved the smallest error rate (4%) when using local band description.

Three different datasets used in Graham et al. (1992): Copenhagen dataset that contains 8106 chromosomes, Edinburgh dataset that contains 4569 chromosomes, and Philadelphia dataset that contains 5817 chromosomes. They investigated two approaches. In the first approach, they used normalized length, centromeric index, density profiles features as input to the MLP classifier. They classified chromosomes into 22 classes and obtained 6.9%, 19.3%, and 25.6% misclassification rates on Copenhagen, Edinburgh, and Philadelphia datasets, respectively. In the second approach, a compound network (pre-classifier) consisting of a two-layer Multilayer Perceptron (MLP) with two input nodes and ten output nodes was introduced. It produced an output approximating the Denver classification based on size and centromeric index. These outputs, together with the density profile, form the inputs to a second stage classifier, also a two-layer MLP. They classified chromosomes into 24 classes and obtained 6.2%, 17.1%, and 23% misclassification rates on Copenhagen, Edinburgh, and Philadelphia datasets, respectively.

After conducting normalizing feature measurement technique, authors in Carothers and Piper (1994) used different classifiers, like transportation procedure, rearrangement classifier RC3, and context-independent machine learning with the following features: size, shape, and banding pattern in order to classify chromosomes into 24 classes. They utilized Copenhagen dataset that contains 8106 chromosomes, Edinburgh dataset that contains 4569 chromosomes, and Philadelphia dataset that contains 5817 chromosomes. Transportation procedure got 4.4%, 15.5%, and 19.9% error rates on Copenhagen, Edinburgh, and Philadelphia datasets, respectively. For rearrangement classifier RC3, they obtained 5.7%, 16.4%, and 20.6% error rates on Copenhagen, Edinburgh, and Philadelphia datasets, respectively. A context-independent ML achieved 6.5%, 18.3%, and 22.8% error rates on Copenhagen, Edinburgh, and Philadelphia datasets, respectively.

The work in Lerner et al. (1995) examined two approaches to the Medial Axis Transform (MAT)-based features; one of them was based on skeletonization and the second was based on a piecewise linear

(PWL) approximation. They acquired the dataset from the Institute of Medical Genetics of Soroka Medical Center, Beer-Sheva, Israel, that contains chromosomes of five different classes (2, 4, 13, 19 and X) from more than 150 different cells. The authors proposed the MLP classifier that used a combination of the features (density profile, centromeric index, and chromosome length) to classify five classes of the chromosomes (2, 4, 13, 19, and X). The approach based on PWL received a classification accuracy greater than 97%, while the approach based on skeleton received over 98% classification accuracy. They concluded that centromere index and chromosome length features are more suitable for classification than the density profile features.

Chromosomes in Badawi et al. (2003) were separated from 20 metaphase images by using OTSU. They extracted the following categories of features: G-Banding greyscale profile features, global chromosome's features, chromosome's centromeric features, number of real bands, real grey level banding, distances between centers of bands, and real bands thickness. They proposed a classification system based on template matching, fuzzy rule, and neural networks. They achieved 93.54% classification accuracy when using fuzzy rule, 94.76% accuracy when using neural networks, and 96.89% accuracy when using template matching.

Also, the authors in Moradi and Setarehdan (2006) obtained the dataset from Cytogenetic Laboratory of Cancer Institute, Imam Hospital, Tehran, Iran, and it consisted of 303 curved chromosomes of classes 16, 17, and 18. They proposed a group of features that consisted of width, position, and intensity of the most important characteristic regions (bands). Width of first and second dark bands, position of first and second dark bands, and the grey level Intensity of first and second dark bands in the chromosomes and applied a three-layer Artificial Neural Network (ANN) to classify each chromosome into one of these classes (16, 17, and 18). The obtained a classification accuracy of 98.6%.

Chromosome images in Mashadi and Seyedin (2007) collected from the Royan Institute include 42,000 chromosome images and they normalized the images in size and intensity. The authors decided to test a support vector machine (SVM) with Gaussian kernel efficacy using pixels as the input to the classifier instead of extracting the banding profiles or other features. They used One-versus-All method to apply SVM to multiclass problem where the number of trained SVMs is equal to the number of classes (24) and by different experiments they got the best classification accuracy of 95.9%.

Around 150 metaphase cells in Wang et al. (2009) were acquired from the genetic laboratory of the University of Oklahoma Health Science Center (OUHSC). They computed 31 features from these feature categories: pixel distribution, centromere index, local band patterns, and processed band patterns. Chromosome classes are differentiated by applying multistep classifier that includes two decision layers with eight ANNs. In the first layer, a single ANN was applied to classify chromosomes to seven classes then in the second layer, seven ANNs were employed for classifying the seven classes to identify individual chromosomes. To find the optimal topology of the ANN by selecting an optimal set of features from a pool of 31 features and determine the appropriate number of hidden neurons, the genetic algorithm (GA) was used. They achieved 67.5% to 97.5% classification accuracy.

Authors Vantiha and Venmathi (2011) extracted global and textural features from the dataset consisted of 4600 segmented chromosome images. The extracted global features are relative area, relative length, and centromeric index while the extracted textural features are angular second moment, contrast, correlation, inverse difference moment, variance, and homogeneity. Two stage neural network classifiers were proposed. The first stage classified chromosomes into seven groups (A–G) based on chromosomes morphological features using Self Organising Map Neural Network. Then, in the second stage, authors developed a hybrid neural network approach that combined the LVQ, K- Mean and Naïve Bayes in conjunction with a serial fusion for classifying the seven groups to twenty-four classes and obtained 98% accuracy.

Biomedical Imaging Laboratory (BioImLab) dataset was employed in Poletti et al. (2012) and contains 5474 chromosome images which are publicly available. They extracted area, chromosome length, perimeter, and 64 samples each for the density and contour profile features. The classification was accomplished by two-layer ANN classifier, followed by reassignment of the classes to rearrange the 24 classes. They achieved 94% average accuracy.

Gagula-Palalic and Can (2014), utilized Sarajevo dataset that has chromosome features (chromosome length, length of short p-arm, and ten principal components obtained from band pattern vectors) for 3300 chromosomes obtained from the Clinical Center of the University of Sarajevo. A Competitive Neural Network Teams (CNNTs) was proposed that ensemble of ANN and nearest neighbor classifiers. It consisted of 462 simple perceptrons where each perceptron was trained to distinguish between two classes to form 22×21 learning machines. They obtained 1.73% error rate in classifying chromosomes to 22 classes.

In Saranya et al. (2015), the separation of chromosomes was done by using Fuzzy-C Mean method. Prior to that, preprocessing was carried out by median filter. The features were extracted based on grey level co-occurrence feature extraction algorithm and using SVM as a classifier and got 95.89% accuracy in classifying 24 classes. They didn't mention the source and size of the dataset.

The research work in Sharma et al. (2017) utilized a non-expert crowd from CrowdFlower to segment chromosomes in the metaphase images collected from a hospital and contains 400 healthy patient metaphases. They preprocessed the images by straightening of bent chromosomes, finding bending orientation, finding the bending center of curved chromosomes, stitching of the two arms of the chromosome, reconstruction, and normalizing the length. For feature extraction and classification, they built their own Convolutional Neural Network (CNN) compromised of four blocks. Every block contained two convolutional layers, one dropout, and one maxpooling layer. The final section of CNN contained two fully connected layers and a softmax layer with 24 units according to the number of chromosome classes. They got 68.5% accuracy without preprocessing steps and 86.7% accuracy with preprocessing steps.

Authors in Somasundaram et al. (2018) preprocessed the images by using median filter and morphological operations, then applied MOGAC method to segment the chromosomes from dataset containing 1000 touching chromosomes, 1000 overlapping, and 500 multiple overlapping chromosomes with normal and abnormal cases. For separating overlapping and touching chromosomes, they applied hypothesis analysis. The SVM and PNN were used for classification of chromosomes into 24 classes. With SVM they extracted chromosome length, centromere index, and similarity index features and obtained 97% accuracy. While with PNN, they extracted chromosome length, centromere position, and perimeter features and achieved 96% accuracy.

Table 7.2 presents a summary of some studies regarding the chromosome classification systems. This is done with respect to segmentation technique, selected features, classification methods, datasets used, number of classified classes, and outcomes.

As can be observed from previous studies, none of them recognized the most common trisomy (13, 18. 21, x, and Y) caused by all five chromosomes classes. Therefore, in this work, we train a special type of multi-layer neural network (known as an Autoencoder) because it gave the best results in the classification of chromosomes. This followed by calculating the number of chromosomes to determine whether there is a trisomy or not and identify the extra chromosome's class.

7.3 METHODS

Figure 7.2 shows a block diagram of the proposed automated chromosome classification system. The chromosome segmentation and feature extraction processes are explained in the following sections. Before any feature extraction, in order to achieve higher accuracy, we need to enhance the true color (RGB) input image and convert it to a greyscale version.

TABLE 7.2 Summary of reviewed past chromosome classification systems

RESEARCH	SEGMENTATION TECHNIQUE	FEATURE EXTRACTION	CLASSIFICATION	DATASET	NO. OF CLASSES	CLASSIFICATION PERFORMANCE
Tso & Graham, 1983	NA	Areas and centromeric indices	Optimal maximum-likelihood	90 human cells	10 groups ((1), (2), (3), (4–5), (6–12, X), (13–15), (16), (17–18), (19–20), (20–21,Y))	Misclassification rate: 3.3%
(Piper & Granum, 1989)	Threshold	Size, area, density, length, centromere index, global band pattern and global shape features	Maximum likelihood	• Copenhagen (180 metaphase) • Edinburgh (125 metaphase) • Philadelphia (130 metaphase)	24	Error rate: • Copenhagen: 8.4% • Edinburgh: 19.6% • Philadelphia: 27%
(Keller et al., 1995)	Connected Component Analysis and Thinning Objects to One Pixel Wide 'Skeletons'	Centromeric index, relative length, and banding pattern features	Fuzzy logic	23,400 of 16 and 18 chromosomes	Classes: 16, 18	Classification accuracy: • Chromosome 16: 100% • Chromosome 18: 87%
(Popescu 1999)	• Kittler-Illingworth and Otsu • Pale path and CSSG	Relative length, area, and 18 features based on weighted density distribution (WDD)	• Benchmark for non-overlapped chromosomes • Mathematical programming for overlaped chromosomes	219 metaphase images from the University of Missouri Ellis Fischel Cancer Center	24	Accuracy: 87%
(Rungruangbaiyok & Phukpattaranont, 2010)	Thresholding and Otsu's	Area, band's area, perimeter, band profile, and singular value decomposition	PNN	60 metaphase images	24	Accuracy: • Female: 68.18% • Male: 61.30%
(Roshtkhari & Setarehdan, 2008)	NA	Density profile features, centromeric	three-layer feed-forward PNN		Classes: 16, 17 and 18	Accuracy: 99.3%

(Continued)

TABLE 7.2 (Continued) Summary of reviewed past chromosome classification systems

RESEARCH	SEGMENTATION TECHNIQUE	FEATURE EXTRACTION	CLASSIFICATION	DATASET	NO. OF CLASSES	CLASSIFICATION PERFORMANCE
		index, and relative length		303 chromosome images of classes 16, 17 and 18		
(Sathyan et al., 2016)	Four-connectivity labelling algorithm and bounding box	• Chromosome global features (medial axis length, area, contour length, and average gray value of each chromosome) • Chromosome centromeric features (contour length p/q ratio and medial axis length ratio of p/q) • Textural features (homogeneity, correlation, contrast, energy, and entropy)	Two feed-forward ANN	1628 chromosome images	24	Accuracy: 75%
(Groen et al., 1989)	NA	Length, centromere index, and 8 weighted density distribution (WDD) functions	• Non-parametric Bayes rule • Nearest Neighbor Linear Fisher discriminant	• Copenhagen (7000 chromosomes) • Leyden (920 chromosomes)	Classes: 4, 5, 7, 9-12, X	Error rate: • Copenhagen: 2.1% • Leyden: 4%
(Graham et al., 1992)	NA	Normalized length, centromeric index, density profiles	Two approaches: 1. MLP 2. Compound network (two-layer MLP)	• Copenhagen (8106 chromosomes) • Edinburgh (4569 chromosomes) • Philadelphia (5817 chromosomes)	24	misclassification ratesFirst approach: • Copenhagen: 6.9% • Edinburgh: 19.3% • Philadelphia: 25.6% Second approach: • Copenhagen: 6.2%

(Continued)

TABLE 7.2 (Continued) Summary of reviewed past chromosome classification systems

RESEARCH	SEGMENTATION TECHNIQUE	FEATURE EXTRACTION	CLASSIFICATION	DATASET	NO. OF CLASSES	CLASSIFICATION PERFORMANCE
(Carothers & Piper, 1994)	NA	Size, shape and banding pattern	• Transportation procedure • Rearrangement classifier RC3 • Context-independent machine learning	• Copenhagen (8106 chromosomes) • Edinburgh (4569 chromosomes) • Philadelphia (5817 chromosomes)	24	• Edinburgh: 17.1% • Philadelphia: 23% Error rates: • Transportation procedure: 4.4%, 15.5%, and 19.9% on Copenhagen, Edinburgh, and Philadelphia respectively. • Rearrangement classifier RC3: 5.7%, 16.4%, and 20.6% on Copenhagen, Edinburgh, and Philadelphia respectively. • Context-independent ML: 6.5%, 18.3%, and 22.8% on Copenhagen, Edinburgh, and Philadelphia respectively
(Lerner et al., 1995)	NA	Density profile, centromeric index, and chromosome length	MLP	5 different classes (2, 4, 13, 19, and X) from more than 150 cells from the Institute of Medical Genetics of Soroka Medical Center, Beer-Sheva, Israel	Classes: (2, 4, 13, 19, and X)	Accuracy: • MAT based on PWL: 97%, • MAT based on skeleton: 98%

(Continued)

TABLE 7.2 (Continued) Summary of reviewed past chromosome classification systems

RESEARCH	SEGMENTATION TECHNIQUE	FEATURE EXTRACTION	CLASSIFICATION	DATASET	NO. OF CLASSES	CLASSIFICATION PERFORMANCE
(Badawi et al., 2003)	OTSU	Feature categories: G-Banding greyscale profile features, global chromosome's features, chromosome's centromeric features, number of real bands, real gray level banding, distances between centers of bands, and real bands thickness	• Template matching • Fuzzy rule • Neural networks	20 metaphase images	24	Accuracy: • Template matching: 96.89% • Fuzzy rule: 93.54% • Neural networks: 94.76%
(Moradi & Setarehdan, 2006)	NA	width, position, and intensity of the most important characteristic regions (bands). width of first and second dark bands, position of the first and the second dark bands, and gray level Intensity of the first and the second dark bands	Three-layer ANN	303 curved chromosomes of classes 16, 17, and 18	Classes: 16, 17, and 18	Accuracy: 98.6%
(Mashadi & Seyedin, 2007)	NA	Using pixels as the input to the classifier instead of extracting the banding profiles or other features	SVM	42,000 chromosome images	24	Accuracy: 95.9%
(Wang et al., 2009)	NA	31 features from these feature	Multistep classifier:	150 metaphase cells from the genetic	24	Accuracy: 67.5%–97.5%

(Continued)

TABLE 7.2 (Continued) Summary of reviewed past chromosome classification systems

RESEARCH	SEGMENTATION TECHNIQUE	FEATURE EXTRACTION	CLASSIFICATION	DATASET	NO. OF CLASSES	CLASSIFICATION PERFORMANCE
		categories: pixel distribution, centromere index, local band patterns, and processed band patterns	1. In the first layer, a single ANN classify chromosomes to seven classes 2. In the second layer, seven ANNs for classifying the seven classes to identify individual chromosomes.	laboratory of the University of Oklahoma Health Science Center.		
(Vanitha & Venmathi, 2011)	NA	• Chromosome global features (relative area, relative length, and centromeric index) • Textural features from GLCM (angular second moment, contrast, correlation, inverse difference moment, variance, and homogeneity)	• Classifying seven groups (A-G) using Self Organising Map Neural Network. • The seven groups were classified to 24 classes using a hybrid neural network approach that combines Naïve Bayes, LVQ, and K- Mean, in conjunction with a serial fusion.	4600 chromosome images	NA	Accuracy: 98%
(Poletti et al., 2012)	NA	Area, chromosome length, perimeter, and 64 samples each for the density and contour profile features	ANN	5474 chromosome images from BioImLab	NA	Accuracy: 94%
	NA				NA	Error rate: 1.73%

(Continued)

TABLE 7.2 (Continued) Summary of reviewed past chromosome classification systems

RESEARCH	SEGMENTATION TECHNIQUE	FEATURE EXTRACTION	CLASSIFICATION	DATASET	NO. OF CLASSES	CLASSIFICATION PERFORMANCE
(Gagula-Palalic & Can, 2014)		Chromosome length, length of short p-arm, and ten principal components obtained from band pattern vectors	Competitive Neural Network Teams (CNNTs) that ensemble of ANN and nearest neighbor classifiers	Features of 3300 chromosomes from the Clinical Center of University of Sarajevo		
(Saranya et al., 2015)	Fuzzy c mean	Texture features from GLCM	SVM	NA	24	Accuracy: 95.89%
(Sharma et al., 2017)	Non-expert crowd from CrowdFlower	CNN of four blocks. Every block contained two convolutional layers, one dropout, and one maxpooling layer. The final section contained two fully connected layers and a softmax layer with 24 unit		400 metaphase images from a hospital	24	Accuracy: 86.7%
(Somasundaram et al., 2018)	MOGAC and Hypothesis Analysis	• For PNN: chromosome length, centromere position, and perimeter features • For SVM: chromosome length, centromere index and similarity index	• PNN • SVM	• 1000 touching chromosomes • 1000 overlapping • 500 multiple overlapping chromosomes	24	Accuracy: • PNN: 96% • SVM: 97%

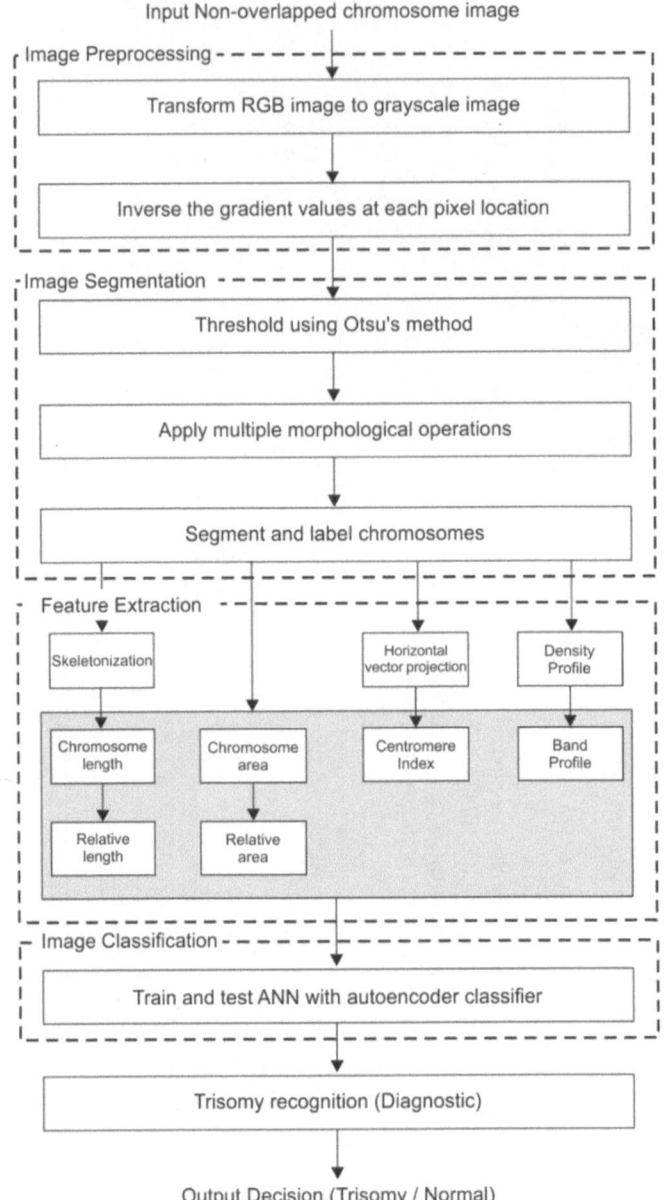

FIGURE 7.2   Block Diagram for Proposed Chromosome Classification.

7.3.1 Image Preprocessing and Segmentation

Human chromosomes contain important information for cytogenetic analysis. Chromosome images are acquired by microscopic imaging of metaphase on specimen slides. Digitized G-banding chromosome images usually show chromosomes in greyscale on white background. Therefore, if we apply segmentation directly, the white areas inside the chromosomes will be considered as background and it will not be included in segmented chromosomes. As a result, the images must be enhanced to increase the accuracy of the segmentation process and minimize the chances of losing internal chromosome parts.

We have adopted the system in Bashmail et al. (2018) in which the image is sharpened by applying the Difference of Gaussian (DoG) filter on a greyscale image before a classic technique of using Otsu's thresholding method followed by dilation, hole filling, and erosion morphological operations. Algorithm 7.1 (Bashmail et al., 2018) shows the preprocessing and segmentation steps. The input for Algorithm 7.1 is represented by an RGB chromosome metaphase image and the output is a segmented and outlined chromosome image. The steps are tested on 130 metaphase images (6011 chromosomes) that are free of overlap and severe bending. The average segmentation accuracy was 99.8% (Bashmail et al., 2018).

ALGORITHM 7.1   PREPROCESSING AND SEGMENTATION

1. Convert input image (RGB) to greyscale.
2. Apply (Difference of Gaussian (DoG)).
3. Apply normalization to a (0, 1) distribution.
4. Increase intensity for each pixel.
5. Find threshold using OTSU's method.
6. Apply Morphological Operations
    a. Filling holes on background pixels (change to 1).
    b. Erosion, using 6 × 6 square structuring element.
7. Remove small objects less than 50 pixels.
8. Label segmented chromosomes.
9. Store segmented chromosomes in a cell array.
10. Apply the final image as a mask to the original image

Figure 7.3 shows the preprocessing and segmentation steps applied to an original metaphase image. Samples of segmented images are displayed in Figure 7.4.

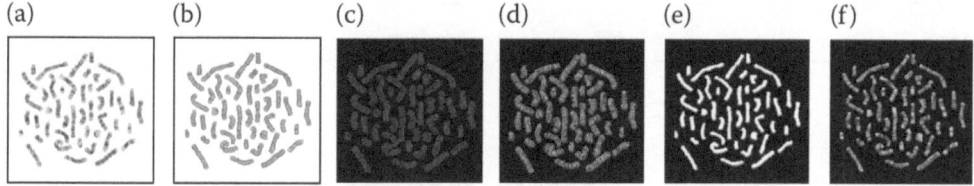

FIGURE 7.3   Preprocessing and Segmentation Steps: (a) Grayscale Image, (b) Inverse Weight, (c) Threshold Using OTSU's Method, (d) Filled Holes, (e) Erosion, (f) Labeled Segmented Chromosomes (Final Result).

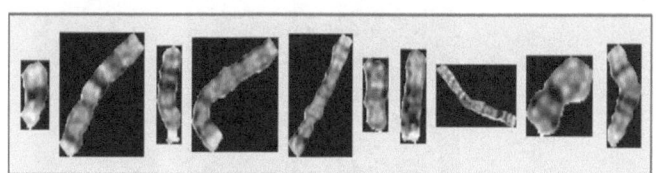

FIGURE 7.4   Sample of Segmented Chromosomes.

7.3.2 Feature Extraction

The quality of a chromosome classification system is mainly determined by its performance in the feature extraction. The main goal of the feature extraction process is to find a small set of characteristics that describe each class in the best possible way. Chromosome features are divided into two groups: morphological-based and texture-based features.

Information about the size and shape of the chromosome is provided by morphological-based features. However, they are not enough to make a complete classification; that is why texture-based features are necessary. The most important morphological features are chromosome length (L), the medial axis, and centromeric index (CI). From the texture-based features of chromosomes, the most popular are: density profile, mean grey value profile, gradient profile, and shape profile. They are used in the classification. Here, based on the previous studies, we select six features to classify the chromosomes. Those features are length, relative length, area, relative area, CI, and density profile.

7.3.2.1 Chromosome Length

Based on their length, human chromosomes are classified into four groups, as shown in Table 7.3. The geometric feature (length) is identified by the length of the medial axis which is determined from the skeleton (based on a distance transform) of each chromosome (Abid & Hamami, 2018), after removing the small branches (Figure 7.5).

7.3.2.2 Relative Length

The length of the chromosome differs from a sample to another according to the level of resolution of the chromosome's image. Therefore, we need to use the relative length because even if the length of the chromosome is different in the same class, the relative length is not. Relative Length is calculated

TABLE 7.3 Chromosomes groups based on length

Chromosome Class	1–5	6–15, X	16–18	19–22, Y
Size	Large	Medium	Short	Shortest

(a) (b) (c) (d)

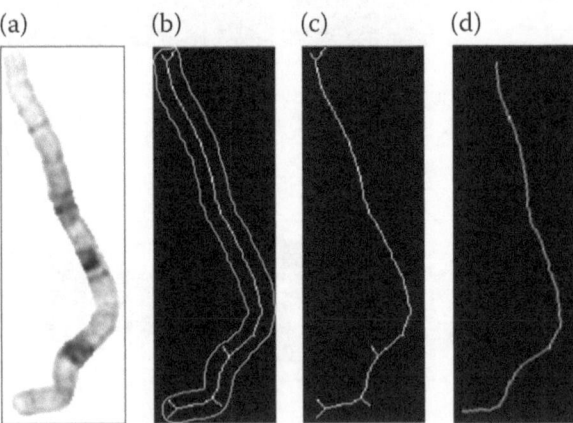

FIGURE 7.5 Chromosome Skeletonization: (a) Segmented Chromosome, (b) Showing Skeletonization and Boundary, (c) Skeletonization Only, and (d) Skeletonization Without the Small Branches.

individually for each chromosome within a metaphase image (Roshtkhari & Setarehdan, 2008) as in Equation (7.1).

$$Chromosome\ relative\ length = \frac{chromosome\ length\ of\ the\ individual\ chromosome}{total\ length\ of\ all\ chromosomes} \quad (7.1)$$

The relative length feature can be found correctly if the chromosome is straight or slightly curved (as in Figure 7.6(a)). However, there are cases where the bending is severe or the chromosome is coiled (Figure 7.6(b)) and the length will be calculated incorrectly. For that reason, we calculate the area and the relative area.

7.3.2.3 Chromosome Area

The chromosome area is calculated by estimating the summation of each white pixel in the thresholding image in the segmentation step (Rungruangbaiyok & Phukpattaranont, 2010) as shown in Figure 7.6(c).

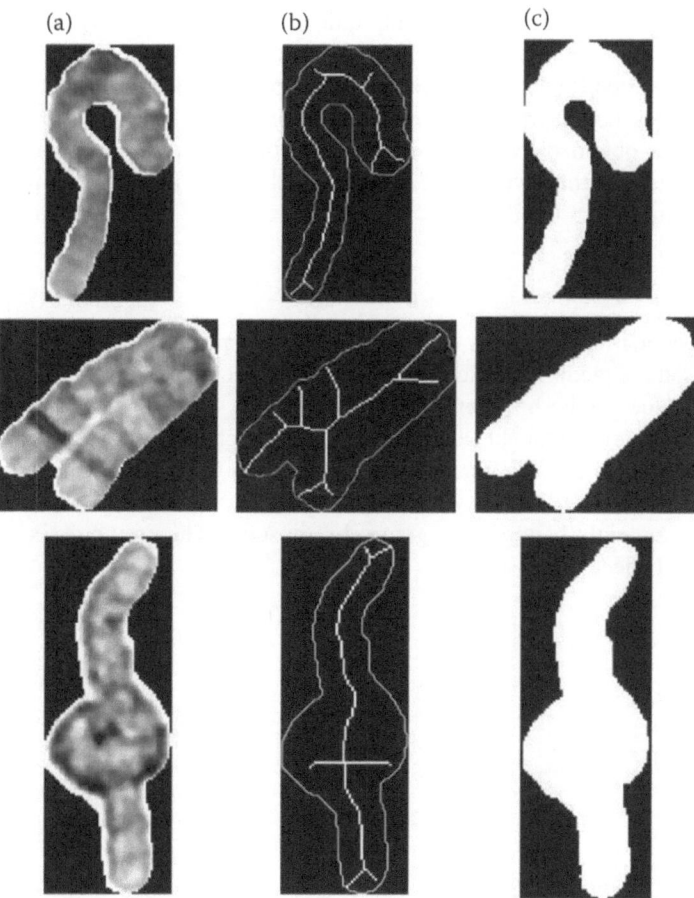

FIGURE 7.6 Samples of Chromosomes and Its Skeleton and Area: (a) (Simply/Severely) Bent and Coiled Chromosomes, (b) Its Skeleton, and (c) Its Area.

7.3.2.4 Relative Area

The area of the chromosome differs from a sample to another according to the level of resolution of the chromosome's image. Therefore, we extract the relative area. The relative area is one of the significant morphological features used to identify a chromosome. The relative area is calculated individually for each chromosome within a metaphase image (Webster, 2007) as in Equation (7.2).

$$Chromosome\ area\ length = \frac{chromosome\ area\ of\ an\ individual\ chromosome}{total\ area\ of\ all\ chromosomes} \tag{7.2}$$

7.3.2.5 Centromere Index (CI)

Due that any one of the 46 chromosomes belongs to the same length group as other chromosomes in the same group, as shown in Table 7.3, we can not use only length or area feature to classify individual chromosomes. For this reason, we also extract the CI.

The centromere is the narrowest part of the chromosome image and is not always located in the central area of a chromosome, as seen in Figure 7.7(a) (learn.genetics). The centromere connects the chromosome's arms, namely the short arm region (p arm) and a long arm region (q arm) as shown in Figure 7.7(b) (Locus (genetics)). The centromeric index is defined as the ratio of the short arm length to the whole chromosome length, To compute this ratio (Lerner et al., 1995), we must first find the centromere location.

The centromere is identified by applying the horizontal projection vectors from a binary image. Because it has only ones and zeros, the binary image is processed for projection vector calculation. The horizontal projection is obtained by adding all the ones of each row pixels. Figure 7.8 shows the horizontal projection vector for the given input image. This helps in identifying the centromere position, a vital feature for chromosome classification.

7.3.2.6 Density Profile

The density profile value is computed as the intensity value of pixel along the medial axis line. Figure 7.9 depicts the chromosome's medial axis and the density profile curve (Abid & Hamami, 2018). They are a sequence and their values range between 0 and 255.

After these values were obtained, and after several experiments of different threshold values, the threshold value (70) was chosen. To get the number of bands of the chromosome, the output values are thresholded to zero (if the pixel intensity value is less than or equal to 70) and one (if it is greater than 70). As shown, there are eight bands in the given chromosome. The previous six features were extracted and stored in an excel sheet.

(a)   (b)

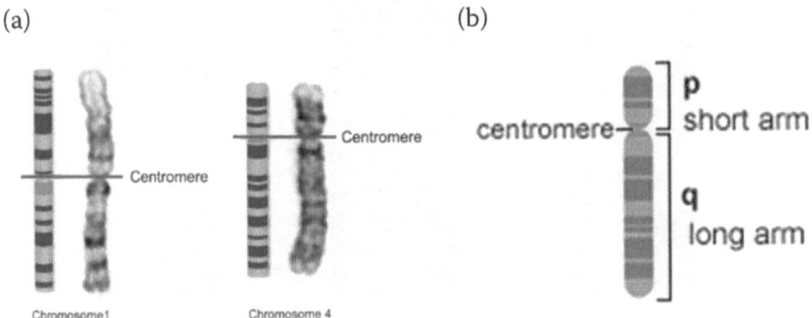

FIGURE 7.7   (a) Chromosome Centromere (learn.genetics); (b) Chromosome (p Arm) and (q Arm) (Locus (genetics)).

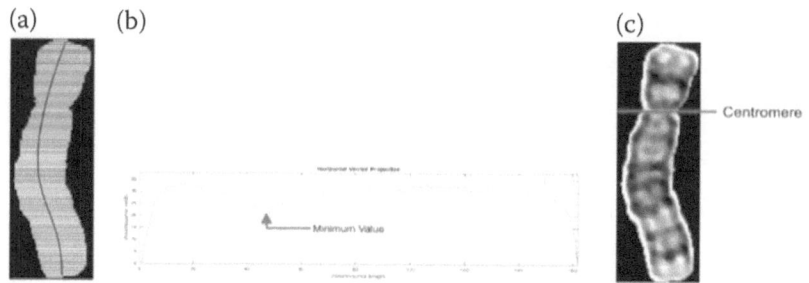

FIGURE 7.8 Chromosome Centromere by Horizontal Projection: (a) Horizontal Projection of Binary Image, (b) Horizontal Projection Results, and (c) Centromere Representation on the Chromosome Image.

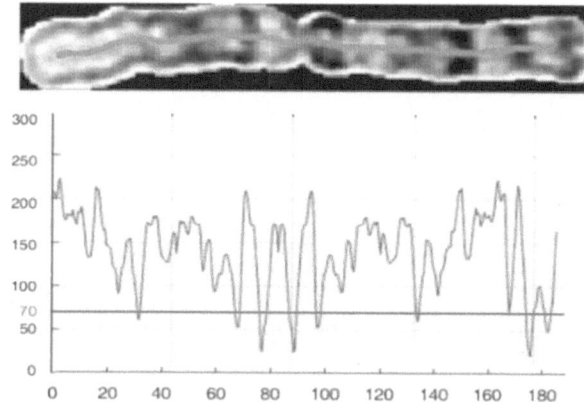

FIGURE 7.9 Extracted Chromosome Density Profile. Green Line Shows the Skeleton, Threshold Line Shown in Red.

7.3.3 Chromosome Classification

The ability of an ANN to achieve chromosome classification has been deeply studied in the past (Abid & Hamami, 2018). It was suggested that ANNs are the best chromosome classifiers. Neural networks with multiple hidden layers are able to solve complex classification problems, such as those relating to images or their features which makes neural networks be considered as black-box models because of the complexity of their underlying structures, therefore the difficulties to explain and provide the information for which reason the output is produced. However, most artificial neural network-based models are designed to provide powerful and accurate predictors and not designed with interpretability constraints. On the contrary, Explainable Artificial Intelligence (XAI) provides a better understanding of how predictions are made by studying methods that explain and interpret artificial neural network-based models (Adadi & Berrada, 2018; Vilone & Longo, 2020). In general, there is a trade-off between explainability and performance. Models that achieved the best accuracy are not very successful in terms of explainability. As well, models whose predictions are easy to understand are usually disadvantageous in terms of accuracy (Evren, 2020; Linardatos et al., 2021). In our case we concern about the accuracy and the final decision will be managed by the lab technician. For that reason, we apply multiple layer neural network as shown in Figure 7.10 (Text Classification Using Neural Networks). However, training neural networks with multiple hidden layers can be difficult in practice. One way to effectively train a neural network

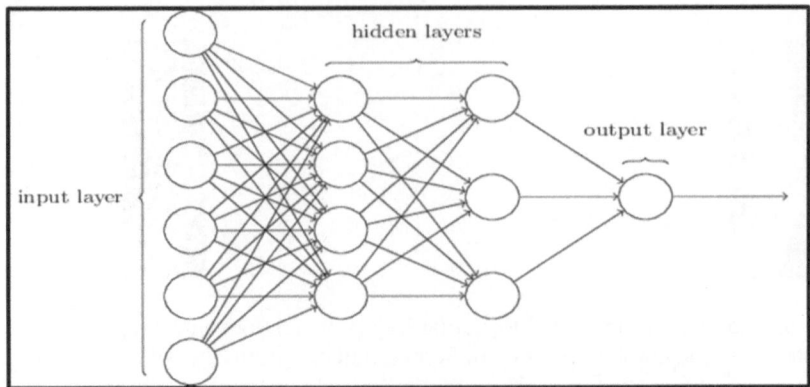

FIGURE 7.10   Multiple Layer Neural Network Architecture (Text Classification using Neural Networks).

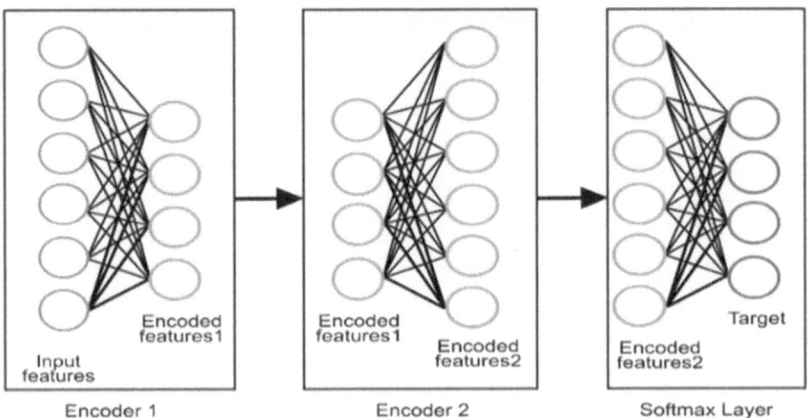

FIGURE 7.11   Multiple Layers Neural Network with Encoder Architecture.

with multiple layers is training one layer at a time by training a special type of network known as an Autoencoder for each desired hidden layer (Karaca et al., 2017) as shown in Figure 7.11.

The autoencoder is typically an unsupervised feedforward neural network trained to 'reconstruct' or 'recreate' its inputs, which forces the hidden layer to try to learn good representations of the inputs. Autoencoders consist of an encoder and a decoder. The algorithm of Autoencoders is as follows (Arpit et al., 2015):

The Autoencoder takes an input vector x and maps it to a hidden representation y through a deterministic mapping:

$$y = f_\theta(x) = s(\mathbf{Wx} + \mathbf{b})$$

where $\theta = \{W, b\}$. W is a weight matrix, b is a bias vector, and s is the sigmoid activation function:

$$S(m) = \frac{1}{1 + e^{-m}}$$

The hidden representation, y, is then mapped back to a reconstructed vector z, where

$$z = g_\theta(y) = s(Wy + b)$$

with $\theta = \{W, b\}$. Thus, each training $x^{(i)}$ is thus mapped to a corresponding $y^{(i)}$ and a reconstruction (of $x^{(i)}$) $z^{(i)}$.

The autoencoder normally has a small number of hidden units to force the network to learn a compressed representation of the input. Here, we will use a large number of hidden units. By doing so, the autoencoder will enlarge the given input's representation. Also, the autoencoder classifier was trained to classify chromosomes into six classes (13, 18, 21, X, Y, and other).

7.3.4 Trisomy Detection

We will use the trained neural network to test a new sample and give the decision whether there is a trisomy or not as in Algorithm 7.2, where the input is the identified targeted chromosome matrix and the output is the decision if there is a trisomy or not. The testing procedure is explained in Figure 7.12.

ALGORITHM 7.2   DECISION MAKING AND DIAGNOSTIC

1. Count no. of chromosomes in each class (13, 18, 21, X, Y)
2. If class13 ==3 Then

   trisomy 13
   Else if class18 ==3 Then
   trisomy 18
   Else if class21 ==3 Then
   trisomy 21
   Else if sex_class ==3 Then
   trisomy XY
   Else
   Case is normal
   End if

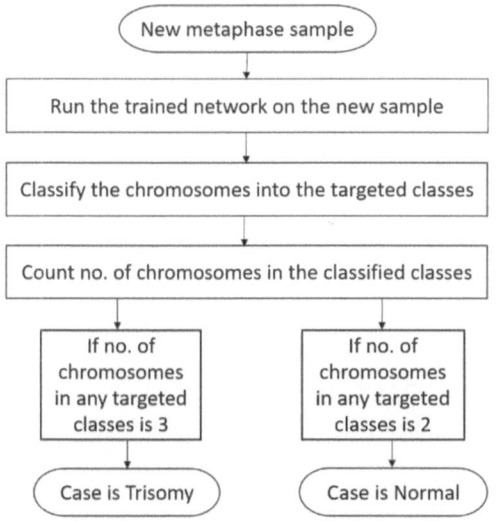

FIGURE 7.12   New Sample Test Procedure.

7.4 EXPERIMENTAL EVALUATION AND RESULTS

The proposed system is implemented using MATLAB® 2018a (Image processing toolbox, Neural network toolbox) on a laptop having a processor Intel (R) Core TM i7- 4770S CPU @ 3.10 GHz 3.09 GHz, 12 GB RAM, and graphics card Intel(R) HD Graphics 3000.

7.4.1 Dataset

A peripheral blood G-banding images are collected from Diagnostic Genomic Medicine Unit (DGMU) laboratory at King Abdulaziz University (KAU) based on a standard protocol. All samples have consent to be used for public research. For this study, the DGMU technicians provide 157 metaphase images obtained from patients containing (7266) chromosomes. These images are free of overlap and have severe bending. From this dataset, 130 images (82%) for training and 27 (18%) for testing were used.

Chromosome digital images are recorded by Olympus DP-10 Digital Camera, which has a unique 1.8-inch liquid crystal (LCD) color monitor. It provides the true color of 24-bit reproduction and sharp 1280 × 1024 pixel quality resolution in optical photomicrography. Digital images JPEG (Joint Photographic Experts Group) are recorded and compressed at a resolution of 72 or 144 dpi (Digital Imaging in Optical Microscopy).

7.4.2 Performance Measurement

The test results are indicated by two parameters, chromosome recognition accuracy and trisomy recognition accuracy. To find the chromosome recognition accuracy, the confusion matrix is used. It contains information about actual and predicted classifications done by the system. The total true positive TTP in the system is obtained through Equation (7.3) and the chromosome recognition accuracy is calculated with Equation (7.4). The trisomy recognition accuracy is calculated with Equation (7.5).

$$TTP_{all} = \sum_{j=1}^{n} \mathbf{x}_{jj} \qquad (7.3)$$

$$chromosome\ recognition\ accuracy = \frac{TTP_{all}}{Total\ Number\ of\ Testing\ Entries} \qquad (7.4)$$

$$trisomy\ recognition\ accuracy_{for\ each\ class} = \frac{TP_{Cases}}{Total\ Number\ of\ Cases\ Testing\ Entries} \qquad (7.5)$$

where TP is total true positive (diagnosed correctly) cases, n is the number of targeted classes.

7.4.3 Experiments and Results

We performed two experiments. The first one, train the network using the input of six classes (13, 18, 21, X, Y, and other) to classify chromosomes according to these six classes. The second experiment, train using the targeted chromosome classes (13, 18, 21, X, and Y). The test results of these two experiments will be presented according to two levels (individual chromosome recognition and trisomy recognition).

The training accuracy for the two experiments was 100%. Details of training, testing, and results will be presented in the next section.

Before performing the experiments, we have listed three combinations of the extracted features that can define the chromosomes.

1. Length, relative length, and centromere index
2. Area, relative area, centromere index, and density profile
3. Length, relative length, area, relative area, centromere index, and density profile

The previous feature sets were chosen because they were combining the basic properties of the chromosome (centromere index, and density profile) and the secondary properties (length, relative length, area, and relative area).

7.4.3.1 Experiment 1: Test (Classic and Multiple Autoencoder) Neural Network to Classify Chromosomes into Six Class

In this experiment, the network was trained using 1339 chromosomes as the input of the six classes that were distributed as shown in Table 7.4. The 'other' class consists of 14 chromosomes of each (1–12, 14–17, 19, 20, 22) classes to make the data balanced.

We trained a feed-forward neural network with two hidden layers (Autoencoder) by back-propagation (bp) learning rule and by benchmarking it against a classic neural network. Classic Neural Network and Autoencoder parameters are shown in Tables 7.5 and 7.6. Multiple layers of neural network with encoder architecture are shown in Figure 7.13. First, we train the two hidden layers individually in an unsupervised procedure by Autoencoder as shown in Figure 7.13. Training the first Autoencoder begins by training data without using the labels. This Autoencoder attempts to replicate its input at its output. Training the second Autoencoder is the same as training the first Autoencoder, but the difference is using the training data for the second Autoencoder – features that were generated from the first Autoencoder.

Then, to represent a categorical distribution over class labels, we train a final softmax layer in a supervised procedure using training data labels as shown in Figure 7.14. The obtained probabilities of

TABLE 7.4 Chromosomes distribution

Chromosome class	13	18	21	X	Y	Other
No. of chromosomes	264	267	279	214	51	264
Total	1339					

TABLE 7.5 Classic feedforward network MATLAB® parameters

PROPERTIES	VALUE
Network type	**Pattern recognition**
Training algorithm	**trainscg** (default)
Size of the hidden layer	**10** (default), **240** (used)
Performance function	**cross entropy** (default)

Bold entities are parameters used during the training of the both networks.

TABLE 7.6 Autoencoder MATLAB® parameters

PROPERTIES	VALUE
Autoencoder type	**sparse Autoencoder** (default)
Training algorithm	**trainscg** (default)
Size of the hidden layer	**10** (default), **240** (used)
Transfer function for the encoder	**logsig** (default)
Weights for the encoder	**0.001** (default)
Bias values for the encoder	**0 or 1**
Loss Function	**Msesparse** (default)
Maximum epochs	**1000** (default) **100000** (used)

Bold entities are parameters used during the training of the both networks.

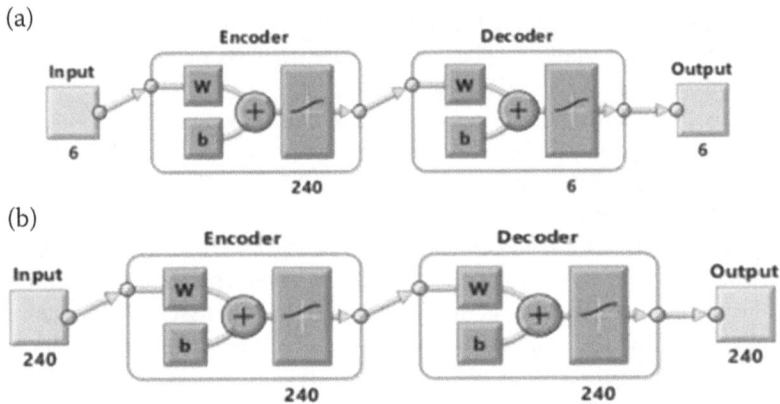

FIGURE 7.13 Training First and Second Autoencoders Neural Network: (a) First and (b) Second Autoencoder.

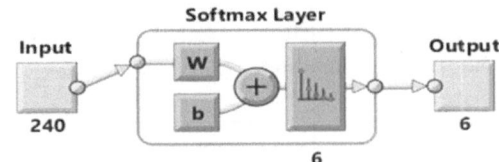

FIGURE 7.14 Softmax Function in the Output Layer.

each input element belong to a label. The next step consists of joining the layers together to form a deep network as shown in Figure 7.15. Then, it was trained in a supervised procedure.

The result for the Multilayer Autoencoder Neural Network can be improved by fine-tuning which implements backpropagation on the whole multilayer network by retraining network on the training data in a supervised procedure.

Table 7.7 shows a comparison of the average chromosome recognition accuracy test results (27 samples) between the Classic Neural Network and Multilayer Autoencoder Neural Network using three feature combinations.

As shown in Table 7.7, the best performance of the system is 84.6%. It is the best performance for area, relative area, CI, and DP set using the Multiple Autoencoder Neural Network. The result indicates

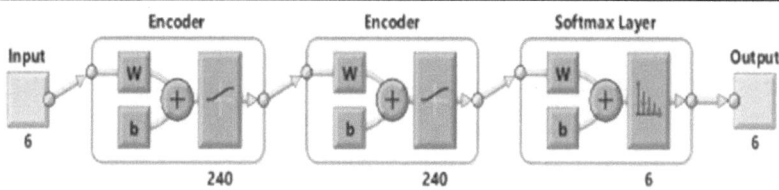

FIGURE 7.15   Stacked Network Formed by Encoders and Softmax Layer.

TABLE 7.7   Average chromosome recognition accuracy

	FEATURE COMBINATIONS	CLASSIC NEURAL NETWORK	MULTIPLE AUTOENCODER NEURAL NETWORK
1	Length, relative length, and Centromere Index (CI)	74.4 %	78.6 %
2	Area, relative area, Centromere Index (CI), and Density Profile (DP)	78.6 %	84.6 %
3	Length, relative length, area, relative area, Centromere Index (CI), and Density Profile (DP)	74.8 %	81.2 %

TABLE 7.8   Multiple autoencoder performance (average trisomy recognition accuracy) details

SAMPLE TYPE	NO. OF SAMPLES	CORRECTLY DIAGNOSED	%	AVERAGE TRISOMY RECOGNITION ACCURACY
Normal	3	3	100%	**84.6%**
Trisomy 13	6	5	83%	
Trisomy 18	7	4	57%	
Trisomy 21	5	5	100%	
Trisomy XY	6	5	83%	

that this combination leads to more robust classification performance. Likewise, the more complex model mostly gives better accuracy as multilayer autoencoder neural network outperform classic neural network.

Table 7.8 shows the details of the average of testing result (trisomy recognition accuracy) for 27 samples using the Multiple Autoencoder Neural Network where it was achieved 84.6% average trisomy recognition accuracy. The average computation time for the (preprocessing, segmentation, features extraction, and classification) steps for these samples was 34.32 s. Table 7.9 shows samples of chromosome recognition accuracy testing results presented by the confusion matrix using the Classic Neural Network and Multiple Autoencoder Neural Network.

In Table 7.9, the diagnosis process is done by counting the number of chromosomes in the targeted classes (13, 18, 21, X, and Y). If there are three copies, then the case is trisomy and if there is two or less, then the case is normal. Accordingly, in the presented test samples, when the Classic Neural Network used, cases 131 and 140 are diagnosed correctly (normal, trisomy 21), respectively. And cases 139, 134, and 137 are wrongly diagnosed (diagnosed as normal). However, when using the Multilayer Autoencoder Neural Network, all test samples are diagnosed correctly.

TABLE 7.9 Chromosome recognition accuracy testing results

SAMPLE NO.	SAMPLE IMAGE	DIAGNOSIS	CLASSIC NEURAL NETWORK	MULTIPLE AUTOENCODER	CLASSIC NEURAL NETWORK PREDICTION	MULTIPLE AUTOENCODER PREDICTION
131		Normal			Normal	Normal
139		Trisomy 13	80.9%	95.7%	Normal	Trisomy 13
140		Trisomy 21	85.1%	87.2%	Trisomy 21	Trisomy 21

(Continued)

TABLE 7.9 (Continued) Chromosome recognition accuracy testing results

SAMPLE NO.	SAMPLE IMAGE	DIAGNOSIS	CLASSIC NEURAL NETWORK	MULTIPLE AUTOENCODER	CLASSIC NEURAL NETWORK PREDICTION	MULTIPLE AUTOENCODER PREDICTION
134		Trisomy 18	68.1%	76.6%	Normal	Trisomy 18
137		Trisomy XY	80.9%	91.5%	Normal	Trisomy XY

Based on previous results, it is clear that the results of using Multilayer Autoencoder Neural Network are better than the classic neural network. So, in the next experiment, we will use the Multilayer Autoencoder Neural Network.

7.4.3.2 Experiment 2: Test Multiple Autoencoder Neural Network to Classify Chromosomes into Five Classes

In this experiment, the network was trained using the set (area, relative area, CI, and DP) features of the targeted classes (13, 18, 21, X, and Y) to compare the results with the state of the art that worked also on classifying groups of the chromosome. Here, we present the results of Individual chromosome recognition.

Table 7.10 shows the average of testing results (chromosome recognition accuracy) for 27 samples using the Multiple Autoencoder Neural Network for the targeted classes (13, 18, 21, X, and Y).

This result is comparable to the other methods reported in Table 7.11. Although we acknowledge that, this comparison may only be indicative, because of the different datasets used and their specification. All listed studies in Table 7.11 used a multi-layer neural network to classify curved and no overlapped chromosomes. There targeted chromosome groups do not contain Y class due to the small number of this type. In the proposed system, Multilayer Autoencoder Neural Network is used to classify five chromosomes classes (from (C, D, E, G) Denver groups) containing Y chromosome (the most common trisomy chromosomes) which increase the complication of the classification process. In addition, the chromosomes images dataset, which we used, containing severely curved, coiled chromosomes from different G-banding resolutions.

7.5 CONCLUSION AND FUTURE WORK

Karyotyping of human chromosomes is a laboratory technique that produces an image of an individual's chromosomes. It is an essential part of trisomy recognition. A trisomy is a type of abnormality in which there are three copies of a particular chromosome, instead of the normal two. Karyotyping is expensive and time-consuming when performed manually. Therefore, efforts to automate some or all of the procedures have continued for decades. We developed an automated karyotyping (chromosome recognition) system to recognize the most common trisomies (13, 18, 21, X, and Y), through four main processes: image enhancement, chromosome segmentation, features extraction, and classification. To do this, we used an assembled local images database which consisted of 157 images that involved 7266 chromosomes.

We trained a multi-layer neural network classifier with Autoencoder to classify chromosomes into targeted chromosomes classes. The classification accuracy for the training was 100% and the average of trisomy recognition test accuracy was 84.6% while the average of chromosome recognition test accuracy

TABLE 7.10   Experiment 2 performance accuracy (average chromosomes recognition accuracy)

CHROMOSOME TYPE	NO. OF CHROMOSOMES	NO. OF RECOGNIZED CHROMOSOMES	CHROMOSOMES RECOGNITION %	AVERAGE CHROMOSOMES RECOGNITION ACCURACY
Chromosome 13	60	57	95%	95%
Chromosome 18	61	57	98.36%	
Chromosome 21	59	55	93.22%	
Sex Chromosomes XY	61	57	93.44%	

TABLE 7.11 Comparison with other methods

REFERENCE	SELECTED FEATURES	CLASSIFICATION	DATA SETS COMPOSITION	INPUT/ OUTPUT CLASSES	ACCURACY (%)
(Lerner et al., 1995)	1. Density profile 2. Centromeric index 3. Chromosome's length	Multilayer perceptron neural network (MLP)	150 cells (622 chromosomes) No overlapped chromosome included	• Input: Features dataset of (2, 4. 13. 19. X) • Output: 5 Class	98%
(Moradi & Setarehdan, 2006)	1. Width, 2. Position 3. Intensity of the most important characteristic regions (bands). • Width of first and second dark bands • Position of the first and the second dark bands • The grey level Intensity of the first and the second dark bands	Three-layer artificial neural networks	Cytogenetic Laboratory of Cancer Institute 303 chromosomes No-overlapped	• Input: Features dataset of (16, 17, 18) group E • Output: 3 Class	98.6%
Proposed	1. Area 2. Relative area 3. Density profile 4. Centromeric index	Multiple autoencoder neural network	239 severely curved and coiledchromosome-No-overlapped	5 Classes (13,18,2-1, X and Y)	95%

was 95%. We found this result to be comparable with other previous methods with Note that studies used a multi-layer neural network to classify curved and no overlapped chromosomes. Their targeted chromosome groups do not contain Y class due to the small number of this type. While in our system, Multilayer Autoencoder Neural Network is used to classify five chromosomes classes (from (C, D, E, G) Denver groups) containing Y chromosome (the most common trisomy chromosomes) which increase the complication of the classification process.

For future investigation, we can try to work on the overlapped, touched, and curved chromosomes (two or more). In addition, we can try to predict the sample's G-banding resolution to reduce the dataset complexity. Moreover, increasing the size of the dataset (number of chromosome images) can be improve the accuracy. On the other hand, improvements of feature selection methods can reflect positively on the overall performance of the ANN classifier. It will provide a better result for the classification of an individual chromosome. Additionally, we will examine different methods to open the 'black box' of a neural network and provide a deeper understanding of its functions targeting the highest level of both explainability and performance.

REFERENCES

Abid, F., & Hamami, L. (2018). A survey of neural network based automated systems for human chromosome classification. *Artificial Intelligence Review*, *49*, 41–56.

Adadi, A., & Berrada, M. (2018). Peeking inside the black-box: A survey on explainable artificial intelligence (XAI). *IEEE Access*, *6*, 52138–52160.

Arpit, D., Zhou, Y., Ngo, H., & Govindaraju, V. (2015). Why regularized auto-encoders learn sparse representation? *arXiv preprint arXiv:1505.05561*.

Badawi, A. M., Hasan, K. G., Aly, E.-E., & Messiha, R. A. (2003). *Chromosomes classification based on neural networks, fuzzy rule based, and template matching classifiers*. 2003 46th Midwest Symposium on Circuits and Systems, Cairo, Egypt, 27th–30th December 2003 (pp. 383–387).

Bashmail, R., Elrefaei, L. A., & Alhalabi, W. (2018). *Automatic segmentation of chromosome cells*. International Conference on Advanced Intelligent Systems and Informatics, Springer, Cairo, Egypt, 29 August 2018 (pp. 654–663).

Carothers, A., & Piper, J. (1994). Computer-aided classification of human chromosomes: A review. *Statistics and Computing*, *4*, 161–171.

Digital Imaging In Optical Microscopy - Olympus DP-10 Digital Camera. https://www.olympus-lifescience.com/en/microscope-resource/primer/digitalimaging/olympusdp10/.

El-Khateeb, M. (2013). The steps in the process of creating a karyotype for chromosome analysis. [Online]. http://studylib.net/doc/5783675/ppt

Gagula-Palalic, S., & Can, M. (2014). *Human chromosome classification using competitive neural network teams (CNNT) and nearest neighbor*. IEEE-EMBS International Conference on Biomedical and Health Informatics (BHI), Valencia, Spain, 1–4 June 2014 (pp. 626–629).

Evren, D. (2020). Explainable artificial intelligence (xAI) approaches and deep meta-learning models. *Advances in deep learning*. IntechOpen.

Graham, J., Errington, P., & Jennings, A. (1992). A neural network chromosome classifier. *Journal of Radiation Research*, *33*, 250–257.

Groen, F. C., Ton, K., Smeulders, A. W., & Young, I. T. (1989). Human chromosome classification based on local band descriptors. *Pattern Recognition Letters*, *9*, 211–222.

Karaca, Y., Cattani, C., & Moonis, M. (2017). Comparison of deep learning and support vector machine learning for subgroups of multiple sclerosis. In *Computational science and its applications – ICCSA 2017* (pp. 142–153). Springer International Publishing.

Keller, J. M., Gader, P., Sjahputera, O., Caldwell, C. W., & Huang, H.-M. (1995). *A fuzzy logic rule-based system for chromosome recognition*. Proceedings Eighth IEEE Symposium on Computer-Based Medical Systems, Lubbock, TX, USA, 9–10 June 1995 (pp. 125–132).

Learn.genetics. *How do scientists read chromosomes?* http://learn.genetics.utah.edu/content/basics/readchromosomes/.

Lejeune, J., Levan, A., Böök, J., Chu, E., Ford, C., Fraccaro, M., Harnden, D. G., Hsu, T. C., Hungerford, D. A., Jacobs, P. A., Makino, S., Puck, T. T., Robinson, A., Tjio, J. H., Catcheside, D. G., Muller, H. J., &Stern, C. (1960). A proposed standard system of nomenclature of human mitotic chromosomes. *The Lancet*, *275*, 1063–1065.

Lerner, B., Guterman, H., Dinstein, I., & Romem, Y. (1995). Medial axis transform-based features and a neural network for human chromosome classification. *Pattern Recognition*, *28*, 1673–1683.

Linardatos, P., Papastefanopoulos, V., & Kotsiantis, S. (2021). Explainable AI: A review of machine learning interpretability methods. *Entropy*, *23*(1), 18.

Locus (genetics). https://familypedia.wikia.org/wiki/Locus_(genetics)#.

Markou, C., Maramis, C., Delopoulos, A., Daiou, C., & Lambropoulos, A. (2012). *Automatic chromosome classification using support vector machines* [Online]. https://olympus.ee.auth.gr/~chmaramis/website/documents/iConceptPress2012.pdf.

Mashadi, N. T., & Seyedin, S. A. (2007). *Direct classification of human G-banded chromosome images using support vector machines*. 2007 9th International Symposium on Signal Processing and Its Applications, Sharjah, United Arab Emirates, 12–15 February 2007, pp. 1–4.

Moradi, M., & Setarehdan, S. K. (2006). New features for automatic classification of human chromosomes: A feasibility study. *Pattern Recognition Letters*, *27*, 19–28.

Piper, J., & Granum, E. (1989). On fully automatic feature measurement for banded chromosome classification. *Cytometry Part A*, *10*, 242–255.

Piper, J., Granum, E., Rutovitz, D., & Ruttledge, H. (1980). Automation of chromosome analysis. *Signal Processing*, *2*, 203–221.

Poletti, E., Grisan, E., & Ruggeri, A. (2012). A modular framework for the automatic classification of chromosomes in Q-band images. *Computer Methods and Programs in Biomedicine*, *105*, 120–130.

Popescu, M., Gader, P., Keller, J., Klein, C., Stanley, J., & Caldwell, C. (1999). Automatic karyotyping of metaphase cells with overlapping chromosomes. *Computers in Biology and Medicine*, *29*, 61–82.

Rayan, Z., Alfonse, M., & Salem, A. B. M. (2019). Machine learning approaches in smart health. *Procedia Computer Science*, *154*, 361–368. 2019/01/01/.

Roshtkhari, M. J., & Setarehdan, S. K. (2008). *Linear discriminant analysis of the wavelet domain features for automatic classification of human chromosomes*. 2008 9th International Conference on Signal Processing, Beijing, China, 26–29 October 2008 (pp. 849–852).

Rungruangbaiyok, S., & Phukpattaranont, P. (2010). Chromosome image classification using a two-step probabilistic neural network. *Songklanakarin Journal of Science & Technology*, *32*(3): 255–262.

Saranya, S., Loganathan, V., & RamaPraba, P. (2015). *Efficient feature extraction and classification of chromosomes*. International Confernce on Innovation Information in Computing Technologies, Chennai, India, 19–20 February 2015 (pp. 1–7).

Sathyan, N., Remya, R., & Sabeena, K. (2016). *Automated karyotyping of metaphase chromosome images based on texture features*. 2016 International Conference on Information Science (ICIS) (pp. 103–106).

Sharma, M., Saha, O., Sriraman, A., Hebbalaguppe, R., Vig, L., & Karande, S. (2017). *Crowdsourcing for chromosome segmentation and deep classification*. 2017 IEEE Conference on Computer Vision and Pattern Recognition Workshops (CVPRW) (pp. 34–41).

Shoblak, D., & Baraka, D. M. (2015). *Trisomy* [Online]. http://www.slideshare.net/palpeds/trisomy-07042015.

Somasundaram, D., Kumaresan, N., Subramanian, V., & Sacikala, S. (2018). Structural similarity and probabilistic neural network based human G-band chromosomes classification. *International Journal of Human Genetics*, *18*(3), 228–237.

Text Classification Using Neural Networks. https://machinelearnings.co/text-classification-using-neural-networks-f5cd7b8765c6

Tso, M., & Graham, J. (1983). The transportation algorithm as an aid to chromosome classification. *Pattern Recognition Letters*, *1*, 489–496.

Vanitha, L., & Venmathi, A. R. (2011). Automatic abnormality detection in chromosomes using hybrid multi-layer neural network. *International Journal of Neural Networks and Applications*, *4*, 37–45.

Vilone, G., & Longo, L. (2020). Explainable artificial intelligence: A systematic review. *arXiv preprint arXiv:2006.00093*.

Wang, X., Zheng, B., Li, S., Mulvihill, J. J., Wood, M. C., & Liu, H. (2009). Automated classification of metaphase chromosomes: Optimization of an adaptive computerized scheme. *Journal of Biomedical Informatics*, *42*, 22–31.

Wang, X., Zheng, B., Wood, M., Li, S., Chen, W., & Liu, H. (2005). Development and evaluation of automated systems for detection and classification of banded chromosomes: Current status and future perspectives. *Journal of Physics D: Applied Physics*, *38*, 2536.

Webster, J. (2007, May–June). Third Kuala Lumpur International Conference on Biomedical Engineering 2006 [Around the World]. *IEEE Engineering in Medicine and Biology Magazine*, *26*(3), 7.

Smart Learning Solutions for Combating COVID-19

Jaspreet Kaur[1] and Harpreet Singh[2]

[1]PG Department of Commerce & Management, Hans Raj Mahila Maha Vidyalaya, Jalandhar, India
[2]PG Department of Bioinformatics, Hans Raj Mahila Maha Vidyalaya, Jalandhar, India

Contents

DOI: 10.1201/9781003172772-8

8.1 INTRODUCTION

8.1.1 Background: Smart Education an Essential Component of Smart Cities

For the last few decades, there has been an upward trend of migration of people to urban areas. This shift of population is creating stress in urban areas in the form of various problems like water, sanitation, energy, etc. So, the governments are working to integrate various technologies with the help of connectivity and the Internet of Things (IoT) to convert the cities into smart cities. The word smart refers to the use of intelligent IoT solutions to enhance the standard of public life, that is, government has established detectors, transducers, sensors, and connected these with the help of software and programmes to provide real-time information about the level of traffic, pollution, energy, waste, etc.[1] In addition to this, the information generated on a real-time basis is openly accessible. So, it can be used by the general public for their benefit. For example, by getting traffic-related information, they can plan their routes. Many other components of smart cities include smart street lighting to save energy, smart waste management, smart water conservation, etc.[2]

Also, for the burgeoning of smart cities, smart education is a must, that is, education having the power to develop students with contemporary capabilities, experiential abilities, and synergetic orientation. These qualities can be imbibed among students with kicky digital technologies.[3] Moreover, their performance can also be gauged on a real-time basis by their mentors and a specific approach for their improvement can be applied to enhance their performance. The use of these technologies will lead to enhancement in student's engagement and improvement in their attendance, thereby resulting in better learning outcomes.[4] In addition, smart technologies will ensure the safety of students while also increasing the trust of the parents. Keeping in view the concept of smart cities and sensing smart education as one of its major components, the current chapter is planned to give an overview of the teaching-learning process, its transformation from traditional to digital, and the emergence of smart technologies, including Artificial Intelligence (AI) that is going to govern the future education. In addition, the role of smart learning technologies and other strategies in combating the effects of the COVID-19 pandemic on the education system has also been discussed.

8.1.2 Teaching-Learning Process

The process of learning begins with the birth of a baby and he keeps on learning throughout his lifetime. His experience of life makes him learn how to behave if the same situation arises again. Learning involves transformation, transformation in the potential, proficiency, knowledge, orientation, and confidence of a student. This transformation is escorted due to the grouping of new concepts, skills, and laws. Educators perceive that education focuses on the needs, requirements, abilities, and interests of learners and it also involves a check on them in the form of exams to probe to what extent they have assimilated and grasped the subjects. Persistent and cautious endeavours are required on the part of virtuous mentors to nourish their understanding, apprehension, expertise, and ability. In-depth apprehension of the teaching and learning process is the sole indispensable requirement for becoming a virtuous mentor. Extensive cognizance of the teaching-learning procedure is essential to expedite the procedure of disseminating knowledge. An individual gains knowledge and understanding by watching, listening, and performing what is being taught by the mentor.

The mere presence of the learner does not guarantee that he or she is learning. A learner should put in considerable interest and strong emphasis while learning so that he or she can gain an understanding of the concepts and can apply the acquired knowledge as per his or her needs and requirements. Teaching and learning can be done effectively if there is the presence of five elements, that is, mentors, mentee, content, teaching facilities, and physical resources. The mentor should have a pellucid purpose, wholehearted understanding of the concepts to be taught. Moreover, he or she should be passionate, fervent, and keen and should ask his or her pupils to share their ideas and understanding with him or her. He or she should be quick, inclined, affectionate, and polite. Moreover, a mentor should plan his or her teaching and be able to communicate with his or her mentees. He or she should be a role model for his or her mentees.

Mentees on the other hand should have the want and desire to learn and they should be proficient enough to grasp the concepts and apply the gained knowledge and understanding when in need. The content or subject matter that is required to be taught should be as per the need of the hour and be relevant, factual, realistic, and practical. In addition to this, content should be methodical and be presented analytically and intellectually in order to satiate and assuage the mentees. The physical decorum of the place should be calm, congenial, and work oriented. The place should be spacious and adequately furnished. The teaching aids should fulfil the requirements constructively. The teaching and learning process involves the following steps:

i. Creating a learning environment: The environment should be conducive for learning. A congenial and pleasant environment makes the process of learning easy and smooth while a bad and negative atmosphere prevents the learner from learning. A constructive, positive, supportive, and optimistic learning environment is key to the academic, spiritual, emotional, and social growth of a learne[5]

ii. Assessment of learning needs: If the subject matter is as per the needs of the mentee then the mentee is more engrossed in the process of getting knowledge or skill. So, a mentor should make an assessment of the educational requirements of mentees in order to attract their attention in the process of learning.

iii. Developing curriculum: After assessing the learning needs of the mentees, the mentor is required to develop a curriculum as per the learning curve of the mentees. If the curriculum is as per the needs of mentees they will be motivated to acquire knowledge, skill, and understanding.

iv. Teaching methods: The mentor should teach the students as per their learning curve. He should act as a facilitator as well as a guide for them. He should focus on the needs of the students and use the means to grasp the attention of the students. The method should be such that whatever is being taught should be retained by the students.

v. Follow-up: This step is very essential as the mentor gets to know whether mentees have understood the concepts which have been taught. For this, a mentor should interact with mentees, or conduct tests, etc. to know to what extent students have grasped the subject matter.

8.1.3 Teaching Pedagogies

The inherent part of teaching and learning is the teaching Pedagogy. Pedagogy can be simply defined as the method and practice of teaching.[6] The teacher uses different and innovative ways to plan and deliver their content depending upon their experience, context of teaching, level of students, and availability of infrastructure. Teaching pedagogy has been constantly evolving with time. Its three major components, namely, Teaching styles, Teaching Theory, and Feedback & Assessment, have undergone a major makeover with the increasing use of ICT tools in teaching. The COVID-19 pandemic has further triggered the adoption of novel and diverse teaching pedagogies, particularly those based on the internet and

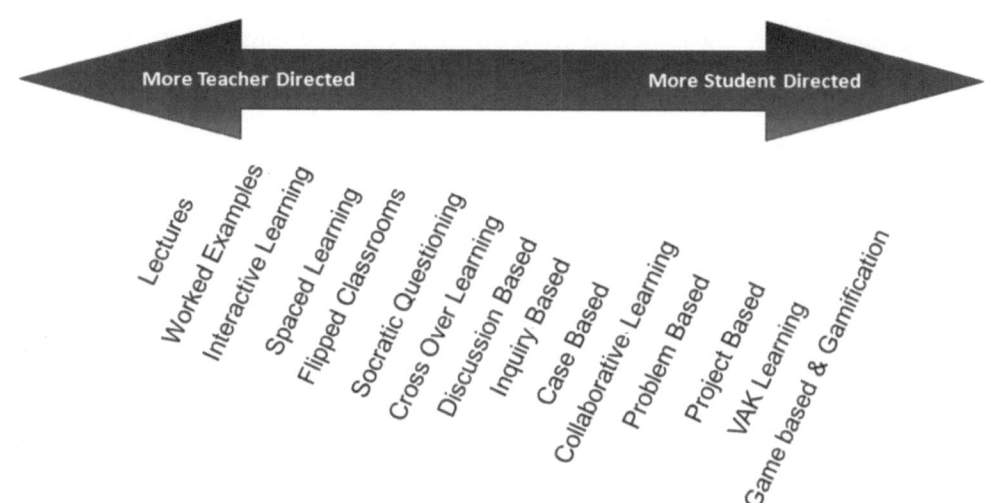

FIGURE 8.1 A Spectrum of Teaching Pedagogies from Teacher Centric Towards Student Centric. Figure Adapted from https://fctl.ucf.edu/teaching-resources/teaching-strategies/teaching-methods-overview/.

information technology to make teaching interesting as well as fruitful even during stressful conditions. Some of the major teaching pedagogies have been depicted in Figure 8.1 and are summarized below.

8.1.3.1 Lectures

The lecture method has been one of the most used teaching pedagogy across the world. It has been a widely used communication mode in the classrooms for many years, even with the introduction of many other teaching strategies. This method gives an opportunity to directly share a large amount of information with a variety of listeners while providing maximum control to the instructor. The instructor has the freedom to cite suitable examples, guided practice, and assigning small group-based assignments in the class. Moreover, it provides an excellent way to get constructive feedback. The lecture method can be effectively combined with other active teaching strategies to achieve learning goals. The lecture method has some inherent disadvantages including the assumption of an unrealistic level of understanding, frequent disengagement with the students, and getting minimal feedback from students.

8.1.3.2 Worked Examples

Worked examples refer to a step-by-step demonstration of a process. During the initial steps, the audience is introduced to a concept in its simplest form. In subsequent steps, more complex aspects of the given process or problem are highlighted. It can be considered as an extended form of a lecture method, particularly suitable to problems related to STEM fields of which the students are little or not aware of.

8.1.3.3 Interactive Learning

Interactive learning attempts to create more dialogue between the instructor and the student. In interactive learning, lectures are built around questions that students try to answer using coloured flashcards or various polling options. Interactive learning also includes other strategies such as group discussions, writing short exercises, debates, extempore, model making, drawing, and individual or collaborative problem solving. The recent use of ICT has further diversified the ways by which various stakeholders can interact to enhance learning. In STEM fields such as computer science, where coding is required for giving live programming assignments or live debugging is an excellent way to foster interaction among students and the faculty.

8.1.3.4 Spaced-Learning

Spaced learning is an excellent modern way to achieve learning goals in which the entire lecture is divided into short modules, which are repeated several times until the students understand these modules completely. However, in between each repetition, a small interval (space) is given to refresh the mind of the learner by involving them in some physical activities or other mindfulness techniques so that they can be refreshed to focus on the next concept. This method enables students to learn concepts step-by-step with increasing order of complexity thereby creating more interest and ultimately leading to a better learning experience.

8.1.3.5 Flipped Classrooms

Flipped classroom method of teaching received its name as it reverses the conventional way of classroom teaching. In conventional teaching, the content is provided to the students at schools in which they practice and revise at their homes. However, in flipped classrooms, students are motivated to search and read content on their own at their homes, which is then discussed and practiced at schools in a guided manner. This method gives more time to students to go through the content rather than getting very little time at school. Therefore, students can learn at their own pace and convenience. Flipped classrooms sensitize students towards a particular concept well before that concept is actually discussed in the classroom thus giving them an opportunity to ask more questions and dig deeper into the concept. Flipped classrooms give students a chance to explore many more dimensions of a given concept making them aware of some entirely new aspects of a given concept.

8.1.3.6 Socratic Questioning

Socratic questioning is one of the most powerful strategies to foster critical thinking among students.[7] Socratic teaching challenges the students with questions rather than providing answers with the aim of developing inquiring and probing minds. It facilitates the inculcation of logical thinking and the ability to explore ideas in depth with the help of very carefully designed questions around a specific concept.[8] More importantly, Socratic questioning promotes independent thinking among the students while giving them ownership of what they have learned.

8.1.3.7 Discussion-Based Learning

Discussion-based learning is another way of developing critical thinking and logical reasoning among students. In this type of teaching pedagogy, students are given a topic and are motivated to discuss its various aspects. In this strategy, students apply their learning by engaging themselves in fruitful discussions via logical reasoning. Discussion-based learning leads to real-time interactions giving students an opportunity to put forward their viewpoints while listening to and learning from the viewpoints of other group members. The major challenge of Discussion-based learning is to facilitate the participation of all students as well as to ensure that students can express their ideas with proper reasoning rather than merely exchanging their viewpoints.

8.1.3.8 Case-Based Learning

Case-based learning is one of the widely used and most effective ways to disseminate and integrate knowledge. Case-based learning engages students in instructional activities as well as discussions on scenarios based on various real-world problems. This strategy is learner-centric which creates the ideal environment for interaction among participants by working together in groups, leading to their knowledge building. The scenarios used can be as simple as addressed in a single setting in a short duration of time or can be as complex as sequential cases requiring multiple settings. The students work together collaboratively to analyze problems that often have multiple solutions, thereby highlighting critical issues or dilemmas. In this way, case-based learning stimulates critical thinking and makes students aware of multiple perspectives of a problem. A case-based approach is often used by instructors to teach content by connecting students to real-life data by putting them in the decision maker's shoes by role-playing. Case-based learning makes the students aware of how (positively or negatively) a decision will impact different participants. Students are thus exposed to the complexity of solving critical socio-economic problems. In the process, they acquire substantive knowledge and develop analytic, collaborative, and communication skills. The content of the cases can be obtained from a variety of sources. With the easy availability of validated cases over the Internet, case-based learning has further widened the range of disciplines it covers. Many online cases also include additional resources including handouts, readings, assignments, etc. Apart from social and natural sciences, case-based learning is an excellent teaching tool for various STEM fields where cases are used to impart relevance and aid in connecting theory with practice.

8.1.3.9 Collaborative Learning

Cooperative learning is a teaching pedagogy in which small groups of students work together to achieve a common goal. Although Collaborative and cooperative terms are used as synonyms, the latter covers a much broader range of group interactions namely, developing learning communities, stimulating student/ faculty discussions, and encouraging electronic exchanges (Bruffee, 1993). Both approaches encourage the involvement of faculty as well as students in the learning process. However, to achieve success, these strategies require careful planning and implementation in terms of group formation, maintaining individual accountability while ensuring positive interdependence, resolving group conflicts, developing appropriate assignments, and managing active learning environments. In a more planned group-based assignment, students are often given specific roles in rotation so that they can learn different aspects of a problem through involvement in different activities one after the other. Cooperative Learning is very suitable in STEM subjects.

8.1.3.10 Enquiry-Based Learning

Enquiry-based learning uses question-driven strategies to inculcate critical-thinking and problem-solving capabilities among students. This strategy works by gradually increasing student's self-direction. In the initial stages, the instructor guides the students towards a learning goal using strategies such as worked scenarios, process worksheets, group discussions, and challenges. Subsequently, the instructor decreases guidance, allows students to take greater responsibility and lead. Students are finally allowed to practice, fail, and learn to improve from the mistakes or improve their performance based on feedback from their mentors or peers. In this mode of learning, the cognitive capabilities of students are nurtured as they enquire deeper into the problems in an attempt to solve them. The success of enquiry-based learning depends on the assessment of prior knowledge of students and recognizing the stages as well as the extent of scaffolding by the instructors.

8.1.3.11 Problem-Based Learning

Problem-based learning is an educational approach in which the learning process starts with a real-life problem such as in case studies. However, the details of the problem, process, and outcomes are kept ambiguous to make learners work outside their comfort zone (Allen et al., 2011). The real-life problems are often intentionally edited to make them ambiguous and more challenging In addition to real-life problems, hypothetical problems can also be formulated to meet the educational objectives. In problem-based learning, a highly motivational environment is provided where students work in small groups and learn by resolving complex, realistic problems. In this way, in addition to acquiring knowledge, the students also develop generic skills related to problem-solving, self-directed learning, teamwork, negotiation, writing, and communication (De Graaf & Kolmos, 2003). This learning strategy has been originally introduced in the 1950s for medical education but has now been very effectively used in many other disciplines, particularly in higher education (Hung et al., 2008; Kilroy, 2004). Its implementation, however, poses major challenges to change the way educators conceive, design, deliver, and assess the curriculum (Hung, 2013).

8.1.3.12 Project-Based Learning

In project-based learning (PBL) students works independently or in groups to solve problems that are authentic, curriculum-based, and often interdisciplinary (Kokotsaki et al., 2016). PBL is an active student-centric approach that engages students in an investigation of authentic problems. PBL motivates learners to select a specific problem as well as to decide specific approaches and methodologies to solve it under the guidance of a teacher (Blumenfeld et al., 1991). To achieve these goals students gather information from a variety of sources and synthesize, analyze, and derive knowledge from it thereby making this strategy a very effective tool for knowledge integration. Towards the end of the project, students get a chance to demonstrate their newly acquired knowledge followed by an evaluation of how much they've learned and how well they can communicate it. PBL is an excellent way of learning by experience. During the entire process of PBL, students develop skills of self-learning, collaboration, communication, goal-setting, constructive investigations, and reflection (Scarbrough et al., 2004). In addition to learning core content and getting awareness, students also learn to solve complex global issues by breaking them into smaller, easily achievable steps (Solomon, 2003).

8.1.3.13 Self-Learning

In self-learning, students explore subjects or topics of their interests through various means such as the internet, library, seminars, conferences, and webinars. The most challenging aspect of this strategy is to make students curious about certain aspects of a topic such that they are motivated to explore them further on their own. In this way, self-learning is very useful in making students self-dependent and it also develops a deeper understanding of the subject among learners.

8.1.3.14 Game-Based Learning and Gamification

In the last 15 years, evolution in games has made it a very useful tool to engage and educate students across various disciplines. Game-based approaches help to achieve educational objectives while making the learning process easier, more student-centred, funny, interesting, and more effective (Lepper & Malone, 1987; Papastergiou, 2009; Prensky, 2001; Rieber, 1996; Rosas et al., 2003). Games promote multisensory, active, experiential, and problem-based learning (Papastergiou, 2009). Games are very important to recall pre-existing knowledge in the sense that players use previously gained knowledge to improve their score in the gaming process. Games also provide very quick feedback thereby allowing players to test different hypotheses and making them learn from their actions (Pesare et al., 2016). In addition to imbibing knowledge, game-based learning helps in promoting logical-mathematical and critical thinking while helping in the development of personal and social skills, language abilities, communication and collaboration skills, creativity, and problem-solving capabilities (McFarlane et al., 2002). Due to the involvement of digital technologies, game-based learning has also been named digital game-based learning (Prensky, 2001). Games are very important in support pedagogical principles such as individualization, feedback, active learning, motivation, social, scaffolding, transfer, and assessment (Oblinger, 2004).

In recent years, much interest has been growing in Gamification, which deals with the use of game mechanics and rules in non-gaming contexts (Cózar-Gutiérrez & Sáez-López, 2016; Deterding et al., 2011; Zichermann & Cunningham, 2011). The major focus of gamification is using game dynamics and thinking to better engage users and stimulate them to actively participate, thereby leading to improved outcomes. Gamification is being widely used in the e-commerce domain, where it impacts the "FOUR Is" (Epps, 2009) that is, Involvement, since it improves the active participation; Interaction, since it leads to a high level of interactivity; Intimacy, since it stimulates the familiarity with the brand; Influence, as it allows the expansion of the brand and the products. Similar to e-commerce, the above-mentioned outcomes are very much desirable in education as well. In education, active participation, interactivity, and competitiveness are essential to motivate and engage the students. However, to effectively apply gamification in education, it is critical to understand the game elements that can be used and the way these elements can enhance the learning process.

The main elements that can be inherited from the game and used in the learning context, are (Bunchball, 2010; Simões et al., 2013):

- **Status:** The acknowledgement of the user's reputation leads to her or his fame and prestige in the community.
- **Recognition of Results:** The use of points and levels to keep track of achievement and progress is useful to maintain interest and encourage a greater commitment to higher goals.
- **Competition:** The comparison of the players' performances is a motivational element that can be achieved for example with charts that allow the player to view the results and the winners to celebrate.
- **Ranks:** The measurement of all participants' progress and their achievements can be used to encourage players to do better, driven by the desire to improve their position.
- **Social Dimension:** The activities of friends can influence those of other users, in real life as well as in virtual communities. Gamification initiatives must therefore be able to create a strong sense of community.
- **Customization:** The game elements and feedback can be customized, thus promoting a sense of belongingness to the 'game' and the community.
- **Scores and Levels:** Scores can be used as rewards for users' progress and for achieving the objectives in various stages. Collecting points will allow access to higher levels, defining the degree of skills achieved by every single player.
- **Reward:** The reward for the obtained results can be real or virtual and be used to increase player satisfaction for having achieved the desired objective. It triggers the motivational mechanisms necessary to reach new and more ambitious goals. This requires an increased involvement of the users who will be encouraged to offer even better performance.

All the above-mentioned elements can be utilized in improving the engagement of participants in learning activities, thereby motivating them to acquire new knowledge. However, in order to transform Gamification into an effective learning experience, it is necessary to investigate how it can be combined with the dimensions involved in the learning process, that is, cognitive, emotional, and social (Illeris et al., 2002; Poscente, 2006; Lee & Hammer, 2011).

8.1.3.15 VAK Teaching

VAK teaching is a teaching pedagogy based on the category of the learner. According to the way the learners prefer to receive and process information, they can be categorized as Visual, Auditory, or Kinaesthetic learners (Sarasin & Celli, 1999). Visual learners are those who can learn effectively by seeing the materials, an Auditory learner likes to gather information by hearing, while Kinaesthetic learners learn best by feeling the data or by doing activities (Surjono, 2011). Learning strategies need to be fine-tuned to match the individual learning styles of the students so as to optimally utilize the learner's potential (Bhattacharyya & Sarip, 2014). Introduced in 1920 to help children with dyslexia, the VAK teaching strategy has become very important in the modern era, particularly with the increasing use of the Internet. Students grasp rapidly when they see, hear, or feel instead of just reading the content. Therefore, in the modern era of the internet, watching and learning through animations and videos has become one of the most attractive medium of teaching. Besides, the learner has the flexibility to choose any combination of or all of the three types of media.

8.1.3.16 Cross-Over Learning

The cross-over learning method uses both formal as well as informal teaching and learning environments. Cross-over learning effectively engages learners giving rise to authentic and innovative learning outcomes (Panke, 2017). In the formal component of cross-over teaching, the traditional classroom setting is used to deliver the content while many other informal or incidental ways, such as seminars, workshops, interactions, field visits, museums, and excursions are used to add new dimensions to the learning experience. The informal component of cross-over teaching is particularly useful in generating curiosity and interest while increasing understanding by giving more opportunities for interaction and asking questions. The learners can use various means to collect data and acquire information which can then be shared with other students as well as teachers invoking discussions, leading to further acquisition of knowledge.

8.2 FROM FACE-TO-FACE TEACHING TO ONLINE TEACHING

In the recent past, there has been a gradual transition in the teaching and learning process from physical classrooms to virtual environments. Initiated by computer-aided teaching, digitization of education has led to a revolution where students across the globe can access quality education through online MOOCs developed by masters in their respective fields, thereby opening a world of new opportunities for the learners. This revolution in online teaching further proved as a gift to combat the impact of COVID-19 in the education sector and enable students to achieve their educational goals. The following section highlights various components of this digital transition in teaching and learning.

8.2.1 Computer-Aided Teaching-Learning

Computer-aided teaching-learning means the use of a computer as an aid in the teaching-learning process. Aid means assistance and support. So, the computer is not used in place of a mentor rather it is used to assist the mentor.[9] It is a device-based learning methodology that makes the teaching-learning process more engrossing, exciting, enthraling, effervescent, and liveable. Learning becomes tranquil if it is an amalgamation of sane, delineation, graphics, and expression. Mentors can teach with the help of audio lectures, video lectures, images, media, graphs, and animation effects which make concepts more gripping and interactive for learners and mentees (Cingi, 2013). So, all the schools, colleges, and universities have been using computer-assisted methodologies for teaching. In fact, computers have been used for teaching in many colleges and universities for more than 15 years to enhance the grasping power of mentees and to multiply the potency and efficacy of the mentor (Külür, 2001). Computer-aided teaching has immense advantages, some of which are discussed below:

- Makes the teaching environment lively with colours, graphics, and animations as compared to the traditional teaching-learning environment which has gradually become dull, intricate, complex, and knotty due to the usage of monochronous teaching materials such as books.
- Helps in economizing paper.
- Makes it easy to streamline and revise the content as per the new amendments and needs of the mentees.
- Enhances the apprehension of concepts through charts, figures, animations, images, videos, and media clips.
- Mentors can easily explain with visual aids strenuous and arduous topics.
- In STEM and some other disciplines including practical education, mentors can explain the theory and practicals simultaneously.

8.2.2 Digital Education

Digital education means the transfer of information, knowledge, skills, and education with the help of online media and digital media. It not only uses electronic devices like computers, laptops, mobiles, and tablets but also Internet connectivity. In digital learning knowledge and information are transmitted to all students of the class simultaneously or allochronically using emails, discussion boards, forums, wikis, blogs, etc. A teacher can take live lectures with the help of online apps like Google Meet, WebEx Meet, Zoom, Microsoft Teams or can record a lecture and share it with students with the help of WhatsApp group, Telegram group, or email or post in Google Classroom, etc. Digital learning is the same as that of classroom learning in which there is a mentor teaching mentees but the difference is that there is no physical classroom, teaching is done in a virtual classroom. Moreover, it also involves interactive two-way communication. In addition to this students can submit their assignments via email, WhatsApp, or Google Classroom. Digital education is also referred to as ICT (information and communication technology) enabled learning, electronic learning, e-learning, Internet-based learning, web-based learning, online education, virtual learning, and synchronous learning (Guragain, 2016). Due to the COVID-19 Pandemic, there have been gigantic transformations in the sphere of teaching and learning. All over the globe, every educational institution is using an online mode of knowledge delivery. This is all due to digitalization. Now, teaching and learning are not confined to divine educational institutions but it is in the fist of the mentor and mentee. The present form of digital education has not evolved easily. It has gone through many transformations; some of them have been discussed below.

8.2.2.1 Time Line of Digital Education

No doubt, the advent of the Internet is recent. But education has been imparted via wireless connectivity using radiotelegraphy since the year 1919. A transmission post named 'amateur' has been established by the University of Wisconsin for knowledge delivery. Trails have been started in the year 1930 by the University of Lowa to impart knowledge via T.V. Federal Communications Commission had booked a T.V. channel for imparting education through television programmes in the year 1950.[10] In 1960 professors of the University of Illinois, USA taught students with the help of computers which have been interconnected with each other.[11] The timeline of digital education is summarized below in Figure 8.2.

Western Behavioural Sciences Institute, California, had initiated matchbook courses (correspondent courses) in the year 1982 by conducting meetings using computer-based telecommunications. While the absolute digital course had been initially started at the University of Toronto in the year 1984. Moreover, the EUN (electronic university network) had been fabricated in the year 1984 as well for providing ingress to digital education. The promulgation of open networking had been done by U.S. research and education network in 1985 by creating NSFNET for exploration and experimentation. In 1986, the EUN initiated its online curriculum exploiting DOS and Commodore 64 computers.[12] Phoenix University initiated a digital course of study in 1989.[13] Then in 1990, Open University of Britain started matchbook

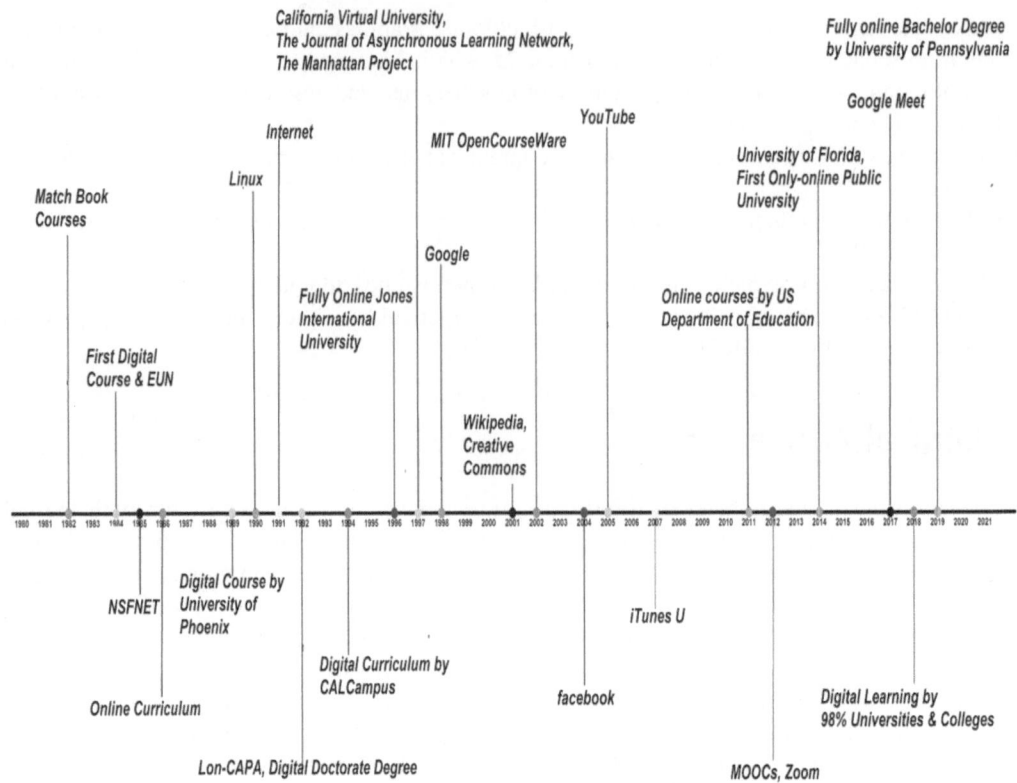

FIGURE 8.2 Agraphical Depiction of Major Milestones During the Evolution of Online Education, 1982 Onwards. Image Created Using ggplo2 Library (Wickham, 2016) in R (R Core Team, 2020). The Details Have Been Mentioned in the Text.

courses imparting knowledge via digital technology, and in the same year software named Linux was developed by Linus Torvalds which is being used to date for imparting digital education. The Internet was innovated by Al, Gore in the year 1991 permitting people to avail the benefits of the Internet and to use it in digital learning. In the year 1992, Michigan State University invented the technique of pre-programmed machine evaluation referred to by the name of LON-CAPA i.e. the learning online network computer-assisted personalized approach and the EUN started doctorate degree digitally using the Internet connection of America. In the year 1994 CALC (computer-assisted learning centre) expanded its operations and renamed it as CALCampus and it started its initial digital curriculum using the World Wide Web with the aid of contemporaneous directions and intercommunication.[14] Jones International University was started by Glan Jones and Bernand Luskin in the year 1996. This university has been fully online.[15]

In the year 1997 California Virtual University (CVU) was formulated for furnishing details of digital education offered by various institutes of California. Moreover in this year a journal titled *Journal of Asynchronous Learning Network* was brought under publication for the first time for encouraging, exploration, and investigation in the field of digital learning. In addition to this blackboard LMS (learning management system) has provided web-based education using blackboard course management operating system[16] and Steven Narmontas developed a software named 'The Manhattan Project'. It has provided the mentors a platform to share notes with their mentees with the help of an internet site[17] and in the years 1998 and 2001 Google and Wikipedia were invented, respectively. Moreover, in the year 2001 Creative Commons was developed as a platform for the researchers or inventors to share their copyrighted or patented research without any cost with the general public for their use.[18] This platform has been used by The Massachusetts Institute of Technology (MIT) to share digital educational information and knowledge with its own development MIT OpenCourseWare in the year 2002.[19] Facebook was developed by Mark Zuckerberg with his fellow beings in the year 2004 as a Harvard mentees social educational website.[20] In 2005, YouTube was developed for making public useful videos[21] and from 2005 Liberty University, Virginia had been providing digital learning programmes for grown-ups.[22] In the year 2007, iTunes U was created for sharing knowledgeable material of various universities in the form of presentations, talks, discourse, oration, online books, etc.[23] Salman Khan, in the year 2008, fabricated Khan Academy for making available digital material to the mentees without any cost.[24]

In 2011, the US Department of Education issued a mandate specifying conditions for educational institutes for digital delivery of education to ensure the efficiency and effectiveness of online classes. In 2012, Udacity formulated MOOCs (Massive Open Online Courses) in collaboration with MIT and Harvard for providing education electronically.[25] In April 2013 a bill was passed granting permission to establish the University of Florida by January 2014 as a digital learning institute by the government of Florida[26] and by the year 2018, almost 98% of the universities and colleges all over the world started digital learning.

All around the world Zoom app, Google Meet, WebEx Meet, Microsoft Teams, etc. are being used for teaching during the time of the COVID-19 Pandemic. Zoom was developed by Eric Yuan in 2012. Zoom has been at the forefront of providing a platform to professionals all over the world to meet their commitments during the COVID-19 pandemic.[27] In the year 2017, Google Meet was invented by combining Hangout meet and Google Chat features in one app and during this pandemic of the Corona virus, this app has been greatly used for the purpose of teaching.[28]

8.2.2.2 Advantages of Digital Learning

Digital learning comes with a series of advantages due to which it has become acceptable, affordable, and the favourite among learners as compared to the traditional methods (Table 8.1). These advantages made Digital learning a most potent tool to fill the vacuum created in the education world during the COVID-19 pandemic.

TABLE 8.1 Comparison of traditional and e-learning environments (Adapted from Zhang et al., 2004)

	TRADITIONAL CLASSROOM LEARNING	DIGITAL/E-LEARNING
Advantages	• Learning and immediate feedback • Cultivation of a social community • Immediate example citing and guided practice	• Learner-centred • Self-paced • Flexibility in time and location • Cost-effective for learners • Potential for global audience • Unlimited access to knowledge • Archival capability for knowledge reuse and sharing • More lively teaching environment using colours, graphics, animations • Better assessment
Disadvantages	• Instructor-centred • Pace of learning mainly depends on the instructor • Time and Location constraints • More expensive to deliver • Limited audience • Difficult to achieve knowledge for future use • Unrealistic level of understanding • Frequent disengagement	• Lack of immediate feedback in asynchronous e-learning • Increased preparation time for the instructor • Not comfortable to some people • Potentially more frustration, anxiety, and confusion • Influenced by internet connectivity and power shortage, particularly in remote areas • Influenced by unaffordability of digital devices by some sections of the society

Some of the major advantages of digital learning are:

- *All-time Availability:* As online delivered lectures can also be recorded, so it will be available for mentees all the time. If mentees find any issues, they can re-listen to the recorded lecture and resolve their queries by themselves. No doubt mentees can ask their problems to the mentor but it is an amazing pleasure to resolve one's queries by oneself. Moreover, if the mentees want to study at odd times, that is, at night they can also have access and even if they are busy at the time of the lecture, they need not worry because they will have a digital record and can access it when they have time.

- *Associated Study:* Mentees can associate with their classmates any time with the help of apps like zoom, Google Meet, WebEx, Microsoft teams and they can also be associated with a mentor when they have time and can resolve their queries. Mentees can also send messages on WhatsApp, email their queries to the mentor, and can get their queries resolved before the next lecture. Even though mentees are bodily apart from their classmates and mentors they can still contact them and stay in touch with them.

- *Autodidactic Learning:* Self-determined learning is highly significant in higher education because the mentee can concentrate on the concepts in which he is weak. Moreover, he can contact his mentor when he finds any problem. This also enhances the morale, inner strength, confidence of the mentee.

- *Enhance Writing and Speaking Skills:* In the physical classroom teaching, a mentor can understand the facial expressions of mentees while in the case of online teaching that is not possible. If the mentee has any query in understanding any concept, he would have to ask his mentor either verbally or by writing in the chatbox. So, this will enhance his speaking and writing skills as well, that is, communication skills of mentees.

- *Exposure to Technology:* As mentees are using online communication apps, they will not find any problem, when they will be in jobs. Because as the world is going global, so companies are having their subsidiaries in various countries, and employees of these subsidiaries can communicate and hold meetings with the head office with the help of these online apps.

- *Engaged Learning:* In traditional classroom teaching, there have been mentors, mentees, and books while in the case of online teaching-learning methodologies, mentors use a variety of ways like online video clips, images, online content to make their lecture interesting and interactive. All this enables the mentees to better understand the concepts.

- *Better Assessment:* With the help of digital means mentors can assess the performance of mentees by taking online tests on a real-time basis. It provides feedback through auto-generated reports about the performance of the mentee.[29]

- *Updated Content:* While teaching online, mentors can access appropriate and pertinent matters through digital media and can share updated knowledge and information with his mentees. Updating ameliorates the precision and validity of matter.

- *Mentor's Acquisition of Knowledge:* Mentors also acquire knowledge by becoming part of various online teaching communities.[30] Teachers can ask their queries in communities as well as share tips to enhance their teaching ability.

- *Information Sharing:* Information can be shared easily with the help of digital media. Mentors or educators can work in collaboration. They can share Google Docs and make changes or improvements in these docs and that document can be assessed by all those to whom access has been provided. This enhances co-operation, coordination, and collaboration skills among the working group members.

- *Expanded Learning Opportunities:* Digital media has expanded learning opportunities for students. They can learn from where they are having the facility of easy access.[31]

8.3 INITIATIVES TAKEN BY GOVERNMENTS OF VARIOUS COUNTRIES

The governments of all the countries have taken various initiatives in the time of the COVID-19 Pandemic so that education should be imparted to the students without any problem. The initiatives taken by the governments of some of the developed, developing, and underdeveloped countries are as follows.

8.3.1 Afghanistan

'Alternative Education Scheme for Persistence of Corona Virus in the Country' has been launched by the Directorate of Technical and Vocational Education. This scheme has provided three sets of schemes for online teaching and learning, the first one for one to two months, the second for three to six months, and the third one for beyond six months. This scheme of alternate education has guided that country should rely on its resources and should not depend on other nation's support. Schools should use various modes of sharing of educational videos via websites, social sites (Facebook, YouTube), school portals, memory cards, CDs and broadcasting through television, sharing of audios via radio, mobile phones. In addition to this books have also been published and distributed among students.

8.3.2 Argentina

The Ministry of Education of Argentina has developed a web portal 'educ.ar' for educational purposes. A programme named 'Seguimos Educando' has been created for open access to digital education material from April 1, 2020. Seguimos Educando broadcasts educational programmes on T.V. in the presence of a teacher and a conductor (journalist or artist, etc.) for fourteen hours a day and on the radio for seven hours a day. Each programme is for one hour. So, seven programmes have been broadcasted on the radio. In addition to this, nine books have been published, two for pre-primary level, four for primary level, two for secondary, and one for family level. These books have been delivered to the homes of the students. All the educational material has been made available at educ.ar web site under the Seguimos Educando section. This platform has also made available self-learning programmes through video-conferencing and social networking sites and also provides links to various online educational programmes conducted which can be accessed by the students during their free time through video-conferencing apps and social networking sites. Moreover, the Ministry of Education of Argentina has also ensured free of cost or no data usage access to digital media for educational purposes. To ensure the availability of Netbooks and tablets to socio-economically vulnerable students, 55000 notebooks and 22000 tablets have been distributed.

8.3.3 Austria

The Ministry of Education of Austria through its website has made available an enormous amount of digital educational material. The educational institutes have been using MOODLE and LMS as well as the cloud platform of Microsoft and Google for teaching purposes. A platform known as EDUTHEK has been developed by the Ministry of Education and through this platform, learning material are being provided to the students. In addition to this a T.V. channel named ORF1 has been broadcasting educational programmes from 6 to 9 am for primary school students and from 9 to 12 the ORF1F reistunde channel has been broadcasting educational programmes for students above ten years of age.

8.3.4 Bangladesh

The government has been broadcasting an educational programme 'MY SCHOOL AT MY HOME' for the students of the sixth standard till the tenth standard through television from 9 am to 12:30 pm. This material can be assessed later through the Bangladesh television YouTube channel. The United Nations Children Fund has been providing support to the Government of Bangladesh for effectively conducting educational programmes through T.V., radio, mobile phone, and internet.

8.3.5 Belize

As the lockdown was announced by the Government on March 18, 2020, from March 20, 2020, for at least two weeks, the primary school exams were suspended which had been scheduled from March 30, 2020, to April 30, 2020. The assessment of high school students was done by conducting exams in the form of MCQs (multiple choice questions) and their performance was also taken from the schools concerned and the final assessment was the sum total of MCQs and school assessment. Education has been imparted to mentees by mentors via virtual classrooms.

8.3.6 Bermuda

The Department of Education of Bermuda has ensured that learning would continue from homes in the time of lockdown due to the COVID-19 pandemic. The department has guided parents and caregivers through its website, e-mail, school portals, and its facebook page. The Department of Education has also provided physical education material to the students in those areas where online learning has not been supported. In addition to this, the Department of Education has also directed the mentors and principals to respond to the queries of students, their parents, and caregivers. Moreover, the Department of Education has formulated a feedback system to have updates from teachers as well as parents and caregivers about their queries, suggestions, and comments concerning online learning or remote learning from home.

8.3.7 China

Due to the outbreak of COVID-19 in China in February 2020 complete a lockdown was announced by the Chinese government. Classes have been hampered but education should not be badly affected, due to it, the government in collaboration with school authorities has directed the mentor to take online classes with the help of mobile apps. Moreover, the telecom sector has also been directed to provide better services in underprivileged areas.[32]

8.3.8 United States

The U.S. is the role model for all other nations with respect to every aspect whether it is development, knowledge, education or entertainment, etc. Online education has been started in the U.S. in the year 2011 and almost six million students have taken admission in some of the online courses. During the times of the COVID-19 pandemic, the lockdown has been announced by the government in the most affected areas of the U.S.[33]

8.3.9 United Kingdom

In the U.K. as well, online education was started in the year 2011 at the boosted level. The government incurred about one hundred million pounds for the growth of e-learning in the U.K.[34] That is why during lockdown online learning has been used at a wider level in the U.K. Similar is the case with Australia.[35]

8.3.10 India

The government of India in collaboration with the Department of Education and Ministry of Human Resources has launched a nationwide online learning programme referred to as NMEICT, that is, National Mission on Education through Information and Communication Technology. In this e-learning programme, the government has taken various initiatives.

- Swayam (https://swayam.gov.in/): Swayam i.e. Study Webs of Active Learning for Young Aspiring Minds is a consolidated manifesto for introducing various e-learning programmes. These courses are being offered to the students of ninth standard and above. Even college-going students, as well as teachers, can also enrol themselves in these courses. About 2769 massive open online courses have been started by Swayam. National Council of Education Research and Training i.e. NCERT has been formulating conspectus for standards 9, 10, 11, and 12in twelve subjects.
- Swayam Prabha (https://www.swayamprabha.gov.in/): To make feasible e-learning in secluded regions where information superhighway services are not at hand, the government has started 30 scholastic channels on television through digital satellite services (DTH) round the clock. In these T.V. programmes syllabus is taught covering various subjects.
- National Digital Library (NDL) (https://ndl.iitkgp.ac.in/): It is an online library with more than three crore electronic data sources for students of all disciplines. About fifty lakh learners have enroled themselves in this library and about twenty lakh students use the resources of this library. It is also accessible in the form of a mobile app.
- Spoken Tutorial (https://spoken-tutorial.org/?trk=profile_certification_title/): It is an e-learning platform that offers ten minutes audio-video tutorial to enhance the skills of learners for grabbing big career opportunities. These are autodidactism tutorials available in twenty-two regional Indian Languages to tutor beginner learners in the absence of an instructor.
- Free and open-source software for education (FOSSEE) (https://fossee.in/about/): fossee.in provides free and open access to various programmes and software like Python, R, Open PLC, Open FOAM, etc. It also provides access to various learning tools for self-learning like TBC i.e. Textbook companion, Lab migration, etc. The experts of this site also respond to the queries of the users as well as to conduct various conferences and workshops to make students, researchers, and academicians aware of its benefits.
- Virtual Lab (https://www.vlab.co.in/): Learners use advanced technology online to conduct experiments, simulations in a virtual environment i.e. online in the artificial environment but get bonafide outcomes in the field of engineering, computer science, biotechnology, physical and chemical sciences, etc.
- E-Yantra (https://www.e-yantra.org/): It is an online platform made available by MHRD in collaboration with IIT Bombay for the development of robotics. The robotics developed can be used in various fields like agriculture, industries, home, service sector, etc. for the support of human beings.
- e-Pathshala (https://epathshala.nic.in/): e-Pathshala is an online educational hub of resources in the form of e-textbooks, audios, videos, etc. These resources can be accessed from computers, laptops, smartphones, tablets, etc. Moreover, these resources are available in various

languages. e-Pathshala is also available in the form of an app that can be downloaded on mobiles. It's another variant e-PG Pathshala (https://epgp.inflibnet.ac.in/) has been made specifically of the students of post-graduate classes. The electronic material can be easily accessed using QR codes.

• National Repository of Open Educational Resources (NROER) (https://nroer.gov.in/home/e-library/): It is a nationwide open repository of educational resources. These e-resources are available in multiple languages of India in various subjects in the form of audio, videos, documents and textbooks, etc.

In addition to these, the ICT in Education Curricula Portal has also been developed by the Government of India to develop digital skills in students as well as teachers. Moreover, SARANSH has also been developed to access the performance of students and improve that as well. Smart classrooms are also being established in Government schools and Government aided schools under classroom-centric digital intervention project.[36] It is also worth mentioning that, during the COVID-19 pandemic schools, colleges and universities have taken various initiatives in order to provide students with undisrupted access to learning platforms. In addition to various online apps like Zoom, Google Meet, WebEx Meet, Microsoft Teams, etc., mentors have tried various innovative means of teaching as per the available infrastructure in their respective institutes. For example, teachers have recorded their audios and videos and shared these with their students in respective WhatsApp groups or other social networking apps. Mentors have also shared online presentations, weblinks, lecture notes, research articles, and documents to provide learning material to students. They have also assessed the performance of their mentees by asking them to post their assignments in Google Classroom created for that particular subject.

8.4 SMART TECHNOLOGIES FOR ONLINE LEARNING

COVID-19 pandemic has witnessed significant transformation in the teaching and learning process with online learning emerging as a saviour. This transformation has only been possible due to a variety of smart technologies which have played an important role during the COVID-19 pandemic (Figure 8.3). These technologies have also been visualized as the future means of education blended with traditional methods during the post-COVID-19 era.

8.4.1 Mobile Learning

Mobile learning is also referred to as M-Learning. It is the learning which takes place with the help of electronically portable devices, which can be a mobile, smartphone, handheld computer, laptops, palmtops, etc. It is not confined to any particular space. It provides the convenience of anywhere and anytime learning. In addition to this mentors can also provide them on the spot guidance, support, and feedback with the help of their mobile phone. Mentees and mentors can also work collaboratively online with the help of various apps like Zoom, Google Meet, Microsoft Teams, etc. using their electronic mobile devices. Mobile learning focuses on the mobility of the learner (Mehdipour & Zerehkafi, 2013). With the help of these mobile devices multiple types of data i.e. text, audio, video can be delivered but there are certain issues like the disruption of students' personal and academic lives (Masters & Ng'ambi, 2007), risk of distraction (Crescente & Lee, 2011), network and battery issues, etc.

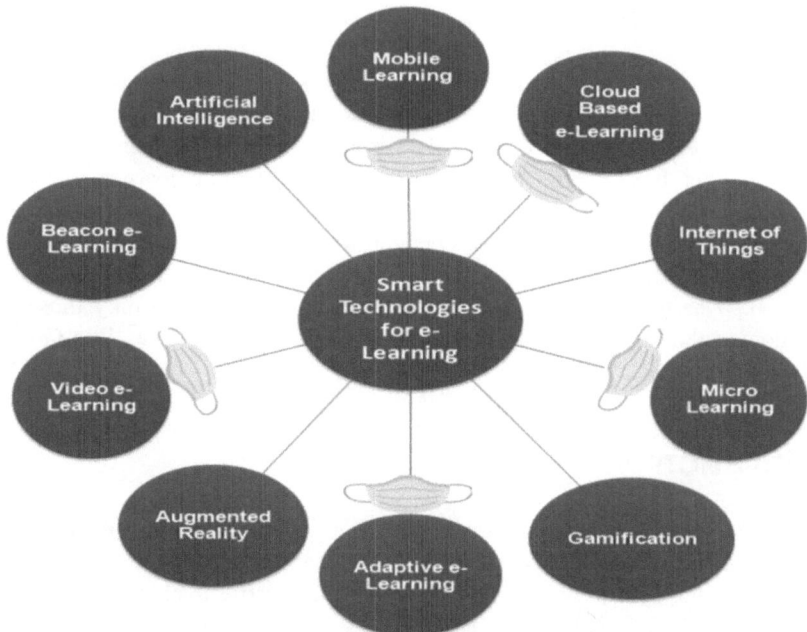

FIGURE 8.3   A graphical summary of various smart technologies used for e-learning. technologies playing a major role during the covid-19 pandemic have been highlighted with masks.

8.4.2 Microlearning

Microlearning means learning in small doses so that it is quite easy for the learner to comprehend. The best example of microlearning is ten to fifteen minutes educational videos on YouTube. Simple concepts can be taught comfortably, but if the concept is quite tedious, this mode of delivery of information is suitable. It has been revealed by the research that the mentee can acquire knowledge in a better way if he is provided with the content in small chunks because if he is made to sit and study for a longer time, he will lack concentration and interest.

8.4.3 Internet of Things

It is an interrelation of computer persons, things, animals, digital machines which transmit information without human and mechanized intervention using a network. A person with a heart monitor implant and an animal with a biochip transponder, a thing with a built-in sensor[37] can transfer data because these devices are assigned unique IP addresses. With an IoT mentor, mentees and parents of mentees can establish contact with each other and can track the progress of the mentee, improve his efficiency and share information. In addition to this, IoT provides digital text, books with 3D pictures and videos, so that mentees can better understand the concepts. Moreover, with the help of IoT technology attendance of the students is taken automatically, mentors can focus on teaching only and parents are also updated with an electronic message by the institute if the mentee is absent. Safety of the children and students is also ensured with the help of IoT as parents can track their children on a real-time basis and ensure their safety[38] and IoT-enabled drones are also available.

8.4.4 Cloud-Based e-Learning

Cloud-based computing means one is using someone else's resources. Let us understand it with the help of an example that one user needs LINUX software that is not on his PC. He or she can work with it, if has Internet connectivity, that is, he or she can use a resource, not of his or her own but someone else's with the help of the Internet. It is referred to as 'The Cloud'. In e-learning, learners use a large number of cloud-based software; some are free and some are paid. For example on Google Drive, one can save one's files in the cloud by using Google Drive. One can work with Google Docs if one has to do one's task in Microsoft word and the task done by the individual will be automatically saved in the cloud and only he can access it. If he gives its access to one or two individuals, then only those individuals can access it. In addition to this, there are other e-learning cloud-based services like Google Classroom, Sheets, Slides, Jamboard, etc.

8.4.5 Gamification

Gamification in learning or education means the application of game elements or rules, characteristics in learning concepts, or in educational concepts in order to enhance the mentees learning capability, their retention, engagement, and productivity.[39] These techniques also enhance competition, team spirit, learning capabilities, socializing aspirations, achievement desire among the participants. Kahoot is an example of gamification. Mentors, while teaching online create a quiz and share it with the help of a web link and mentees respond to the quiz on a real-time basis. In addition to this, mentors use Arcy learning technique, they share YouTube link, and children after watching the video answer the questions, and get an online certificate on the spot[40] and there are other gamification apps as well like Minecraft-Education edition, Math Blaster, Treasure Mountain, Google Read-Along, etc.

8.4.6 Adaptive e-Learning

Adaptive technique is applied in online learning in which the learning is adapted as per the needs and aspirations, abilities of each learner to provide him unique and personal experience.[41] A user is asked before the beginning of the course to select the level of difficulty that is, easy, medium, and hard. The course content is presented to the learner as per the level of difficulty the learner has chosen. The learner cannot move to the next stage before responding to all the questions correctly. Based on the level of learning achieved, the next course is presented to the learner.[42] Thus adaptive learning provides customized tailored programmes for each learner[43] to make him learn as per his capability.

8.4.7 Augmented Reality

Augmented Reality is an amalgamation or unification of cybernated data and media with a natural real-life physical world on a concurrent basis. While wholly or completely unnatural or stilted environment is referred to as virtual reality. Augmented reality utilizes actual existent conditions and surroundings where computerized content is superimposed.[44] With its help mentees can grasp the concepts more knowledgeably by using digital media, graphics, sounds, animations, etc. in real-life environments. For example, students studying surgery sciences can perform surgery digitally without risking any life and money as well. So, in the same way, military learning, space learning, has become quite easy and conceptualized with the aid of augmented reality.

8.4.8 Video e-Learning

Online learning is not possible in the absence of audiovisual aids. Reading something online is quite difficult, it affects the eyes as well as takes more time. While if the concept is presented via video, it can be easily grasped in lesser time. Videos help in better understanding of complex topics, by actually demonstrating the concept visually. Videos shatter the dullness of study by appending graphics, animation, sound, media, etc. for enhancing attentiveness and engrossment.

8.4.9 Beacon e-Learning

Beacon is a compact wireless Bluetooth spammer that sends out gesticulation that can be received by Bluetooth-enabled appliance. These signals can also be received by Bluetooth-enabled mobiles.[45] By structurally splitting an area of a classroom and furnishing appropriate details and facts to the mentees of related subdivided areas (Griffiths et al., 2019) with the aid of audio-video recordings through devices which are connected with the help of Bluetooth is called beacon e-learning.

8.4.10 Artificial Intelligence e-Learning

Artificial intelligence online-learning podium has the power to function as a mentor for mentees by resolving their queries on a real-time basis. The mentee is neither required to wait and ask his mentor about the query nor is he required to type his query on the internet. He can take the help of an AI podium and resolve his problem. Moreover, the mentee is not required to learn any special language, he can simply ask in his language and the AI podium will understand and respond to his problem. It acts as a personal tutor for the learner. It makes changes in the content by apprehending the level of understanding, knowledge, and intelligence of the learner and based on the former learning ability of the learner it also equips him with learning material as per his needs and aspirations.[46]

8.5 SMART RESOURCES FOR ONLINE TEACHING, LEARNING, AND EVALUATION

Having discussed various technologies for online teaching, the current section provides a glimpse of various innovative resources being used to deal with various aspects of online teaching and learning. Some of them have been discussed below.]

8.5.1 Dropbox

Dropbox can store files in an online cloud in a secured and accessible manner. The files in the cloud can be shared with the required number of users. It can be edited by the team members.[47] It can be used for educational purposes as well. It enables the mentors and mentees to share files with the help of Dropbox. It provides 15 GB of storage space to each individual. So, the number of team members space is multiplied which enables the members to store a quite large amount of data in the cloud without affecting their computer storage. Moreover, deleted files can also be restored within 120 days and the admin can check the working of team members and limit their access.[48]

8.5.2 Class Dojo

The mission of Class Dojo is to collaborate the working of mentor, mentee, and parents of mentees and also provide them an excellent experience. It acts as a transmission desk that enables sharing of photos, videos, messages, etc., that is, of what is being learned by mentees in the class, and mentors also share their feedback with the mentee's parents about the performance of their children on a real-time basis.[49] By collaborating in the correct manner teachers and parents can make the student grow excellently.

8.5.3 Edmodo

Edmodo enables online communication between teacher, students, as well as parents of students. It also enables a teacher to post assignments for their students and grade them, after the student has submitted and measure their performance. Moreover, teachers can get feedback from their students and can guide them. The teacher can monitor the content posted by students and control their work. In addition to posting assignments, a teacher can take a quiz, create polls, as well as put the link to make reliable content available to their students.[50]

8.5.4 Educreations

Educreations is a mentors' platform that enables them to prepare captivating, interesting, and knowledgeable lessons for their mentees with the help of a recording option. The mentor can record his audio and can design a visual lesson for his mentees that can be accessed by them at their convenience. Educreations also enables the mentors to import images and documents from Dropbox, Google Drive, etc. into the whiteboard. Moreover, whatever a mentor creates on Educreations is automatically stored in his account and can be shared there. But it can be accessed only by those to whom the right has been given.[51]

8.5.5 TED Ed

It is a platform via which educational videos can be accessed by the students. It has enabled the mentors to create and post videos that can be accessed by curious mentees all over the world.[52]

8.5.6 Unplag

Unplag is used to detect content similarity. When the mentees submit their work with the mentor. The mentor can store the work in his library and check the content similarity of the work with his e-library and over the Internet with the help of Unplag and detect the extent of copied work. The report generated after checking plagiarism can be saved by the mentor and he can also email the report to his mentee.[53]

8.5.7 Slack

Slack is a social media tool that is greatly being used for educational purposes.[54] It is an app that is used for assigning group tasks to students. They can create a slack group and collaborate. They can discuss the topic online by sitting at their places from remote locations (Darvishi, 2020). Slack also has the feature of messenger, Twitter, and Dropbox, which means one can share documents, and all the persons to whom the access has been given can edit the Google Docs file and team members can also arrange an online meeting if required using Google Meet. Slack allows the use of Google Suite as well.[55]

8.5.8 Google Apps for Education

Google Docs, Sheets, and Slides allow the sharing of documents, sheets, slides and these can be accessed, edited by those to whom access has been given. By using Google Drive one need not save anything on a computer. One can simply save one's files, folders on Google Drive without affecting P.C. space. Google Docs, Sheets, and Slides are automatically saved in Google Drive and can be accessed therefrom. Jamboard allows writing text, drawing images and removing them, and even allows a user to add text and image from the web.[56] To communicate for educational purposes Google has provided the facility of Gmail and even Google Meet. Google Meet allows mentors and mentees to meet virtually and mentors can effectively teach their mentees. The Chat feature is also available in Google Meet so that mentors and mentees can communicate with each other. Using Google Classroom mentors can assign tasks to mentees and mentees can submit their tasks personally to their mentors so others could not copy their work. It also enables mentors to check the work submitted and evaluate the performance of their mentees. In order to gather feedback or comments, mentors can use Google Forms. It can also be used for attendance and quiz. Google also allows the creation of assignments. Using Google admin function mentors can handle security issues, mentees, and devices in a secured manner.[57]

8.5.9 Remind

Remind app is used by mentors to remind mentees about some important task. Mentees cannot revert any message. Moreover, the reminders sent to the mentees will be anonymous. They cannot find out who has sent them the reminder. In addition to this, by using this app, a mentor can arrange a virtual class for his or her mentees and can send them the link generated using the app. Mentees can join the class using the link shared by the mentor. Through this app, a teacher can also stay connected with the parents of their mentees.[58]

8.5.10 Edublogs

Edublogs is composed of two words, that is, edu and blog. Edu means educational and blog means a web page written by an individual or a group of individuals. So, Edublogs means a web page written for educational purposes. These blogs can be created by the teacher, researcher, learner, or administrator and can be shared with students. The teacher creates Edublogs to share information with the students. Even students can also create Edublogs individually or as a part of a team to reveal their learning or to share the results of the study undertaken by them.[59]

8.5.11 Socrative

Socrative helps mentors to engage mentees by sending them an online quiz and it also helps mentors to access the performance of the mentees and grade them. The mentor can engage students on laptops, smartphones, PCs, or tablets having Internet access.[60]

8.5.12 Moodle

MOODLE stands for a modular object-oriented dynamic learning environment. It is a vigorous, dependable, and cost-free e-learning platform created by Martin Dougiamas in the year 2002.[61] Lessons can be created and courses can be managed with the help of MOODLE by the educators. Students can submit their tasks, attempt MCQs, access videos and text, and interact with mentors and other students using it. Class executives can tailor-make the MOODLE classroom and course as per need and requirement. To work with MOODLE, it must be downloaded and installed on one's system.[62]

8.5.13 Discord

Discord has been developed in the year 2015 as a group chat app for the gaming community. Servers composed discord and each server is composed of its members, channels, rules, and topics. Using this app, one can share text messages, audio, video, images, music, web links, etc. with the help of the chat function. The mentor can add mentees on discord by forwarding them a link generated by it for adding members. Moreover, the mentor can mute his mentees if they have unmuted knowingly or unknowingly. In addition to this, it allows sharing of the screen.[63]

8.6 ARTIFICIAL INTELLIGENCE-BASED LEARNING AND THE EMERGENCE OF THE INTELLIGENT TUTORING SYSTEM (ITS)

In the previous sections, we had deliberated on various aspects of the teaching-learning process and tried to highlight various smart technologies, which have been used to combat the impact of COVID-19 on the education sector. In the absence of physical classrooms, Internet-based or information-based teaching-learning strategies played a vital role in continuing educational activities during the COVID-19 era. Though information-based teaching has been adopted by many countries in the recent past, the COVID-19 pandemic has resulted in an exponential increase in their usage. In fact, information-based teaching has been the front runner to provide affordable and scalable education to learners around the world. The implementation of Information-based teaching has primarily been dominated by online teaching platforms and multimedia technology (Gu, 2019). However, these two verticals have some inherent problems including the lack of classroom experience, lack of standard curriculum, difficulty to observe student's learning behaviour, etc. (Gu, 2019; Mor et al., 2006). If we can dynamically and scientifically observe, monitor, and track learner's online behaviours, we can provide adaptive feedback, customized assessment, and more personalized attention to them (Kelly and Nanjiani, 2004). Furthermore, online teaching caters to a large number of students from across the world with different time zones, different foundational skills, different progress rates, and different frames of reference. Such a diversity of students often makes it challenging to achieve desired outcomes, assess progress, and provide personalized constructive feedback (Popenici & Kerr, 2017). That is the reason why a new information-based sophisticated technology known as AI is the recent buzzword to make information-based teaching more effective, not only during the current pandemic but also during the post-COVID-19 era.

In simple terms, AI can be expressed as the association of intelligence with machines. It aims to develop machines that can mimic human intelligence and take decisions matching human capabilities. AI complements computer science by creating efficient programmes that help to develop virtual machines with capabilities of reasoning, problem-solving, and learning. It has many applications including pattern recognition, natural language processing, automatic theorem proving, programme automation, intelligent database system, robotics, expert systems, artificial neural network, and intelligent tutoring system (ITS) (Malik et al., 2019). ITS is a rapidly evolving area where AI plays a vital role in developing intelligent tutors or assistants, predicting student behaviors (Huang et al., 2019; Waheed et al., 2020), recommending the most suitable content for the learners (Ammar et al., 2020), and managing huge volumes of data (Daniel, 2017). Recent advances in big data, DL, or ML techniques have significant potential to further strengthen the AI-based teaching-learning process, leading to novel applications, more efficient operations, and more human approaches (García-Peñalvo et al., 2020). Moreover, it helps to design interactive content such as e-books, video lectures, natural games, and individual assessments of teaching agents (Malik et al., 2019). With the application of AI in developing Interactive graphics, enhanced gaming platforms, virtual teaching agents, context-specific feedback systems, and precision curriculum, information-based education is ready for a paradigm shift (Malik et al., 2019; Pareto, 2014). Some examples of intelligent tutor systems include BEETLE II, ASSISTments ecosystem, Reasoning Mind Genie 2, and AutoTutor and family (Dzikovska et al., 2014; Heffernan & Heffernan, 2014; Khachatryan et al., 2014; Nye et al., 2014). These systems include features such as multiple 3D representations with dynamic graphical user interfaces, various degrees of problems (lower, intermediate, and high), logical reasoning platform, analysis of simulated and real data, teaching with virtual agents, and fusion of human and computer intelligence with strong and context-related decision-making qualities (Malik et al., 2019).

Also, AI-based real-time gaming solutions such as BELLA (for primary school children) and Robot laboratory (for undergraduate students of computer science) have been introduced. In these games, problems are represented as adventure games which make students deeply involved in the concept, enhancing their learning capabilities. The ITS has specific applications in STEM-based teaching-learning. During the COVID-19 pandemic, ITS has been smartly utilized in the form of high-fidelity simulations to make medical students learn and practice surgical skills, thereby Re-Envisioning Surgical Education (Mirchi et al., 2020Mirchi 2020).

8.7 EXPLAINABLE ARTIFICIAL INTELLIGENCE (XAI)

AI has gained a lot of momentum in the last decade and has achieved an unprecedented level of learning performance for solving increasingly complex problems. The scope of AI has further widened, making this technology vital for developing human society to its next level (West, 2018). However, a transition from initial transparent interpretable models to the present-day opaque Deep Learning models-powered systems has made AI systems so sophisticated that they merely require any human intervention in their design and deployment (Arrieta et al., 2020). This lack of transparency in many complex black-box model-based AI systems has created a lot of interest in including an explanation in these systems so that their behaviour can be better understood and easily interpreted (Putnam & Conati, 2019). Transparency is critical in building user's trust, especially during AI applications directly influencing humans' lives such as in the fields of medicine, law, defence, transportation, security, and finance.

For example in precision medicine, experts require much more information from an AI model than that generated by a binary prediction model to correctly diagnose a disease condition (Tjoa & Guan, 2020). In the field of education also, AI-based systems have been further advanced using XAI. In ITS, the concept of XAI has been used to design educational systems that can take care of the relevant states, needs, and properties of their learners while interacting with them, thereby giving personalized instructions to best suit their specific requirements (Anderson et al., 1985). As such systems have long-lasting effects on the learners's transparency, and interpretability of the backend algorithm driving becomes extremely important (Conati et al., 2018). A special ITS known as Adaptive Constraint Satisfaction Problem (ACSP) was used to provide personalized instructions to the students using interventions in the form of textual hint messages (Putnam & Conati, 2019). Furthermore, with a paradigm shift in using Virtual Learning Environments (VLEs), the data related to student's interaction with VLEs has increased exponentially, providing a digital footprint of the students' engagement with the learning materials and activities. AI techniques have been commonly used to mine this big data to build models to predict students' outcomes and other applications.

XAI is of great need in the field of education which involves interaction between humans and AI systems to develop more transparent AI systems (Alonso & Casalino, 2019). A system called ExpliClas (Alonso & Bugarín, 2019) has been used for the generation of explanations in the process of education (Alonso & Casalino, 2019). An XAI tool called Virtual Operative Assistant has also been created to impart effective simulation-based training in the field of surgery and medicine (Mirchi et al., 2020Mirchi 2020). Such XAI innovations driven will pave the way to creating more robust and transparent AI-based applications, thereby fulfilling the goal of smart education for smart cities.

8.8 BLENDED LEARNING MODEL FOR FUTURE

The blended learning model is a mix of mentor-led offline learning and online learning to enhance the level of student's understanding. Blended learning amalgamates conventional mentor-mentee classroom teaching with online learning to meet the needs of the individual mentees. There are different methods of blended learning that uses different levels and modes of technology. Due to the use of technology, formal learning is possible using remote access. The various blended learning models are as listed below.

8.8.1 Face-to-Face Driver Model

In the face-to-face learning model teacher teaches students in the classroom mode. In this blend of online-offline modes, technology is used for assistance only.

8.8.2 Rotation Model

In the rotation model, mentees are rotated on a fixed schedule basis or as per the discretion of the mentor under different learning methodologies but one such methodology which is required to be followed with offline learning is e-learning. Other methodologies that can be used are like dividing the class into small groups and assigning group assignments, individual student teaching methods, etc.[64]

8.8.2.1 Station Rotation

In this rotation model, the stations are changed for teaching. Mentees are divided into groups. At one point in time, one group may be learning offline in the class and at the same point of time another group may be learning using online mode in the same class and the third group may be given individual attention by a group of mentors i.e. each member of the group has been assigned with an individual mentor for teaching and learning and that as well in the same classroom.

8.8.2.2 Lab Rotation

In this rotation model students work in both online and offline modes. Students learn the concepts in offline mode in class and then perform that in the online mode. If they find any difficulty they can again ask their mentors in the offline class. The basic difference between station and lab rotation is that under station rotation mentees study at the same place while in a lab rotation mentees can change their place of learning from classroom to lab for online learning.[65]

8.8.2.3 Flipped Rotation

In the flipped rotation model, students go through the new concept at home before the teacher has explained and then in the class they practice the concept in the presence of the the mentor and get their queries resolved on the spot in the classroom. By following this model, each student learns at his or her speed, each one has online access to the content or material and he or she can get the solution to his or her problems from the mentor.[66]

8.8.2.4 Individual Rotation Model

In this rotation model, each mentee is rotated as per the fixed schedule formulated by the mentor or at the discretion of the mentor under different learning modalities specifically from offline to online learning methodology. It allows the mentees to work at their speed and they can use the modality which best suits them, some prefer the online and some prefer the offline learning methodology.[67]

8.8.3 Flex Model

In this model of blended learning, each mentee is assigned a PC in the classroom setting and mentees are required to complete the given task at their own pace without any pressure and mentor is also available in the classroom setting to resolve the queries of mentees on a one-on-one basis or if the whole class is facing the same problem, it can be resolved by the mentor for all of them in the form of group discussion.[68]

8.8.4 Online Lab School Model

In this model, teaching is done online by a mentor and mentees learn by sitting in the computer lab of their school having access to the Internet. Mentees are supervised by non-teaching staff and taught by an online mentor. There is no face-to-face teaching and learning.[69]

8.8.5 Self-Blended Model Evaluation of classification methods and learning logs

In this model of blended learning, the learner himself chooses the online course of his own choice along with face-to-face learning at some educational institute. Like, a student is interested in pursuing French that is not available at his college. He can opt for the subjects of his choice at college and can study French by enroling himself in a French online course. So, the student has blended online and offline learning and that is why this learning would be known by the name of self-learning. It is also known as the À La Carte approach of blended learning.[70]

8.8.6 Online Driver Model

In this model of blended learning a student has online access to the learning material and if he does not understand any concept, he can mail his difficulties or problems to the mentor and the mentor will arrange an online chat or online class to resolve his queries. In addition to this, a student also submits his assignments online using a PC, tablet, laptop, or mobile. Using this model a student can enrol himself in an online course in other countries as well.[71]

8.9 LIMITATIONS AND FUTURE PROSPECTS

Online teaching has increased the efficiency of teachers. Due to online teaching, teachers are using online resources beyond books. It has provided the facility of anytime and anywhere learning. Students can join the class from where they are. They can record the lecture if they are busy at the time of the lecture and can hear the lecture when they have time. Online learning has added comfort due to which student attendance has improved. It is also beneficial for physically challenged students but internet access is essential for it. Moreover, if the mentor is busy at the time of the lecture, he can record his lecture before the class and can share the audio or video with the students at the time of the lecture.[72] In addition to this, experts can also be invited to online classes easily while it is quite difficult for them to come physically.

On the other hand, online learning also has some drawbacks. Students are distracted by social sites and online advertisements while e-learning. So, it is required on the part of mentors to keep the online class more collaborative and engrossing to maintain the concentration of the students. Moreover, while learning online, sometimes teachers and sometimes students face connectivity issues. In addition to this, students also feel isolated, so mentors must allow mentees to have online face-to-face interaction and allow other forms of communication among students to lessen this feeling and it is also creating health issues among students like eyesight problems, posture problem, other physical and psychological issues, etc. It is also imperative on the part of educational institutions to provide training to the teachers for enabling them to take online classes.

During the time of the COVID-19 Pandemic, all the schools, colleges, and universities have been closed and teaching and learning have been done through online mode. Now the situation which is prevailing today is referred to as the new normal. Although governments of various countries have permitted the opening up of educational institutes, some students are coming while some study online using various apps. Moreover, students, parents, and teachers have realized the importance of independent learning during the times of online teaching-learning from home due to the pandemic. Students do not need face-to-face interaction, what they need is effective learning, on-the-spot help, and emotional support. All this is possible in online learning as well. So, in the future, blended learning models will work, in which along with classroom teaching and learning there will be online learning as well.[73]

NOTES

---

1 https://www.asme.org/topics-resources/content/top-10-growing-smart-cities
2 https://blog.bismart.com/en/what-is-a-smart-city#:~:text=In%20general%2C%20a%20smart%20city,%2C%2C%20Othe%20idea%20isn't.
3 https://hub.beesmart.city/en/solutions/smart-people/smart-education/viewsonic-smart-education-for-smart-cities
4 https://www.thetechedvocate.org/how-can-smart-city-technologies-impact-education/
5 https://www.educationcorner.com/building-a-positive-learning-environment.html
6 https://www.tes.com/news/what-is-pedagogy-definition
7 https://www.criticalthinking.org/pages/socratic-teaching/606
8 https://www.intel.com/content/dam/www/program/education/us/en/documents/project-design/strategies/dep-question-socratic.pdf
9 https://www.ukessays.com/essays/it-research/computer-aided-learning.php

10 https://en.wikipedia.org/wiki/Online_learning_in_higher_education
11 http://adamasuniversity.ac.in/a-brief-history-of-online-education/
12 https://thebestschools.org/magazine/online-education-history/
13 https://www.britannica.com/topic/distance-learning
14 https://www.petersons.com/blog/the-history-of-online-education/
15 https://www.fnu.edu/evolution-distance-learning/
16 https://en.wikipedia.org/wiki/Blackboard_Learn
17 https://en.wikipedia.org/wiki/History_of_virtual_learning_environments
18 https://en.wikipedia.org/wiki/Creative_Commons
19 https://en.wikipedia.org/wiki/MIT_OpenCourseWare
20 https://en.wikipedia.org/wiki/Facebook
21 https://en.wikipedia.org/wiki/YouTube
22 https://en.wikipedia.org/wiki/Liberty_University
23 https://www.insidehighered.com/digital-learning/article/2019/06/12/apple-winds-down-itunes-u#:~:text=Apple
    %20eliminated%20the%20iTunes%20U,to%20the%20Apple%20Podcasts%20app.
24 https://en.wikipedia.org/wiki/Khan_Academy
25 https://www.mcgill.ca/maut/current-issues/moocs/history
26 https://www.insidehighered.com/news/2013/10/01/u-florida-races-create-online-campus-jan-1-opening-date-
    approaches
27 https://en.wikipedia.org/wiki/Zoom_Video_Communications#:~:text=Zoom%20was%20founded%20by
    %20Eric,videotelephony%20market%20was%20already%20saturated
28 https://en.wikipedia.org/wiki/Google_Meet
29 https://www.theasianschool.net/blog/what-is-the-digital-education-system-and-its-advantages-for-students/
30 https://www.gettingsmart.com/2019/02/a-teachers-perspective-on-the-importance-of-sharing-students-learning/
31 https://focusband.com/7-benefits-digital-learning/
32 https://www.worldbank.org/en/topic/edutech/brief/how-countries-are-using-edtech-to-support-remote-learning-
    during-the-covid-19-pandemic
33 https://monitor.icef.com/2018/05/continuing-expansion-online-learning-us/
34 https://london.ac.uk/ways-study/distance-learning
35 https://www.weforum.org/agenda/2020/04/coronavirus-education-global-covid19-online-digital-learning/
36 https://pib.gov.in/Pressreleaseshare.aspx?PRID=1577240
37 https://internetofthingsagenda.techtarget.com/definition/Internet-of-Things-IoT
38 https://www.digiteum.com/iot-applications-education/
39 https://en.wikipedia.org/wiki/Gamification
40 https://www.gamify.com/what-is-gamification
41 https://www.iadlearning.com/adaptive-e-learning/
42 https://blog.commlabindia.com/elearning-design/adaptive-learning-in-elearning
43 https://elearningindustry.com/adaptive-learning-for-schools-colleges
44 https://whatis.techtarget.com/definition/augmented-reality-AR
45 https://kontakt.io/what-is-a-beacon/
46 https://elearningindustry.com/artificial-intelligence-based-platform-impact-future-elearning
47 https://www.dropbox.com/?landing=dbv2
48 https://help.dropbox.com/accounts-billing
49 https://www.classdojo.com/about/
50 https://www.emergingedtech.com/2013/12/10-reasons-why-edmodo-is-a-excellent-and-hugely-popular-digital-
    learning-platform/
51 https://blogs.umass.edu/onlinetools/learner-centered-tools/educreations/
52 https://www.ted.com/about/programs-initiatives/ted-ed
53 https://edtechreview.in/trends-insights/insights/2529-student-online-learning-supervision
54 http://www.diva-portal.se/smash/get/diva2:1393381/FULLTEXT01.pdf
55 https://edtechmagazine.com/higher/article/2015/09/4-lessons-college-it-teams-switch-slack
56 https://www.pocket-lint.com/gadgets/news/google/139279-what-is-google-jamboard-how-does-it-work-and-
    when-can-you-buy-it
57 https://edu.google.com/intl/en_in/products/gsuite-for-education/
58 https://www.youtube.com/watch?v=486gZe0wCbU
59 http://desarrolloweb.dlsi.ua.es/blogs/what-is-an-education-blog
60 https://www.definitions.net/definition/socrative#:~:text=Socrative%20helps%20teachers%20engage%20%26%2
    0assess,districts%20personalize%20and%20improve%20learning.
61 https://ethinkeducation.com/what-is-moodle-guide/
62 https://techterms.com/definition/moodle

63 https://www.businessinsider.com/what-is-discord?IR=T
64 https://www.christenseninstitute.org/blended-learning-definitions-and-models/
65 https://www.easygenerator.com/en/blog/blended-learning/lab-rotation-model/
66 https://www.cae.net/flipped-classroom-or-station-rotation-lab-which-blended-learning-model-is-right-for-your-center/
67 https://sites.google.com/site/blendclass/individual-rotation
68 https://study.com/academy/lesson/flex-mode-in-blended-learning-definition-application-examples.html
69 https://study.com/academy/lesson/online-lab-model-in-blended-learning-definition-application-examples.html
70 https://study.com/academy/lesson/self-blend-model-definition-application-examples.html
71 https://study.com/academy/lesson/online-driver-model-definition-application-examples.html
72 https://elearningindustry.com/advantages-and-disadvantages-online-learning
73 https://blogs.ibo.org/blog/2020/09/28/the-future-of-online-learning/

REFERENCES

Allen, DE, Donham, RS, & Bernhardt, SA (2011). Problem-based learning. *New Directions for Teaching and Learning, 2011*(128), 21–29.

Alonso, JM, & Bugarín, A (2019). ExpliClas: Automatic generation of explanations in natural language for WEKA classifiers. In *2019 IEEE International Conferences on Fuzzy Systems*, 1–6.

Alonso, JM, & Casalino, G (2019). Explainable artificial intelligence for human-centric data analysis in virtual learning environments. In D Burgos et al. (Eds.), *Higher education learning methodologies and technologies online. HELMeTO 2019. Communications in Computer and Information Science 1091*. Springer. Retrieved from 10.1007/978-3-030-31284-8_10.

Ammar, WBH, Chaabouni, M, & Ghezala, HB (2020). Recommender system for quality educational resources. In V Kumar & C Troussas (Eds.), Intelligent Tutoring Systems, Proceedings of the 16th International Conference. ITS 2020. Athens, Greece, 8–12 June 2020. Lecture Notes in Computer Science. Springer International Publishing, 327–334.

Anderson, JR, Boyle, CF, & Reiser, BJ (1985). Intelligent tutoring systems. *Science, 228*(4698), 456–462.

Arrieta, AB, Díaz-Rodríguez, N, Del Ser, J, Bennetot, A, Tabik, S, Barbado, A, García, S, Gil-López, S, Molina, D, Benjamins, R, & Chatila, R (2020). Explainable Artificial Intelligence (XAI): Concepts, taxonomies, opportunities and challenges toward responsible AI. *Information Fusion, 58*, 82–115.

Bhattacharyya, E, & Sarip, ABM (2014). Learning style and its impact in higher education and human capital needs. *Procedia - Social and Behavioral Sciences, 123*, 485–494.

Blumenfeld, PC, Soloway, E, Marx, RW, Krajcik, JS, Guzdial, M, & Palincsar, A (1991). Motivating project-based learning: Sustaining the doing, supporting the learning. *Educational Psychologist, 26*(3–4), 369–398.

Bruffee, KA (1993). *Collaborative learning: higher education, interdependence, and the authority of knowledge.* Baltimore, US: John Hopkins University Press.

Bunchball, I (2010). Gamification 101: An introduction to the use of game dynamics to influence behavior. *White paper 9.*

Cingi, CC (2013). Computer aided education. *Procedia - Social and Behavioral Sciences, 103*, 220–229.

Conati, C, Porayska-Pomsta, K, & Mavrikis, M (2018). AI in education needs interpretable machine learning: Lessons from Open Learner Modelling. arXiv preprint arXiv:1807.00154.

Cózar-Gutiérrez, R, & Sáez-López, JM (2016). Game-based learning and gamification in initial teacher training in the social sciences: An experiment with MinecraftEdu. *International Journal of Educational Technology in Higher Education, 13*(1), 1–11.

Crescente, ML, & Lee, D (2011). Critical issues of M-Learning: Design models, adoption processes, and future trends. *Journal of the Chinese Institute of Industrial Engineers, 28*(2), 111–123.

Daniel, BK (2017). Big Data and data science: A critical review of issues for educational research. *British Journal of Educational Technology, 50*, 101–113.

Darvishi, S (2020). The use of Slack as a social media in higher education: Students perceptions of advantages and disadvantage of Slack during learning process. Retrieved from http://lnu.diva-portal.org/smash/get/diva2:1393381/FULLTEXT01.pdf as on 10 December, 2020.

De Graaf, E, & Kolmos, A (2003). Characteristics of problem-based learning. *International Journal of Engineering Education, 19*(5), 657–662.

Deterding, S, Sicart, M, Nacke, L, O'Hara, K, & Dixon, D (2011). Gamification. Using game-design elements in non-gaming contexts. *In CHI'11 extended abstracts on human factors in computing systems*, 2425–2428.

Dzikovska, M, Steinhauser, N, Farrow, E, Moore, J, & Campbell, G (2014). BEETLE II: Deep natural language understanding and automatic feedback generation for intelligent tutoring in basic electricity and electronics. *International Journal of Artificial Intelligence in Education*, 24(3), 284–332.

Epps, S. (2009). *What Engagement Means for Media Companies*, Forrester, New York.

García-Peñalvo, FJ, Casado-Lumbreras, C, Colomo-Palacios, R, & Yadav, A (2020). Smart learning. Retrieved from file:///C:/Users/BIOINFO/Downloads/applsci-10-06964.pdf as on 20 December, 2020.

Griffiths, S, Wong, MS, Kwok, CYT, Kam, R, Lam, SC, Yang, L, Yip, TL, Heo, J, Chan, BSB., Xiong, G, & Lu, K (2019). "Exploring Bluetooth Beacon Use Cases in Teaching and Learning: Increasing the Sustainability of Physical Learning Spaces". Sustainability, 11 (15), 4005 1–17.

Gu, R (2019). Application of Big Data and artificial intelligence technology in the construction of information-based teaching model in colleges and universities. Retrieved from https://webofproceedings.org/proceedings_series/ESSP/ICRTPE%202019/ICRTPE199.pdf as on 10 December, 2020.

Guragain, N. (2016). *E-Learning Benefits and Applications*.

Heffernan, NT, & Heffernan, CL (2014). The Assistments ecosystem: Building a platform that brings scientists and teachers together for minimally invasive research on human learning and teaching. *International Journal of Artificial Intelligence in Education*, 24(4), 470–497.

Huang, AYQ, Lu, OHT, Huang, JCH, Yin, CJ, & Yang, SJH (2019). Predicting students' academic performance by using educational big data and learning analytics: Evaluation of classification methods and learning logs. *Interactive Learning Environments*, 28, 206–230.

Hung, W. (2013). "Problem-based learning: A learning environment for enhancing learning transfer". *New directions for adult and continuing education*, 137, 27–38. 10.1002/ace.v2013.137.

Hung, W., Jonassen, D. H., & Liu, R. (2008). *"Problem-basedlearning. Handbook of research on educational communications and technology"*, 3, 485–506.

Illeris, K, Reader, D, & Malone, M (2002). *The three dimensions of learning: Contemporary learning theory in the tension field between the cognitive, the emotional and the social*. Roskilde University Press.

Kelly, TM, & Nanjiani, NA (2004).*The business case for e-learning*. Cisco Press.

Khachatryan, GA, Romashov, AV, Khachatryan, AR, Gaudino, SJ, Khachatryan, JM, Guarian, KR, & Yufa, NV (2014). Reasoning Mind Genie 2: An intelligent tutoring system as a vehicle for international transfer of instructional methods in mathematics. *International Journal of Artificial Intelligence in Education*, 24(3), 333–382.

Kilroy, DA (2004). Problem based learning. *Emergency Medicine Journal*, 21(4), 411–413.

Kokotsaki, D, Menzies, V, & Wiggins, A (2016). Project-based learning: A review of the literature. *Improving Schools*, 19(3), 267–277.

Külür, S (2001). New possibilities for the photogrammetry education in Turkey. *The International Archives of the Photogrammetry, Remote Sensing and Spatial Information Sciences*, 34(6), 1–3.

Lee, JJ, & Hammer, J (2011). Gamification in education: What, how, why bother? *Academic Exchange Quarterly*, 15(2), 146.

Lepper, MR, & Malone, TW (1987). Intrinsic motivation in the classroom. Research on motivation in education. *Aptitude, Learning and instruction*, 3, 225–286.

Malik, G, Tayal, DK, & Vij, S (2019). An analysis of the role of artificial intelligence in education and teaching. In Recent Findings in Intelligent Computing Techniques. Springer, 407–417.

Masters, K, & Ng'ambi, D (2007). After the broadcast: Disrupting health sciences' students' lives with SMS. In Proceedings of IADIS International Conference Mobile Learning. Lisbon, Portugal, January, 171–175.

McFarlane, A, Sparrowhawk, A, & Heald, Y (2002). Report on the educational use of games. *In TEEM (Teachers evaluating educational multimedia)*, Cambridge.

Mehdipour, Y, & Zerehkafi, H (2013). Mobile learning for education: Benefits and challenges. *International Journal of Computational Engineering Research*, 3(6), 93–101.

Mirchi, N, Bissonnette, V, Yilmaz, R, Ledwos, N, Winkler-Schwartz, A, & Del Maestro, RF (2020). The Virtual Operative Assistant: An explainable artificial intelligence tool for simulation-based training in surgery and medicine. *PloS One*, 15(2), e0229596.

Mirchi, N, Ledwos, N, & Del Maestro, RF (2020). Intelligent tutoring systems: Re-envisioning surgical education in response to COVID-19. *Canadian Journal of Neurological Sciences*, 48(2), 198–200.

Mor, E., Minguilkin, J., & Carb¢, J. M. (2006). Analysis of user navigational behavior for e-learning personalization. *Data mining in e-learning (advances in management information)* (vol. 4, pp. 227–243).

Nye, BD, Graesser, AC, & Hu, X (2014). AutoTutor and family: A review of 17 years of natural language tutoring. *International Journal of Artificial Intelligence in Education, 24*(4), 427–469.

Oblinger, D (2004). The next generation of educational engagement. *Journal of Interactive Media in Education, 2004*(1), 1-18.

Panke, S (2017). Crossover learning. *AACE Review*. Retrieved from https://www.aace.org/review/crossover-learning/ as on 11 December, 2020

Papastergiou, M (2009). Digital game-based learning in high school computer science education: Impact on educational effectiveness and student motivation. *Computers & Education, 52*(1), 1–12.

Pareto, L (2014). A teachable agent game engaging primary school children to learn arithmetic concepts and reasoning. *International Journal of Artificial Intelligence in Education, 24*(3), 251–283.

Pesare, E, Roselli, T, Corriero, N, & Rossano, V (2016). Game-based learning and gamification to promote engagement and motivation in medical learning contexts. *Smart Learning Environments, 3*(1), 1–21.

Popenici, SA, & Kerr, S (2017). Exploring the impact of artificial intelligence on teaching and learning in higher education. *Research and Practice in Technology Enhanced Learning, 12*(1), 1–13.

Poscente, K (2006). The three dimensions of learning: Contemporary learning theory in the tension field between the cognitive, the emotional and the social. Author: Knud Illeris. *The International Review of Research in Open and Distributed Learning, 7*(1), 1–3.

Prensky, M (2001). The digital game-based learning revolution. *Chp1 into from digital game-based training.*

Putnam, V, & Conati, C (2019). March. Exploring the Need for Explainable Artificial Intelligence (XAI). In *Intelligent Tutoring Systems (ITS) in IUI Workshops* 19.

R Core Team (2020). R: A language and environment for statistical computing. In *R Foundation for Statistical Computing*. Vienna, Austria. Retrieved from https://www.R-project.org/.

Rieber, LP (1996). Seriously considering play: Designing interactive learning environments based on the blending of microworlds, simulations, and games. *Educational Technology Research and Development, 44*(2), 43–58.

Rosas, R, Nussbaum, M, Cumsille, P, Marianov, V, Correa, M, Flores, P, Grau, V, Lagos, F, López, X, López, V, & Rodriguez, P (2003). Beyond Nintendo: Design and assessment of educational video games for first and second grade students. *Computers & Education, 40*(1), 71–94.

Sarasin, LC, & Celli, LM (1999). Learning style perspectives: Impact in the classroom. *Atwood Pub.*

Scarbrough, H, Swan, J, Laurent, S, Bresnen, M, Edelman, L, & Newell, S (2004). Project-based learning and the role of learning boundaries. *Organization Studies, 25*(9), 1579–1600.

Simões, J, Redondo, RD, & Vilas, AF (2013). A social gamification framework for a K-6 learning platform. *Computers in Human Behavior, 29*(2), 345–353.

Solomon, G (2003). Project-based learning: A primer. *Technology and Learning Dayton, 23*(6), 20.

Surjono, HD (2011). The design of adaptive e-learning system based on student's learning styles. *International Journal of Computer Science and Information Technologies, 2*(5), 2350–2353.

Tjoa, E, & Guan, C (2020). A survey on explainable artificial intelligence (xai): Toward medical. *In xai. IEEE Transactions on Neural Networks and Learning Systems.*

Waheed, SUH, Hassan, N, Aljohani, R, Hardman, J, Alelyani, S, & Nawaz, R (2020). Predicting academic performance of students from VLE big data using deep learning models. *Computers in Human Behavior, 104*, 106–189.

West, DM (2018). *The future of work: Robots, AI, and automation.* Brookings Institution Press.

Wickham H (2016). *ggplot2: Elegant graphics for data analysis.* Springer-Verlag. ISBN 978-3-319-24277-4. Retrieved from https://ggplot2.tidyverse.org./

Zhang, D, Leon Zhao, J, Zhou, L, & Nunamaker, Jr., JF (2004). Can e-learning replace classroom learning. *Communications of the ACM, 47*(5), 75–79. Retrieved from 10.1145/986213.986216.

Zichermann, G, & Cunningham, C (2011). *Gamification by design: Implementing game mechanics in web and mobile apps.* O'Reilly Media.

An Analysis of Machine Learning for Smart Transportation System (STS)

9

R. Dhaya[1] and R. Kanthavel[2]

[1]*Department of Computer Science, King Khalid University-Sarat Abidha Campus, Abha, Saudi Arabia*
[2]*Department of Computer Engineering, King Khalid University, Abha, Saudi Arabia*

Contents

DOI: 10.1201/9781003172772-9

9.1 INTRODUCTION

In Smart Transport Systems (ST) background, precise calculation of potential circumstances of the traffic is fundamental to relieving gridlock in order to react to passage episodes. Factual machine learning calculations have likewise discovered their approach to supporting brilliant transportation. In any case, ML authorizes a framework to increase the benefits by means of data rather than a set of codes. Be that as it may, ML is certainly not a basic cycle (Mrityunjaya et al., 2017). As the counts ingest planning data, it is then possible to make more correct models reliant on that data. An AI model is the yield created when you train your AI estimation with data. In the wake of planning, when you give model data, we will be given a yield. For example, a farsighted figure will make an insightful model. By then, when you give the perceptive model data, you will get an expectation dependent on the information that prepared the model. ML empowers models to prepare informational indexes prior to being sent. Some ML models are based on the web oriented contents. This iterative cycle of online models prompts an improvement in the kinds of affiliations made between information components. Because of their multifaceted nature and size, these examples and affiliations might have effectively been disregarded by human perception. After a model has been prepared, it may very well be utilized progressively to gain information. The enhancements in precision are a consequence of the preparation cycle and robotization that are essential for ML.

MLg methods are needed to improve the exactness of prescient models. Contingent upon the idea of the business issue being tended to, there are various methodologies dependent on the sort and volume of the information. The four primary classes of machine learning are shown in Figure 9.1 and briefed as follows:

Supervised Learning (SL): It commonly starts with a setup arrangement of information and a specific comprehension of how that information is ordered. Administered learning is planned to discover designs in information that can be applied to an investigation cycle. This information has named highlights that characterize the significance of information. A volume of creatures will be recognized by the ML instantly, based on images and composed portrayals.

Unsupervised Learning (USL): The utilization of this learning starts because the issue needs a gigantic quantity of unrepresented data. For example, in social media applications to understand the importance of using this information that requires the relevant data independently. Solo studies conduct an iterative cycle, breaking down information by not using the intervention of human beings. The important considerations for genuine and phish mail for a researcher to tag natural bulk electronic mail have been the typical task. All things considered, ML representatives, in light of grouping and affiliation, are applied to distinguish undesirable messages (Masaki, 1998).

Reinforcement Learning (RL): The calculation gets input from data analysis, controlling the client to the most excellent product. Reinforcement learning differs from various types of directed studying in light of the fact that the mechanism has not been arranged with the exemplar data guide. Otherwise, the

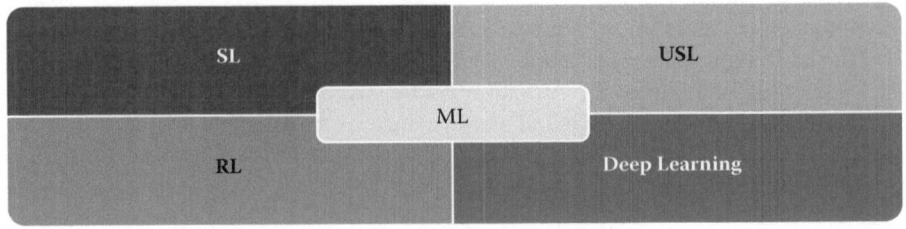

FIGURE 9.1 Main Classes of ML.

TABLE 9.1 Supervised earning, unsupervised learning, reinforcement learning, and deep learning – comparison

SUPERVISED LEARNING	UNSUPERVISED LEARNING	REINFORCEMENT LEARNING	DEEP LEARNING
Works on sample data or existing data	No external data or predefined data	Interacting with the environment	Machine learning based on artificial neural networks
Assets are depreciable	Assets are depreciable	Liabilities are non depreciable	Unstructured or unlabelled data
Regression and classification	Clustering and association	Exploitation or exploration	Classification and clustering task
Operated with interactive and software applications	Operated with interactive and software applications	Supports and works improved in AI and human interaction is common	Operated with interactive software or applications
Many algorithms exist in this learning	Many algorithms exist in this learning	Neither supervised nor unsupervised	CNN, RNN, LSTM
Runs on any platform or applications	Runs on any platform or applications	Runs on any platform or applications	Runs on any platform with any applications (req. large amount of processing power)

mechanism intercepts the trails and, in this manner, an alignment of effective options would bring about the cycle will be equipped, towards the essence of taking care of the present concern (La & Bhatnaga, 2011).

Deep Learning (DL): One of the major machine learning techniques DL combines the concept of the organization of connected neurons layers to get benefits through the data from the experiments. This type of learning is valuable when you're attempting to take in examples from unstructured information. Profound learning complex neural organizations have been proposed to copy the mechanism of studying the human functions of cerebrum in order to enable the systems to arrange and control inadequately characterized outcomes and problems. The neural training samples for young children and learning through virtuality can be connected effectively. Table 9.1 shows the comparison of SL, UL, RL, and DL.

ML had extraordinary appropriateness in the vehicle business. As of late, ML strategies have become a part of brilliant transportation. Through profound learning, ML investigated the perplexing communications of streets, roadways, traffic, natural components, crashes, etc. ML has additional incredible potential in day-by-day traffic board and the assortment of traffic information. ML can help with back-office tasks also. For example, a vehicle organization, every day, can get handfuls if not many requests, contingent upon how huge the organization is (Graefe & Kuhnert, 1991). Envision that each one of those vehicle orders is physically prepared. These activities set aside a colossal measure of effort to do it and, furthermore, is viewed as an exhausting and mistake-inclined assignment. To guarantee the progression of the vehicles, the request handling should be done at a specific time. The advantage is a cycle that can be effectively mechanized with the joined advancements of RPA and ML. The answer to the computerized handling of transport orders is a couple of steps. To begin with, the report should be transferred into a programme from where the boot can get it. Second, the report is perused and grouped. Third, information is extricated and set precisely into fields. Last, the report is traded. In any case, the records change in shape and format, have inadequate information, or need human mediation. RPA, joined with ML, can make a learning cycle that will create exact information, fill in the reports while enhancing time, wiping out the requirement for human intercession for good. The consequence of executing this sort of arrangement would be diminishing the preparing costs, expanding representative fulfilment, excellent outcomes, and a more light-footed organization.

Hence, this chapter is divided into three main sections, namely, Evolution and process of Smart Transport System, Deep learning techniques for Autonomous Vehicle Decision making, Deep Learning Neural Techniques for Autonomous driving and integration of Deep learning with IoT. At the end of this chapter, the important DL techniques are explained.

9.2 EVOLUTION OF SMART TRANSPORT SYSTEM (STS)

STS applications are extensively grouped into three classes, namely, mobility, safety, and environmental. STS mobility applications are proposed to give mobility administrations, for example, the briefest course between beginning objective pair thinking about various variables (e.g., distance, time, energy utilization) in an information-rich travel climate dependent on data gathered by the STS information assortment innovations. By changing traffic lights, progressively overseeing travel activities, or dispatching crisis support benefits, these applications can help the transportation of the board places screen and oversee framework execution (Kanthavel, 2019). The STS safety applications, for example, giving a pace caution at a pointed bend or tricky street, may lessen collides by giving warnings and alerts. The applications incorporate the safety of the vehicles, crisis executives (e.g., crisis steering). The moment gridlock data can help an explorer settle on educated choices that thusly decline the environmental effect of everyday excursions. Explorers can dodge clogs by taking backup ways to go or by rescheduling their outings, which thusly can make the excursions more eco-accommodating.

The improvement of Smart Transportation Systems (STS) innovation started during the 1970s. These frameworks are currently broadly used to screen traffic and to react rapidly to mishaps by dispatching crisis administrations. STS innovation as Advanced Traveller Information Systems is additionally used to educate explorers about movement conditions. Data is made accessible by customary sources, for example, TV, radio, and variable message signs, just as by means of traffic sites. These are available through using smart phones, cell phones, other hand-held gadgets and worldwide situating frameworks recipients. Explorers can get traffic alarms through short message administration, mechanized voice calls, and email. They can utilize the data to change their movement mode, takeoff time, course, and different choices, and they can utilize it as a navigational guide for way-finding and finding administrations (Mannion et al., 2016). Figure 9.2 demonstrates the Priority-Smart Transportation Systems.

Most examinations infer that STS innovation assists with easing gridlock and lessens mishaps, contamination, and carbon dioxide discharges. All things considered, the capability of STS is restricted by the way that drivers can ignore exhortation on the off chance that they trust it isn't to their greatest advantage. And, like the impacts of limited development, better data can instigate individuals to travel

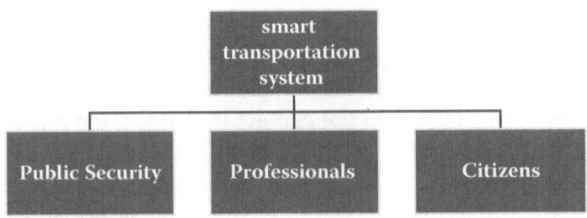

FIGURE 9.2   Priority-Smart Transportation Systems.

more since it diminishes the cost of movement. This deficiency can on a fundamental level be overwhelmed by executing street evaluating in tandem with STS.

The Smart Transportation System has gotten incredible consideration as of late. By integrating data innovation, the IoT, cyber-physical frameworks, and DL, STS offers a promising method to improve the safety, productivity, and manageability of present-day transportation framework. The most key issues in STS, steering issues, target arranging the ways for one or different vehicles so a specific goal is advanced.

9.3 PROCESS OF SMART TRANSPORT SYSTEM

The STS intends to accomplish traffic productivity by limiting traffic issues. It plans to reduce the hours of suburbanites just as it upgrades their security and comfort. The utilization is not simply restricted to gridlock control and data, yet additionally to street well-being and effective foundation use.

A smart transportation system is for identification, investigation management, and exchange advancements from base transport to improve in a secured manner with proficiency. STS includes a broad range of usages that cycle and present information to reduce blockage, enhance traffic administration. Moreover the process is completed by reducing the usual behaviour and increasing the merits for the betterment of the users and general society.

9.3.1 Why Is a Smart Transportation System Required?

- Deficient street advancement
- Speed reduction, prolonged accident ratios
- Ridiculous anticipation of producing sufficient fresh paths or to fulfil the requirement
- Construct transport systems that are more proficient, safe, and reliable by utilizing data information
- Enhance the appeal of public transport
- Facing increasing blockage that expands journey periods and expenses
- Lessen the natural effects of transport
- Points of interest of Smart Transport Scheme
- Decrease in breaks and deferrals at convergences
- Manage speed and its control
- Journey duration enhancement
- Limit the executives
- Episode the executives

9.3.2 How Does a Smart Transportation System Work?

This section of this chapter analyzes the enabling technologies, the process in which the STS undergoes that includes data acquisition, processing, communication, distribution, and utilization (Haydari & Yilmaz, 2020). Figure 9.3 illustrates the working system of the Smart Transportation System.

9.3.2.1 STS Joining Innovations

The specialized centre of an STS has been the use of data and manages innovations to transport scheme jobs and their advancements incorporate correspondence, programmed control, and PC equipment and programming. The transformation of these advancements in transporting needs information of various

FIGURE 9.3   Working System of Smart Transportation System.

designing keys, for instance common, electro mechanical, modern, and their connected controls. Most transport issues are brought about by the absence of opportune and precise data and by the absence of proper coordination between people in the system. Thus, the helpful commitment of data innovation is to present the improved data to assist individuals associated with the scheme to settle on the best choices.

9.3.2.2 STS Technological Facilitation

There is a scope of data and interchange innovations that empowers the advancement of STS. For example, optical cables, Global Positioning System, electromagnetic compasses, Compact Disc Read Only Memory, laser sensors, advanced guide information bases, and show advances. Empowering innovations can be isolated into a few classes, which include the following.

9.3.2.3 Information Acquisition

It is conceivable to screen deals utilizing a few methods, for example, inductive circle identifiers, and sensors used in the traffic. Instances of sensors are radar and ultrasonic, video picture locator and visual pictures from shut circuit TV which give live pictures to assist the traffic community administrator screen confounded traffic circumstances and settle on reasonable choices.

9.3.2.4 Information Processing

Data gathered from the information the board community needed to be prepared, confirmed, and combined into a configuration that was valuable for the administrators. This should be possible utilizing the information combination measure. Further, the Instant Detection System of incidents may likewise be utilized for information preparing. Worldwide situating systems can be utilized on the means of transportation to deal with information.

9.3.2.5 Information Communications

A few different ways can be utilized to pass on directives, for instance wireline or remote, Fibre optics, electronic cost assortment (ETC), business vehicle tasks (CYO), leaving the board, signal seizure, in-vehicle marking, in-vehicle explorer data, and reference point based-course direction systems. A portion of these information correspondence advances are utilized by information the board places while the vehicle can be used by others.

9.3.2.6 Information Sharing

Traffic and added connected information can be appropriated in different manners to develop transport proficiency, security, and ecological excellence, for instance, phones, radio, TV, PCs, fax machinery, and changeable communication symbols, vehicle radios, cell phones, PCs, and hand-held advanced gadgets.

9.3.2.7 Information Exploitation

This includes slope metering to manage the progression of motor vehicles proceeding onto a freeway, and co-appointment of traffic lights inside huge metropolitan territories happens at the traffic executives' place. Notwithstanding powerful course direction which allows the client to settle on essential choices on a moment time premise, and versatile voyage management which permits the rider to naturally decrease vehicle tempo to remain a protected progress from the motor vehicle before.

9.3.3 Intelligent Transportation System User Functions

Smart transportation user functions are given in Table 9.2.

TABLE 9.2   Smart transportation user functions

SL. NO.	IDENTIFY	SERVICES
1	Traveller data	Travel Information, driver data, communal transportation data, individual data administrations, and course direction and route.
2	Traffic organization	Transportation arranging support, traffic light, request the executives, traffic guidelines and, framework upkeep the board.
3	Vehicle systems	Vision Enhancement, computerized vehicle activity, longitudinal impact evasion, sidelong crash shirking, wellbeing status, and pre-crash restriction arrangement.
4	Commercial vehicles	Business vehicle pre-freedom, vehicle managerial cycles, computerized side of the road wellbeing assessment, business vehicle locally available security checking, business vehicle armada the executives.
5	Public transportation	Public transport the board, request responsive transport the executives, shared transport the board.
6	Emergency administration	Crisis warning and individual security, crisis vehicle the executives, perilous materials, and occurrence notice.
7	Electronic payment security	Electronic monetary exchanges, community travel safety, wellbeing development for weak street clients, savvy intersections.

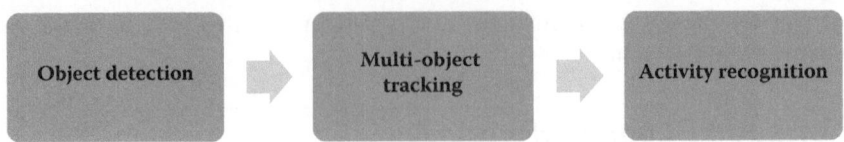

FIGURE 9.4 Primary Use of ML in STS.

9.4 NEED FOR ML TECHNIQUES IN STS

The benefits of ML can be taken to settle on a choice precisely and rapidly even in a congestive climate. The benefit of ML is that it is conceivable to use calculations and models to anticipate results. Try to guarantee that the information researchers accomplishing the work territory utilizing the correct calculations, ingesting the most fitting information (that is precise and clean), and utilizing the best performing models. On the off chance that every one of these components meets up, it's conceivable to persistently prepare the model and gain from the results by learning from the information. The computerization of the way towards displaying, preparing the model, and testing prompts precise expectations to help business change (Ghosh et al., 2018). ML arrangements have just started their promising imprints in the transportation business, where it is demonstrated to try and have a better yield on speculation contrasted with ordinary arrangements. Figure 9.4 explains the primary use of ML in STS. Be that as it may, the transportation issues are as yet well-off in relating and utilizing ML methods and require extra thought (Azgomi & Jamshidi, 2018). The hidden objectives for the mentioned arrangements are to decrease blockage, get better security, and reduce individual blunders, relieve horrible ecological effects, upgrade energy execution, and improve the profitability and productivity of outside transport. Henceforth, there is a significant requirement for ML in STS. Copiousness of confronts while applying ML to ITS include:

- Gathering, cleaning, and naming bulk datasets
- Law-producers and strategy need to stay aware of the tech
- Breakable copies that rupture ,when useful in new spaces
- Protection and confidentiality (Bodhani, 2012)

9.5 DL TECHNIQUES FOR AUTONOMOUS VEHICLE DECISION MAKING

Self-driving vehicles are self-governing dynamic systems that cycle floods of perceptions approaching the various on-board sources, for example, cameras, radars, LiDARs, ultrasonic sensors, Global Positioning Systems elements, and, additionally, sensors. These perceptions are utilized by the vehicle's PC to settle on driving choices. The driving choices are processed either in a particular discernment arranging activity pipeline, where tactile data is straightforwardly planned to control yields (Tizghadam et al., 2019). The segments of the secluded channel can be planned moreover dependent on AI or profound education strategies, or utilizing traditional non-learning draws near. Different stages of learning and non-learning-based parts are conceivable. A security screen is intended to guarantee the well-being of every module. These parts are discernment and Localization, High-Level Path arrangement, performance negotiation, or low-level way of arranging, movement regulator. In view of these various elevated level elements, it has assembled applicable profound learning papers depicting

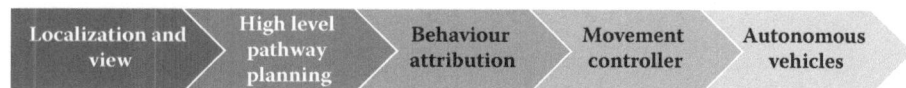

FIGURE 9.5 Sample Pathway – The Deep Learning Supported Self-Driving Vehicle.

techniques created for self-governing driving systems. In the audited calculations, we have additionally gathered significant articles covering the well-being, information sources, and equipment angles experienced when planning profound knowledge components for self-driving vehicles. Specified a course arranged throughout the street organization, the main undertaking of an independent vehicle is to comprehend and confine by self in the general climate. In view of this portrayal, a persistent way is arranged and the future activities of the vehicle are controlled by the conduct intervention system. A sample pathway is given in Figure 9.5 about the DL-based self-driving car. Long last, a movement control system responsively rectifies mistakes created in the execution of the arranged movement.

One self-driving innovation empowers a vehicle to work self-sufficiently by seeing the climate and augmenting an approachable response. Subsequently, we provide an outline of the pinnacle strategies utilized in driving sight thoughtful, taking into account camera supported versus LiDAR climate discernment. We study point discovery and acknowledgement, meaningful division and restriction in self-ruling driving, just as sight thoughtful utilizing inhabitance plans.

9.5.1 DL for Driving Scene Perception and Localization

DL strategies are especially appropriate for distinguishing and perceiving substances in 2D pictures and 3D point mists obtained from camcorders and LiDAR (Light Detection and Ranging) gadgets, separately. In the independent dynamic local area, 3D discernment is basically founded on LiDAR sensors, which give an immediate 3D portrayal of the general climate as 3D point mists. The exhibition of a LiDAR is estimated regarding field of view, reach, and goal, and revolution/outline rate. 3D sensors, for example, Velodyne, ordinarily have a 360∘ flat meadow of vision. To work at lofty paces, a self-ruling vehicle requires at least a 200m territory, permitting the transportation modes to respond to modify in street circumstances as expected. The 3D item discovery exactness is directed by the goal of the sensor, with the most exceptional LiDARs having the option to give a 3cm precision (Yuan et al., 2019). Ongoing discussion started approximately camera versus LiDAR detecting advances. Tesla and Waymo, two of the organizations driving the improvement of self-driving innovation, have various ways of thinking as to their principle insight sensor, just as with respect to the focus on the stage. Waymo structures its transportation straightforwardly as Level 5 systems, in excess of 10 million miles driven self-sufficiently. Then again, Tesla® sends its Autopilot as an ADAS (Advanced Driver Assistance System) part, which clients can go round on or off whenever it might suit them. The information base has been obtained by gathering information from client-possessed vehicles. The primary detecting innovations vary between the two organizations. Tesla attempts to use its camera schemes, though Waymo's driving innovation depends extra on Lidar sensors.

9.5.2 DL Neural Techniques for Autonomous Driving

In this part, we depict the premise of profound learning advances utilized in self-sufficient means of transportation and remark on the capacities of every worldview. We center on Convolutional Neural Networks (CNN), Recurrent Neural Networks (RNN), and Deep Reinforcement Learning (DRL), which

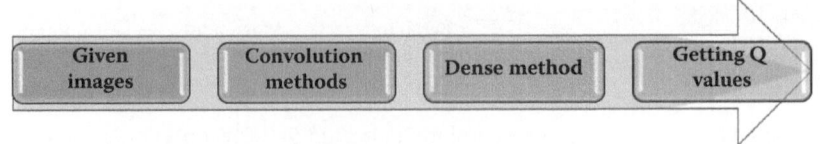

FIGURE 9.6   CNN Method – Autonomous Driving.

are the most widely recognized profound knowledge systems useful for self-sufficient driving methods (Sorin Grigorescu et al., 2019).

Concurrent Neural Network (CNN): CNNs are essentially utilized for handling spatial data, for example, pictures, and can be seen as picture highlights extractors and general non-straight capacity approximately. Prior to the ascent of profound learning, PC vision systems used to be executed depending on high-quality highlights, for example, HAAR, Local Binary values (Dey et al., 2020), or Histograms of Oriented Gradients. In contrast with these conventional handmade highlights, convolutional neural organizations can consequently gain proficiency with a portrayal of the component gap determined in the preparation set. CNN's can be inexactly perceived as surmised similarities to various pieces of the mammalian visual cortex. A picture framed on the eye part-retina is shipped off the image cortex through the thalamus. Every cerebrum half of the globe has its own visual cortex. The visual data is gotten by the visual cortex in an annoying way: the left visual cortex gets data from the right eye, while the privilege visual cortex is taken care of with visual information from the left eye. Figure 9.6 elaborates the CNN Method – Autonomous driving.

The data is handled by the double motion hypothesis, which expresses that the illustration flow follows two fundamental transitions: a type of motion, liable for image identification and item acknowledgement, and another type of motion utilized for construction of the spatial relationships among items. A CNN mirrors the working of the ventral transition, wherein various territories of the cerebrum are delicate to explicit highlights in the illustration part. The previous synapses in the illustration parts are enacted by sharp advances in the illustration part of view, similarly in which an edge indicator features, elaborating the advances among the adjoining picture representations in a picture. The mentioned limits are additionally utilized in the mind to inexact article fractions, lastly to assess unique portrayals of items. A CNN is defined by its loads vector $\theta = [W, b]$, where W is the arrangement of loads overseeing the between neural associations and b is the arrangement of neuron inclination esteems. The arrangement of loads W is coordinated as picture channels.. Convolutional layers inside a CNN abuse neighbourhood spatial relationships of picture pixels to learn interpretation invariant convolution channels, which catch the distinguished picture highlights.

Recurrent Neural Network (RNN): Among profound learning strategies, RNN is particularly acceptable in handling fleeting arrangement information, for example, text, or video transfers. Not the same as regular neural organizations, an RNN contains a period subordinate criticism circle in its memory cell. Given a period of subordinate information succession [s,..., s] and a yield arrangement [z,..., z], an RNN can be 'unfurled' $\tau i + \tau o$ periods to produce a reliable organization design coordinating the information time-span. Such neural organizations are additionally experienced under the label of grouping to-succession representations (Gu et al., 2020). An unfurled system comprises $\tau i + \tau o + 1$ indistinguishable partitions, and each level has similarly educated loads. Once unfurled, an RNN can be prepared utilizing the back spread through time calculation. When contrasted with a regular neural organization, the solitary distinction is that the scholarly loads in each unfurled duplicate of the organization are found the middle value of; along these lines. Figure 9.7 confirms the RNN Method – Autonomous driving.

The fundamental test in utilizing essential RNNs is the disappearing slope experienced during the preparation. The angle signs can wind up being duplicated on countless occasions, the same number as

FIGURE 9.7 RNN Method – Autonomous Driving.

the quantity of an era spans. Subsequently, a customary RNN isn't reasonable for catching long-haul conditions in grouping information. In the event that an organization is profound or measures long successions, the angle of the organization's yield would struggle to engender back to influence the loads of the prior layers. Under slope evaporating, the loads of the organization won't be adequately refreshed, winding up with exceptionally little weight. Repetitive layers abuse worldly relationships of arrangement information to find out time span-subordinate neural network organization. Believe the storage condition and the yield condition in an LSTM organization, examined at time step, just as the information is at time. In repetitive neural organization wording, the advancement system is commonly utilized for preparing many to many RNN models. This advancement issue is usually addressed utilizing slope-based strategies, such as Stochastic Gradient Descent, along with the backpropagation during time span calculation for ascertaining the organization's inclinations (Chopra & Roy, 2020).

DRL: Profound learning is characterized as 'a class of AI methods that misuse numerous layers of non-direct data handling for managed or unaided element extraction and change, and for design examination and arrangement'. Other definitions are given, yet we can notice similar key ideas across every one of these definitions: Profound learning utilizes different layers of nonlinear preparing units and 26 depend on the administered or solo learning of highlight extractions in each layer, with a progressive system from low-level to significant level highlights in the layers. All things considered, complex neural organizations fit into this definition. Profound Reinforcement Learning alludes to the use of profound learning in a support-picking-up setting (Schlichtkrull et al., 2017). The profound support learning field's ubiquity began with the presentation of Deep Q-Networks (DQ). DQ are profound learning models that consolidate profound convolutional neural organizations with the Q-learning calculation. Prior to this, utilizing a nonlinear capacity, for example, a neural organization as a capacity approximate for the Q work was known to be shaky. A significant commitment made by DQNs is the utilization of involvement replay and an objective organization to balance out the preparation of the Q activity esteem work guess with profound neural organizations. Q-network in this way alludes to a neural organization working roughly of the Q-work: Q (s, a; θ) ≈ Q ∗ (s, a), θ be the organization load. Experience Replay Catastrophic failing to remember or disastrous impedance indicates the propensity of neural organizations to abruptly fail to remember data that they recently taught (Jo et al., 2015). Luckily, there are approaches to conquering this issue. Experience Replay consists in keeping a cushion of the old encounters and training the organization against them. During the preparation, we utilize a support where past encounters are put away. An encounter consists of a noticed state and activity pair, the quick prize acquired and the following state noticed. At the point when the cushion is full, new encounters, as a rule, supplant the most seasoned ones in the support. Figure 9.8 demonstrates theDL method of Autonomous driving.

FIGURE 9.8 Deep Learning Method – Autonomous Driving.

To prepare the organization, we test a group of encounters from the cradle and use them to apply the typical backpropagation calculation. The benefit of this methodology is that, by inspecting these clusters of encounters, we break the connection between the information that we get when the organization is prepared in the typical online style. The profound Q-learning calculation utilizes experience replay. A specialist's involvement with a period step t is indicated by et and is a tuple (st, at, rt, st+1) comprising of the present status st, the picked activity at, the prize rt, and the following state st+1. The encounters for constant steps are put away in a replay memory, over numerous scenes.

9.5.3 DL for Passage Preparation and Performance Calculation

A self-governing vehicle to discover a course among its focuses, which has a beginning point for an ideal area, speaks to a way of arranging. As indicated by the arranging measure, a self-driving vehicle ought to consider all potential deterrents that are available in the general climate and ascertain a direction along an impact-free course (Lod, 2020). As expressed in, self-sufficient dynamic is a multimode specialist surrounding wherever the congregation automobile must pertain refined arrangement abilities by means of extra street clients when surpassing, creating path, consolidating, using both left and right tracking, all while exploring shapeless metropolitan streets. The writing discoveries highlight a nontrivial strategy that should deal with the well-being in riding. In view of a prize capacity R (s⁻) = −r for a mishap occasion that ought to be stayed away from and R (s⁻) ∈ [−1, 1] for the remainder of the directions, the objective is to figure out how to carry out troublesome moves easily with security.

9.6 INTEGRATION OF ML WITH IOT IN AUTONOMOUS VEHICLES

AI conveys experiences covered up in IoT information for quick, robotized reactions and improved dynamics. AI for IoT can be utilized to project future patterns, recognize abnormalities, and expand knowledge by ingesting pictures, videos, and sound (Khayyam et al., 2020). One of the fundamental errands of any AI calculation in the self-driving vehicle is a ceaseless delivering of the general climate and the forecast of potential changes to those environmental factors. These assignments are essentially partitioned into four subtasks:

- Object discovery
- Object identification or acknowledgement object grouping
- Object confinement and forecast of development

AI calculations can be inexactly partitioned into four classifications: relapse calculations, design acknowledgement, bunch calculations, and choice lattice calculations. One classification of AI calculations can be utilized to execute at least two diverse subtasks. For instance, relapse calculations can be utilized for object recognition just as for object confinement or forecast of development (McCorduck, 2004). AI calculations are presently utilized widely to discover answers for various provokes going from monetary market forecasts to self-driving vehicles. With the incorporation of sensor information handling in a concentrated electronic control unit (ECU) in a vehicle, it is basic to expand the utilization of AI to perform new undertakings. Potential applications incorporate driving situation characterization or driver condition assessment through information combination from various inside and outside sensors, for example, cameras, radars, Lidar, or the IoT.

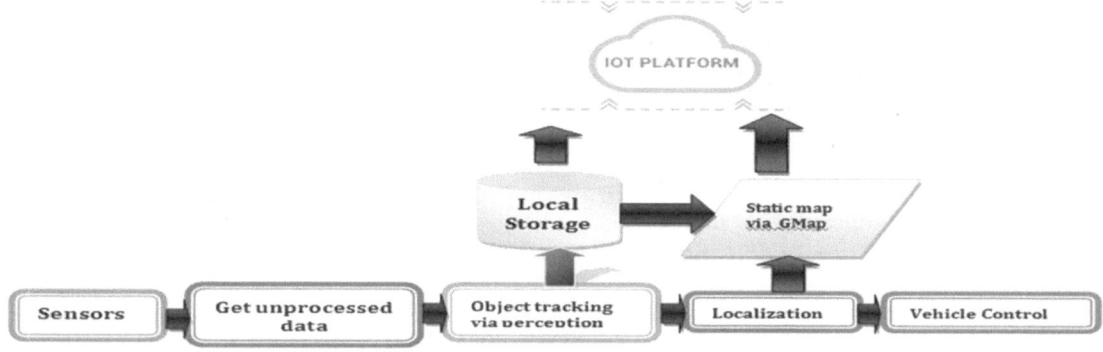

FIGURE 9.9   AI and IoT integration in Autonomous Vehicles.

Why Use AI for IoT?: AI can help demystify the shrouded designs in IoT information by investigating huge volumes of information utilizing complex calculations. It can enhance or supplant manual cycles with robotized systems utilizing measurably inferred activities in basic cycles. Figure 9.9 shows the AI and IoT integration in Autonomous Vehicles.

Sample Use Cases: Organizations are using AI for IoT to perform prescient abilities on a wide assortment of utilization cases that empower the business to acquire new experiences and progress computerization capacities. With AI for IoT, it is possible to:

- Ingest and change information into a reliable configuration
- Build an AI model
- Deploy this AI model on cloud, edge, and gadget

For instance, utilizing AI, an organization can robotize quality investigation and imperfection following its sequential construction system, track action of resources in the field, and estimate utilization and request designs.

Benefits of AI for IoT: The stage comes all set with the devices you need for quick outcomes: gadget network and the executives, application enablement and coordination, just as streaming examination, AI, and AI model organization. The stage is accessible on the cloud, on-premises, as well as at the edge. Extraordinarily, with Complicity IoT, independent, edge-just arrangements are additionally upheld.

Simplify AI Model Preparing: Complicity IoT Machine learning is intended to assist you with building AI models in a simple way. Auto ML uphold permits the correct AI model to be picked depending on the information provided, regardless of whether that be operational gadget information caught on the Complicity IoT stage or chronicled information put away in huge information files. Smart transportation chattels are given in Table 9.3.

Flexibility to Utilize Your Information Science Library of Decision: There is a wide assortment of information science libraries accessible (e.g., Tensorflow, Keras, Scikit-learn) for creating AI models. IoT Machine Learning permits models to be created in information science structures of your decision. These models can be changed into industry-standard configurations utilizing open source apparatuses and made accessible for scoring inside IoT.

Rapid Model Organization to Operationalize AI Rapidly: Regardless of whether made inside IoT Machine Learning itself or imported from other information science systems, model arrangement into creation conditions is conceivable at any place required in a single tick, either in the cloud or at the edge. Operationalized models can be effortlessly observed and refreshed if hidden examples move.

TABLE 9.3 Smart Transportation Chattels

Value Proposition of Smart Transportation

- Safety
- Environmental friendliness
- Productivity and efficiency
- Better life quality

Main concern part and priority measures

- Optimal utilization of street, traffic, and travel information
- Continuity of traffic and cargo the board ITS administrations
- ITS street well-being and security applications
- Connecting the transportation vehicles with the vehicle foundation

Managing the smart transformation journey

- Policy and regulations
- Customer support
- Technology support

Relevant areas of Smart Transport System

- Improved traffic administration scheme
- Improved traveller data scheme
- Improved vehicle manage scheme
- Improved communal transport scheme
- Improved rustic transport scheme
- Improved profit-making vehicle process scheme

Legal limitation for ITS deployment

- Personal data protection
- National security

Moreover, prepared and confirmed models are accessible for sure-fire model sending to quicken reception.

Prebuilt Connectors for Operational and Verifiable Data Stores: Complicity IoT ML gives simple admittance to information living in operational and authentic information stores for model preparing. It can recover this information on an occasional premise and course it through a computerized pipeline to change the information and train an AI model. Information can be facilitated on Amazon S3 or Microsoft Azure Data Lake Storage, just as nearby information stockpiling, and recovered utilizing prebuilt IoT Data Hub connectors.

Integration with IoT Streaming Analytics: IoT Machine Learning empowers superior scoring of constant IoT information inside IoT Streaming Analytics. IoT Streaming Analytics gives an 'AI' building block in its visual examination manufacturer that permits the client to summon a predefined AI model to score continuous information. This gives a no-code climate to incorporate AI models with streaming examination work processes.

9.7 CONCLUSION

This chapter presented an analytics on Machine learning for STS which include evolution and process of Smart Transport System, DL techniques for Autonomous Vehicle Decision making, DL Neural Techniques for Autonomous driving, and integration of DL with IoT. Finally, the important DL techniques have also been explained. From the studies, it is inferred that the smart transportation structure is the exploitation of identification, investigation, organize, and interchange advancements to land transportation to develop well-being, portability, and usefulness. The transportation structure includes an extensive range of utilizing the cycle, tender information to effortlessness obstruction, develop transfer decision-making, boundary normal results, and increase the merits of transportation to trade customers. AI is a better way to generate clever machines that can replicate human thought ability and performance, whereas, ML is a relevance or subset of AI that permits machines to be taught from data without being programmed clearly. It is also studied that ML utilizes algorithms to parse data, learn from that data, and formulate learned assessments based on what is well-read. DL constructs algorithms in layers to generate an 'artificial neural network' that can learn and formulate smart assessment by its individual processes.

REFERENCES

Azgomi, HF, & Jamshidi, M (2018). A brief survey on smart community and smart transportation. *IEEE 30th International Conference on Tools with Artificial Intelligence (ICTAI)*, 932–939.

Bodhani, A (2012). Smart transport. *Engineering & Technology*, 7(6), 70–73, 10.1049/et.2012.0611

Chopra, R, & Roy, SS (2020). End-to-end reinforcement learning for self-driving car, *Advanced Computing and Intelligent Engineering*, 1, 53–61.

Dey, S, Singh, AK, Prasad, DK, & McDonald-Maier, KD (2019). IRON-MAN: An approach to perform temporal motionless analysis of video using CNN. *IEEE Access*, 8, 137101–137115.

Ghosh, R, Pragathi, R, Ullas, S, & Borra, S (2017). Intelligent transportation systems: A survey. *2017 International Conference on Circuits, Controls, and Communications (CCUBE*, 160–165.

Graefe, V, & Kuhnert, K (1991). Vision-based autonomous road vehicle. *Vision-based vehicle guidance*. Springer Series in Perception Engineering (pp. 1–29).

Grigorescu, S, Trasnea, B, Cocias, T, & Macesanu, G (2019). A survey of deep learning techniques for autonomous driving. *Journal of Field Robotics*, 37(3), 362–386. 10.1002/rob.21918.

Gu, Z, Li, Z, Di, X, & Shi, R (2020). An LSTM-based autonomous driving model using a Waymo open dataset. *Applied Sciences*, 10(6), 2046, 10.3390/app10062046.

Haydari, A, & Yilmaz, Y (2020). Deep reinforcement learning for intelligent transportation systems: A survey. *IEEE Transactions on Intelligent Transportation Systems*, 1–22, 10.1109/TITS.2020.3008612.

Jo, K, Kim, J, Kim, D, Jang, C, & Sunwoo, M (2015). Development of autonomous carpart ii: A case study on the implementation of an autonomous driving system based on distributed architecture. *IEEE Transactions on Industrial Electronics*, 62(8), 5119–5132.

Kanthavel, R, Dhaya, R, Devi, M, Algarni, F, Dixikha, P(2019). Assessment on recurrent applications of machine learning and its behaviors. *International Journal of Engineering and Advanced Technology*, 8(6S3), 1174–1180.

Khayyam, H, Javadi, B, Jalili, M, & Jazar, RN (2020). Artificial intelligence and Internet of things for autonomous vehicles. In *Nonlinear approaches in engineering applications* (pp. 39–68). 10.1007/978-3-030-18963-1_2

La, P, & Bhatnaga, S (2011). Reinforcement learning with function approximation for traffic signal control. *IEEE Transactions on Intelligent Transportation Systems*, 12(2), 412–421.

Lod, LL (2020). A survey of deep learning techniques for autonomous driving, http://leonlloyd.co.uk/pkoj2d/a-survey-of-deep-learning-techniques-for-autonomous-driving-b60669

Mannion, P, Duggan, J, & Howley, E (2016). An experimental review of reinforcement learning algorithms for adaptive traffic signal control. In *Autonomic road transport support systems* (pp. 47–66). Springer International Publishing.

Masaki, I (1998). Machine-vision systems for intelligent transportation systems, *IEEE Intelligent Systems*, 13(6), 24–31.

McCorduck, P (2004). *Machines who think: A personal inquiry into the history and prospects of artificial intelligence*. A.K. Peters Ltd.

Mrityunjaya, DH, Kumar, N, Laxmikant, Ali, S & Kelagadi, HM (2017). Smart transportation. *International conference on I-SMAC (IoT in Social, Mobile, Analytics and Cloud) (I-SMAC 2017)*, 1–6,978-1-5090-3243-3/17/$31.00.

Schlichtkrull, M, Kipf, TN, Bloem, P, van den Berg, R, Titov, I, & Welling, M (2017). Modeling relational data with graph convolutiona networks. *European semantic web conference*, 563–607.

Stanford University (2016). Artificial intelligence and life in 2030. One hundred year study on artificial intelligence (AI100), 2016. Retrieved from https://ai100.stanford.edu

Tizghadam, A, Khazaei, H, Moghaddam, MHY, & Hassan, Y (2019). Machine learning in transportation. *Journal of Advanced Transportation Systems, 2019*(4359785), 1–3, 10.1155/2019/4359785

Yuan, T, da Rocha Neto, WB, Rothenberg, C, Obraczka, K, & Barakat, C (2019). Harnessing machine learning for next-generation intelligent transportation systems: A survey. *HAL Project*, hal-02284820.

Classification of Kinematic Data Using Explainable Artificial Intelligence (XAI) for Smart Motion

10

C.Y. Yong

University College of Technology Sarawak, Sarawak, Malaysia

Contents

10.1 INTRODUCTION

There are many ways to capture human motion in the digital domain, the common ways are using mechanical, electro-magnetic, optical, and video-based. Explainable Artificial Intelligence (XAI) is one of the methods to fulfil smart cities environment (Samek & Müller, 2019). The identification of human motion can be done just by focusing on a set of dots on a motion (Song et al., 2000). Gypsy system and Physilog system are commercial motion capture systems that use mechanical ways to capture human motion (Chen & Yuan, 2015). Liberty mocap system from Polhemus, Cabled Flock of Birds system from Ascension, wireless electro-magnetic mocap system Motion Star from Ascension and the motion capture for Lara Craft movie was performed by Motion Star using electro-magnetic motion capture system (Markkanen, 2016). Optical motion capture system uses optical markers as human models for motion estimation. The previous systems are very expensive and thus a technique was proposed to capture human motion by using a wearable sensor (Han et al., 2013). The most challenging elements of the techniques are poor imaging or occlusion and need large samples for all possible poses (Mai et al., 2018). We specifically analyze the primary origins of motion, objectively evaluate the various methods of reconstructing, analyzing, and modelling natural motion signals, and briefly summarize existing efforts to define coding schemes, matched filters, and enhancement of elements involved in information processing (Yong et al., 2011). We conclude the chapter with a study of adaptations to demonstrate the various stages of processing at which motion filters have evolved in response to natural motion signals.

10.2 RESEARCH REVIEW

Signal processing is basically divided into four categories: signal acquisition, quality improvement, signal compression, and feature extraction (Eckert & Zeil, 2001). Signal acquisition or signal collection is the first step of the study, where raw data is stored for later processing (Kirianaki et al., 2002). Sampling and quantization are the typical original signal rebuilding approaches in the digital system (Krishnan & Athavale, 2018). Noise reduction and cancellation are improvement procedures to enhance the signal quality. The main issue in signal processing is the data transmission procedure (Elhoseny & Hassanien, 2019). Large data samples and long-term data collection periods are required a number of time in signal transmission. Therefore, signal compression is performed to reduce the sample size for faster speed (Han et al., 2015). Finally, information is retrieved by extracting the message from the signal for data understanding (Tur & De Mori, 2011).

The implementation of the study on movement or locomotion of animal to human motions is widely used in medical purposes. The study is usually called gait analysis to analyze the locomotion of certain living things due to their habits (Bennett et al., 2016). The implementation aims to illustrate that the different body dynamics depend on certain conditions.

Human motion implementation is normally performed in two approaches, they are vision based and wearable sensor based (Chen et al., 2015). The vision-based approach is an analysis of streams of video from camera or video recorder by using a computer (Prates et al., 2018). The second approach is an analysis of sets of data generating from few mounting sensors on the subject's body part. This project is focusing on the availability and effectiveness of the second approach by using statistical classifiers as analysis instruments, to be the main objective to create smart cities using Explainable Artificial Intelligence (XAI) (Gunning, 2017).

Identification of a new observation to the set of classes it belongs is defined as classification in statistics with machine learning (Tajbakhsh & Suzuki, 2017). The membership of an observation to a set of classes is assigned using explanatory quantifiable properties such as variables and features. Classifier is a mathematical algorithm to perform the concrete class membership implementation for categorization (Russell & Norvig, 2016).

In machine learning, the terminology of classification is divided into supervised learning and unsupervised learning (or normally called clustering or cluster analysis) (Berry et al., 2019). Supervised learning required learning where training is performed on the algorithm for correct observation identification (Nasteski, 2017). Unsupervised learning involves observation grouping based on the measure of inherent similarity (Caron et al., 2018). Classification is performed with logistic regression or a similar procedure (Breiman, 2017).

The paradigms underlying this problem are part of the so-called XAI area, which is commonly recognized as a necessary function for the realistic implementation of AI models (Yong et al., 2013a). This chapter provides a summary of the current literature and contributions in the field of XAI, as well as a look ahead to what is yet to come. We summarized previous attempts to describe explainability in Artificial Intelligence for this reason, creating a novel description of explainable AI that encompasses certain prior philosophical propositions with a significant emphasis on the audience for which explainability is pursued (Arrieta et al., 2020).

10.3 MATERIALS AND METHODS

10.3.1 Study Sample

Five healthy persons were volunteer themselves in taking part of the data collection section in university campus. Their ages are between 20 and 25 years old with normal limbs movement and advocate healthy lifestyles. They are having significant mobility in their everyday routine independent of any walking aids and without serious medical record.

10.3.2 Experimental Set-Up

A wireless 3-axis device was used in the study. The device employs a 3-space sensor breakout board for a 3-axial gyroscope, a 3-axial accelerometer, and a 3-axial compass. The device is mounted in an enclosure with 60 mm × 35 mm × 15 mm measurements, connected to a processor using wireless asynchronous serial real-time transmission. The device is equipped with Kalman filtering for quaternion orientation processing (Chui & Chen, 2017).

The experimental setup of the study started from the transmission of sensing data from the wearable sensor to the processor unit. The next step followed by data managing for saving and retrieval in future need, then preprocessing and back end data processing were performed to retrieve the information from the raw signal. The flow was stopped until good feedback and evaluation were achieved (Ravi & Ravi, 2015).

The device was attached firmly above the right arm of every subject using a designed holder as shown in Figure 10.1. The device holder was specially designed to attach the sensor firmly on the arm to avoid any disturbance and jolting when performing any movement.

(a)   (b)

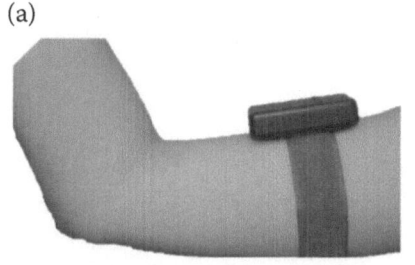
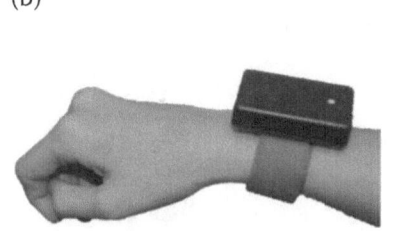

Whole body movement activity   Upper-limb movement activity

FIGURE 10.1 Diagram of Sensor Attachment for Whole Body Movement Activity (a) and Upper-Limb Movement Activity (b).

10.3.3 Data Acquisition

The study was started by collecting data from two types of activities with different motions. The two main activities are categorized to whole-body movement and upper-limbs movement.

    a. Whole-body movement activities include:
- Jogging
- Walking
- Throwing

    b. Upper-limbs movement activities include:
- Square shape drawing
- Circle shape drawing
- Triangle shape drawing

The whole-body movement activity was performed on a treadmill with a regular motor while upper-limbs movement activity was performed with sitting posture. The normal jogging and walking speed of every subject were pre-defined by conducting a pre-test. The subjects were asked to perform normal jogging and walking motions pleasantly at their own pace and convenience along a 10-metre-long track. Their normal jogging and walking speeds were calculated by dividing the distance by the total time taken. The speed was recorded as 6.3–6.7 ft/s (1.92–2.04 m/s) for jogging and 3.5–3.9 ft/s (1.07–1.19 m/s) for walking. Measurement of speed was in feet per second due to the default setting of the treadmill.

Both jogging and walking motions were conducted on a treadmill by adjusting the motor speed according to the subject's normal speed. Activities were carried out in a supervised environment for start and end time stamping. The process was repeated 5 times for every motion with distance 10 metres, 20 metres, 30 metres, 40 metres, 50 metres, 100 metres, 150 metres, and 200 metres.

For throwing motion, the subjects were required to sit on a static chair and perform the throwing activity as shown in Figure 10.2. The throwing activity was a targeting movement for both hands by throwing a paper ball with radius 3 cm in length and 2 g in weight toward four assigned locations (50 cm, 100 cm, 150 cm, and 200 cm) away from the chair. Throwing movements were tested by both hands five times, respectively, for every single assigned location. The motion of subjects was restricted on the static chair in order to reduce or eliminate undesirable motion of subjects, thus provide an ideal phenomenon for motion sensing from sensors.

Figure 10.2 shows the standard of operating procedure in conducting the throwing motion signal capture experiment. X-axis represents the vertical segment while y-axis represents the horizontal segment, or the distance between the landed paper ball from the subject.

Meanwhile for upper-limbs movement activity, the subjects who are right-handed required to draw the three assigned shapes randomly at their own paces following the direction as shown in Figure 10.3. The subjects need to perform the drawings firstly using their left hands then followed by the right hands. The starting and ending points of drawing are essential since the proposed recognition system is required

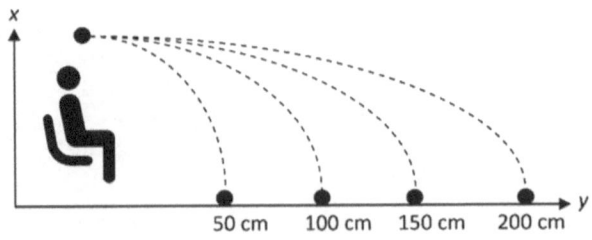

FIGURE 10.2   Throwing Process for Four Assigned Locations.

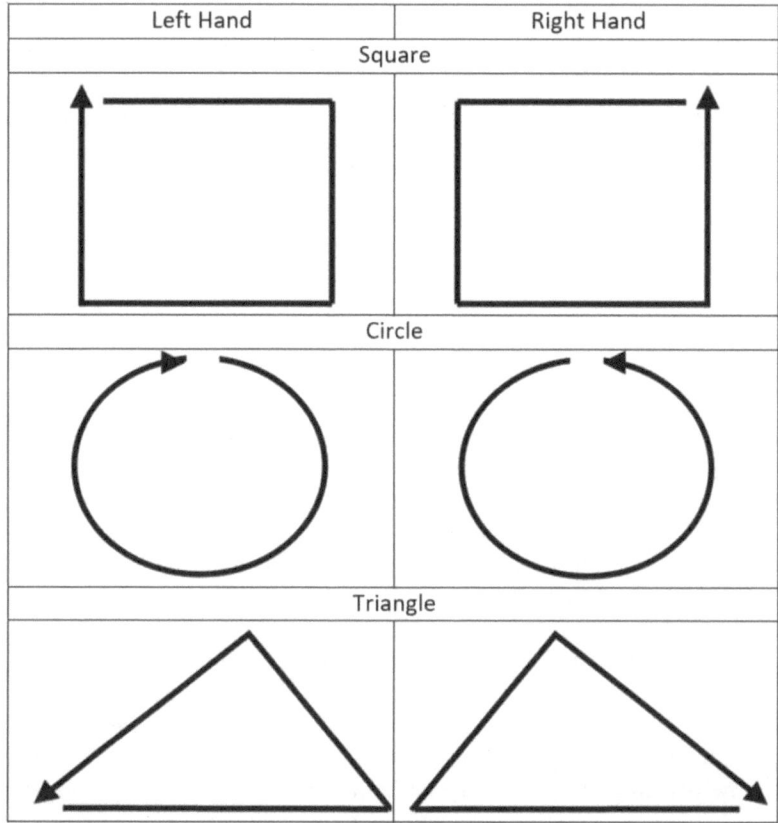

FIGURE 10.3 Direction of Square, Circle and Triangle Drawings for Left Hand and Right Hand.

to represent the drawing motion in 2D animation form starting from the first stroke of the drawing until the end.

Data logged by the sensor were saved in a text file. The files documented gyroscope, accelerometer, and compass orientation in 3-axis. The raw files were then processed into matrix files for MATLAB analysis. The file consisted of nine columns with the first three columns representing gyroscope data in x, y, and z axes, the next three are accelerometer in 3 axes and the last three are compass data.

10.3.4 Data Management

Data collection is a process to prepare and collect information on record. The purpose is to make decisions and retrieve important information issues. In order to fulfil its intended purpose, verification and validation are performed to evaluate the system requirements and specifications. Verification and validation are independent of the interest of third parties (Sargent, 2010). The process is done externally disregarding involvement in recognition analysis.

Cross-validation is a common system result assessment method. Data were divided into several partitions with equal size subsamples for multiple rounds of cross-validation. In k-fold cross-validation, k is defined as the number of subsamples (Rohani et al., 2018). Cross-validation process is performed by selecting a partition among the subsamples and repeated k times. The result of every k subsamples is averaged for the final result.

An amount of 70% of total data is used for system training; this is to ensure the system undergoes a long period and large data sample for learning and become skilled. The remainder is divided into two portions for cross-validation and testing (Berner et al., 2019). Data distribution is very important for artificial intelligence analysis, the proposed system need not be exposed to the repeating data (Yong et al., 2013b). It is an effective way to verify and introduce the proposed system with better performance from deviation or bias (Camilli et al., 1994).

There are two types of data that were being prepared for the proposed system analysis, they are whole-body movement activity (jogging, walking, and throwing) and upper-limbs movement activity (square, circle, and triangle drawings).

10.3.5 System Evaluation

Confusion matrix or contingency table is a specific table layout that allows the presentation of system performance. Performance visualization is used to investigate if the system is confusing among the classes (Verma & Khanna, 2013). The confusion matrix summarizes the results among jogging, walking, and throwing motions and upper-limbs movement for square, circle, and triangle drawings. In all confusion matrix sections, error were inspected: jogging motion was wrongly predicted to walking motion. All correct predictions by the system are located at the diagonal of the confusion matrix. Any value with non-diagonal is categorized as an error (Yong et al., 2013c).

Receiver operating characteristic (ROC) is a curve plotting to illustrate the performance of the system (Carter et al., 2016). ROC is a graph plotting by true positive rate versus false positive rate. A diagonal line divides ROC plot into two equal spaces (Tripepi et al., 2009). Plotted graph above the line represents good classification result; curve below the line represents poor result. Area below the ROC graph also indicates the system performance; the larger the area the better the performance and vice versa (Liu & Wu, 2003). The proposed system achieved good classification.

10.4 RESULT

Figure 10.4 shows the raw data collected from the wearable sensor. The pattern and statistical characteristic of the data does not differ significantly between both jogging and walking motions. It is difficult to distinguish between both motions. Therefore, clustering is needed to differentiate the respective motions.

10.4.1 Whole-Body Movement

Table 10.1 shows the computation time of every classifier proposed in the study. QDA is the fastest and shortest time taken for the whole clustering process. It took 0.093 s followed by proposed PCA-K-Means 0.094 s, KNN 0.182 s, K-Means statistical toolbox 0.223 s, NB-G 0.427 s, NB-KD 1.009 s, LDA 2.175 s, DT 2.183 s, ANN 7.608 s and SVM 17.166 s. The PCA-K-Means classifier is 0.001 s or 1% longer in time computation compared with QDA. It is the most potential classifier for analysis since the following are more than 96% or 2-folds which are very time-consuming.

Table 10.2 shows the resubstitution error, cross-validation error, accuracy, sensitivity, and specificity. SVM achieved 100% accuracy, sensitivity, and specificity. This implies that SVM is the best classifier for motion recognition follow by ANN and the proposed PCA-K-Means.

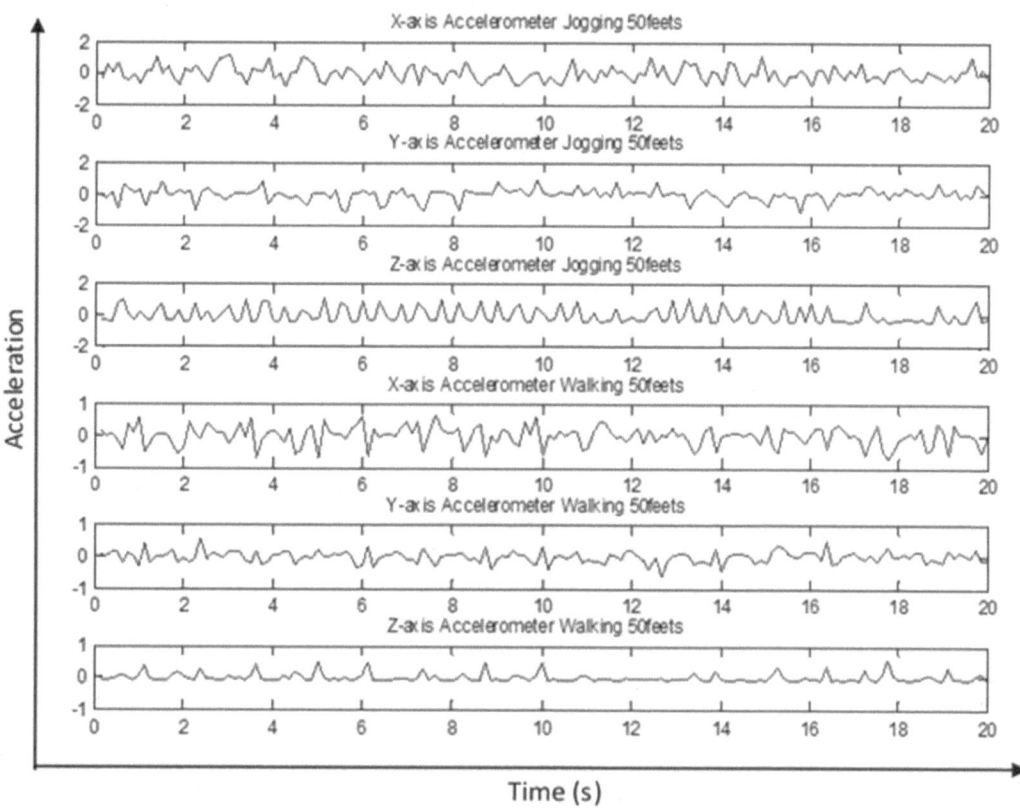

FIGURE 10.4 Raw Data of Walking Motion.

TABLE 10.1 Computation time of every classifier used for whole-body movement analysis

CLASSIFIER	PROCESSING TIME (S)	DELAYED (S) WITH PREVIOUS	DELAYED (S) WITH QDA	DELAYED (%) WITH QDA
Quadratic Discriminant Analysis	0.093	–	–	–
PCA-K-Means (Proposed)	0.094	0.001	0.001	1
K-Nearest Neighbours	0.182	0.088	0.089	96
K-Means Statistical Toolbox	0.223	0.041	0.130	140
Naïve Bayes Gaussian	0.427	0.204	0.334	359
Naïve Bayes Kernel Density	1.009	0.582	0.916	985
Linear Discriminant Analysis	2.175	1.166	2.082	2239
Decision Tree	2.183	0.008	2.090	2247
Artificial Neural Network	7.608	5.425	7.515	8081
Support Vector Machine	17.166	9.558	17.073	18358

TABLE 10.2   Resubstitution error, cross-validation error, accuracy, sensitivity, and specificity of every algorithm for whole-body movement analysis

ALGORITHMS	RESUBSTITUTION ERROR (%)	CROSS-VALIDATION ERROR (%)	ACCURACY	SENSITIVITY	SPECIFICITY
			(NORMALIZED)		
Support Vector Machine	0.000	0.393	1.00	1.00	1.00
Artificial Neural Network	0.010	0.423	0.99	0.99	1.00
PCA-K-Means (Proposed)	0.047	0.347	0.95	1.00	0.93
Naive Bayes Kernel Density	0.140	0.183	0.86	0.64	0.97
Decision Tree	0.053	0.323	0.84	0.73	0.89
Quadratic Discriminant Analysis	0.230	0.253	0.77	0.51	0.90
Naive Bayes Gaussian Kernel	0.250	0.260	0.75	0.51	0.87
Linear Discriminant Analysis	0.357	0.423	0.64	0.54	0.70
K-Means Statistical Toolbox	0.410	0.423	0.59	0.40	0.69
K-Nearest Neighbours	0.693	0.423	0.31	0.29	0.32

Table 10.3 shows the confusion matrix and true positive of every classifier. SVM, ANN, and proposed PCA-K-Means classifiers achieved 100%, 99.67%, and 95.33%, respectively. True positive values are computed by averaging the diagonal true classification of confusion matrix. The higher the percentage of true positive, the higher the classifier accuracy on recognition of data belong to their groups.

10.4.2   Upper-Limbs Movement

Table 10.4 lists ascending orderly from the least expensive on time to the largest. Regarding small sample size, the proposed PCA-K-Means is capable of computing the analysis result in 0.023 s which is 270% faster than the QDA (second fastest), 11739% and 21983% faster than SVM and ANN, respectively. The proposed approach has established a light-speed platform for rapid processing.

Table 10.5 shows the resubstitution error, cross-validation error, accuracy, sensitivity, and specificity of every classifier. Both SVM and ANN classifiers achieve 100% accuracy followed by Naive Bayes kernel density with an accuracy of 84% and proposed PCA-K-Means of 80%.

Table 10.6 shows the confusion matrix and true positive of every classifier. Once again, SVM and ANN classify every single shape drawings 100% correctly. The proposed PCA-K-Means classified 74 correctly out of 93 samples. The extended approach from ordinary K-means statistical toolbox to a PCA integrated algorithm has increased the true positive rate from 27.96% to 79.57%.

TABLE 10.3   Confusion matrix of every algorithm used for upper-limb movement analysis

ALGORITHMS	CONFUSION MATRIX			TRUE POSITIVE (%)
Support Vector Machine	100	0	0	100
	0	100	0	
	0	0	100	
Artificial Neural Network	99	0	0	99.67
	1	100	0	
	0	0	100	
PCA-K-Means (Proposed)	89	11	0	95.33
	0	97	3	
	0	0	100	
Decision Tree	94	4	2	94.67
	3	96	1	
	6	0	94	
Naive Bayes Kernel Density	64	27	9	86.00
	0	99	1	
	3	2	95	
Quadratic Discriminant Analysis	51	33	16	77.00
	13	86	1	
	5	1	94	
Naive Bayes Gaussian Kernel	51	32	17	75.00
	11	81	8	
	7	0	93	
Linear Discriminant Analysis	54	26	20	64.33
	42	49	9	
	4	6	90	
K-Means Statistical Toolbox	40	47	13	59.00
	39	54	7	
	11	6	83	
K-Nearest Neighbours	29	33	38	30.67
	35	33	32	
	36	34	30	

10.5 DISCUSSION

Experimentally, QDA took the shortest time for classification computation while SVM and ANN achieved the highest accuracy and number of true positive; however, the proposed PCA-K-Means classifier achieved more than 95% of true positive and accuracy in 0.094 s of elapsed time for whole-body movement activity. QDA is fast in computation but with low accuracy, SVM and ANN are of high accuracy but slow in computation. Hence, neither QDA nor SVM nor ANN are inappropriate and not preferential for analysis and recognition of large datasets.

The proposed PCA-K-Means extended from an ordinary K-Means classifier is capable of analyzing and recognizing the input data in the shortest time with high accuracy. Results showed that the proposed

TABLE 10.4   Computation time of every classifier used for upper-limb movement analysis

CLASSIFIER	PROCESSING TIME (S)	DELAYED (S) WITH PREVIOUS	DELAYED (S) WITH PCA-K-MEANS	DELAYED (%) WITH PCA-K-MEANS
PCA-K-Means (Proposed)	0.023	–	–	–
Quadratic Discriminant Analysis	0.085	0.062	0.062	270
K-Means Statistical Toolbox	0.134	0.049	0.111	483
K-Nearest Neighbours	0.177	0.043	0.154	670
Naive Bayes Gaussian	0.21	0.033	0.187	813
Naive Bayes Kernel Density	0.649	0.439	0.626	2722
Linear Discriminant Analysis	1.216	0.567	1.193	5187
Decision Tree	1.538	0.322	1.515	6587
Support Vector Machine	2.723	1.185	2.7	11739
Artificial Neural Network	5.079	2.356	5.056	21983

TABLE 10.5   Resubstitution error, cross-validation error, accuracy, sensitivity, and specificity of every algorithm for upper-limb movement analysis

ALGORITHMS	RESUBSTITUTION ERROR (%)	CROSS-VALIDATION ERROR (%)	ACCURACY	SENSITIVITY	SPECIFICITY
			(NORMALIZED)		
Artificial Neural Network	0.000	0.301	1.00	1.00	1.00
Naive Bayes Kernel Density	0.161	0.409	0.84	0.61	0.95
PCA-K-Means (Proposed)	0.204	0.333	0.80	0.94	0.72
Quadratic Discriminant Analysis	0.237	0.333	0.76	0.52	0.89
Linear Discriminant Analysis	0.269	0.301	0.73	0.52	0.84
Naive Bayes Gaussian Kernel	0.280	0.398	0.72	0.45	0.86
Decision Tree	0.118	0.247	0.52	0.39	0.58
K-Nearest Neighbours	0.645	0.301	0.36	0.29	0.39
K-Means Statistical Toolbox	0.720	0.301	0.28	0.52	0.16
Artificial Neural Network	0.000	0.301	1.00	1.00	1.00

TABLE 10.6   Confusion matrix of every algorithm used for upper-limb movement analysis

ALGORITHMS	CONFUSION MATRIX			TRUE POSITIVE (%)
Support Vector Machine	31	0	0	100
	0	31	0	
	0	0	31	
Artificial Neural Network	31	0	0	100
	0	31	0	
	0	0	31	
Decision Tree	21	0	10	88.17
	0	31	0	
	1	0	30	
Naive Bayes Kernel Density	19	1	11	83.87
	1	30	0	
	2	0	29	
PCA-K-Means (Proposed)	31	0	0	79.57
	2	29	0	
	0	4	14	
Quadratic Discriminant Analysis	16	0	15	76.34
	0	31	0	
	7	0	24	
Linear Discriminant Analysis	16	3	12	73.12
	0	31	0	
	10	0	21	
Naive Bayes Gaussian Kernel	14	0	17	72.04
	1	30	0	
	8	0	23	
K-Nearest Neighbours	9	13	9	35.48
	13	10	8	
	9	8	14	
K-Means Statistical Toolbox	16	7	8	27.96
	0	3	28	
	20	4	7	

approach is adequate for motion analysis and thus further experiments will be carried out to evaluate and assess the technique with other bio-signals such as EEG, ECG, EOG, and EMG.

10.6 CONCLUSION

Nine existing classifiers and a proposed integrated classifier are proposed. The recognition process is presented in a developed GUI as shown in Figure 10.5. The proposed method fully satisfies the specifications required for smart cities.

An Explainable Artificial Intelligence (XAI) namely PCA-K-Means classifier was proposed in the study for motion recognition analysis. PCA was integrated with K-Means to further complement each

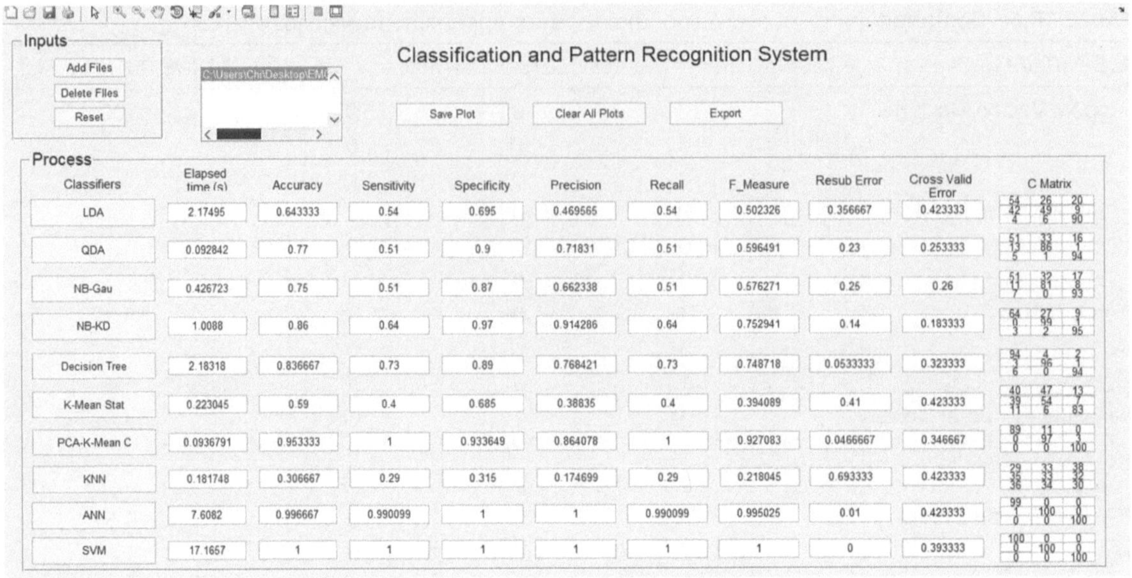

Classification and Pattern Recognition System

Inputs: Add Files, Delete Files, Reset

C:\Users\Chi\Desktop\EMG

Save Plot | Clear All Plots | Export

Process

Classifiers	Elapsed time (s)	Accuracy	Sensitivity	Specificity	Precision	Recall	F_Measure	Resub Error	Cross Valid Error	C Matrix
LDA	2.17495	0.643333	0.54	0.695	0.469565	0.54	0.502326	0.356667	0.423333	54 26 20 / 42 49 9 / 4 6 90
QDA	0.092842	0.77	0.51	0.9	0.71831	0.51	0.596491	0.23	0.253333	51 33 16 / 11 88 1 / 5 1 94
NB-Gau	0.426723	0.75	0.51	0.87	0.662338	0.51	0.576271	0.25	0.26	51 32 17 / 11 81 8 / 7 0 93
NB-KD	1.0088	0.86	0.64	0.97	0.914286	0.64	0.752941	0.14	0.183333	64 27 9 / 0 54 1 / 3 2 95
Decision Tree	2.18318	0.836667	0.73	0.89	0.768421	0.73	0.748718	0.0533333	0.323333	94 4 2 / 6 90 4 / 0 6 94
K-Mean Stat	0.223045	0.59	0.4	0.685	0.38835	0.4	0.394089	0.41	0.423333	40 47 13 / 39 54 7 / 11 6 83
PCA-K-Mean C	0.0936791	0.953333	1	0.933649	0.864078	1	0.927083	0.0466667	0.346667	89 11 0 / 0 97 0 / 0 0 100
KNN	0.181748	0.306667	0.29	0.315	0.174699	0.29	0.218045	0.693333	0.423333	29 33 38 / 36 33 30 / ...
ANN	7.6082	0.996667	0.990099	1	1	0.990099	0.995025	0.01	0.423333	99 0 0 / 0 100 0 / 0 0 100
SVM	17.1657	1	1	1	1	1	1	0	0.393333	100 0 0 / 0 100 0 / 0 0 100

FIGURE 10.5   GUI of the Proposed Classification and Pattern Recognition System.

other and push forward the wide integration. The proposed classifier attains high accuracy and sensitivity with low computation time. Even though SVM and ANN achieve almost 100% of accuracy, yet SVM and ANN take 82 and 185 times longer than the proposed classifier. They are not appropriate for large dataset analysis since the recognition system is necessary to fulfil large data training for fine performance.

The proposed PCA-K-Means classifier successfully classified all three motions data (jogging, walking, and throwing) and three shape drawing motions data (square, circle, and triangle) with lower time consumption, lower errors, higher accuracy, and higher sensitivity.

The initial testing of this study was conducted in a laboratory environment. In future, sample size will be widen with different dynamic and transition activities. Data collection will be carried out in distint environment.

ACKNOWLEDGEMENTS

A project of this magnitude depends on the hard work and commitment of many professionals, and we are pleased to acknowledge their contributions. We are deeply indebted and would like to express our gratitude to the University College of Technology Sarawak for supporting and funding this study under University Research Grants:

- UCTS/RESEARCH/4/2016/14
- UCTS/RESEARCH/3/2019/07
- UCTS/RESEARCH/4/2020/10

REFERENCES

Arrieta, AB, Díaz-Rodríguez, N, Ser, JD, Bennetot, A, Tabik, S, Barbado, A, García, S, et al. (2020). Explainable Artificial Intelligence (XAI): Concepts, taxonomies, opportunities and challenges toward responsible AI. *Information Fusion, 58*, 82–115.

Bennett, TR, Wu, J, Kehtarnavaz, N, & Jafari, R (2016). Inertial measurement unit-based wearable computers for assisted living applications: A signal processing perspective. *IEEE Signal Processing Magazine, 33*(2), 28–35.

Berner, C, Brockman, G, Chan, B, Cheung, V, Dębiak, P, Dennison, C, Farhi, D, et al. (2019). Dota 2 with large scale deep reinforcement learning. arXiv preprint arXiv:1912.06680.

Berry, MW, Mohamed, A, & Yap, BW (Eds.). (2019). *Supervised and unsupervised learning for data science.* Springer Nature.

Breiman, L (2017). *Classification and regression trees.* Routledge.

Camilli, G, Shepard, LA, & Shepard, L (1994). *Methods for identifying biased test items.* Vol. 4. Sage.

Caron, M, Bojanowski, P, Joulin, A, & Douze, M (2018). Deep clustering for unsupervised learning of visual features. In *Proceedings of the European Conference on Computer Vision (ECCV)*, pp. 132–149.

Carter, JV, Pan, J, Rai, SN, & Galandiuk, S (2016). ROC-ing along: Evaluation and interpretation of receiver operating characteristic curves. *Surgery, 159*(6), 1638–1645.

Chen, C, Jafari, R, & Kehtarnavaz, N (2015). Utd-mhad: A multimodal dataset for human action recognition utilizing a depth camera and a wearable inertial sensor. In *2015 IEEE International Conference on Image Processing (ICIP)*, pp. 168–172. IEEE.

Chen, I, & Yuan, Q (2015). Method and apparatus for calibrating a motion tracking system. U.S. Patent 9,119,569, issued September 1.

Chui, CK, & Chen, G (2017). *Kalman filtering* (pp. 19–26). Springer International Publishing.

Eckert, MP, & Zeil, J (2001). Towards an ecology of motion vision. In *Motion vision*, pp. 333–369. Springer.

Elhoseny, M, & Hassanien, AE (2019). Secure data transmission in WSN: An overview *Dynamic Wireless Sensor Networks.* Studies in Systems, Decision and Control (vol. 165, pp. 115–143). Cham: Springer. https://doi.org/10.1007/978-3-319-92807-4_6

Gunning, D (2017). Explainable artificial intelligence (xai). *Defense Advanced Research Projects Agency (DARPA), nd Web 2*, no. 2.

Han, J, Shao, L, Xu, D, & Shotton, J (2013). Enhanced computer vision with microsoft kinect sensor: A review. *IEEE Transactions on Cybernetics, 43*(5), 1318–1334.

Han, S, Mao, H, & Dally, WJ (2015). Deep compression: Compressing deep neural networks with pruning, trained quantization and huffman coding. arXiv preprint arXiv:1510.00149.

Kirianaki, NV, Yurish, SY, Shpak, NO, & Deynega, VP (2002). *Data acquisition and signal processing for smart sensors.* Wiley.

Krishnan, S, & Athavale, Y (2018). Trends in biomedical signal feature extraction. *Biomedical Signal Processing and Control, 43*, 41–63.

Liu, H, & Wu, T (2003). Estimating the area under a receiver operating characteristic (ROC) curve for repeated measures design. *Journal of Statistical Software, 8*(12), 1–18.

Mai, T, Woo, MY, Boles, K, & Jetty, P (2018). Point-of-care ultrasound performed by a medical student compared to physical examination by vascular surgeons in the detection of abdominal aortic aneurysms *Annals of Vascular Surgery, 52*, 15–21.

Markkanen, K (2016). TOMB RAIDER REBOOT as Reviewed through Joseph Campbell's Monomyth.

Nasteski, V (2017). An overview of the supervised machine learning methods. *Horizons. b 4*, 51–62.

Prates, PA, Mendonça, R, Lourenço, A, Marques, F, Matos-Carvalho, JP, & Barata, J (2018). Vision-based UAV detection and tracking using motion signatures. In *2018 IEEE Industrial Cyber-Physical Systems (ICPS)*, pp. 482–487. IEEE.

Ravi, K, & Ravi, V (2015). A survey on opinion mining and sentiment analysis: Tasks, approaches and applications. *Knowledge-Based Systems, 89*, 14–46.

Rohani, A, Taki, M, & Abdollahpour, M (2018). A novel soft computing model (Gaussian process regression with K-fold cross validation) for daily and monthly solar radiation forecasting (Part: I). *Renewable Energy, 115*, 411–422.

Russell, SJ, & Norvig, P (2016). *Artificial intelligence: A modern approach.* Pearson Education Limited.

Sargent, RG (2010). Verification and validation of simulation models. In *Proceedings of the 2010 Winter Simulation Conference*, pp. 166–183. IEEE.

Samek, W, & Müller, K (2019). Towards explainable artificial intelligence. In *Explainable AI: Interpreting, explaining and visualizing deep learning*, pp. 5–22. Springer.

Song, Y, Feng, X, & Perona, P (2000). Towards detection of human motion. In *Computer Vision and Pattern Recognition, 2000. Proceedings. IEEE Conference on*, vol. 1, pp. 810–817. IEEE.

Tajbakhsh, N, & Suzuki, K (2017). Comparing two classes of end-to-end machine-learning models in lung nodule detection and classification: MTANNs vs. CNNs. *Pattern Recognition, 63*, 476–486.

Tripepi, G, Jager, KJ, Dekker, FW, & Zoccali, C (2009). Diagnostic methods 2: Receiver operating characteristic (ROC) curves. *Kidney International, 76*(3), 252–256.

Tur, G, & De Mori, R (2011). *Spoken language understanding: Systems for extracting semantic information from speech*. John Wiley & Sons.

Verma, VK, & Khanna, N (2013). Indian language identification using k-means clustering and support vector machine (SVM). In *2013 Students Conference on Engineering and Systems (SCES)*, pp. 1–5. IEEE.

Yong, CY, Sudirman, R, & Chew, KM (2011). Motion detection and analysis with four different detectors. In *2011 Third International Conference on Computational Intelligence, Modelling & Simulation*, pp. 46–50. IEEE.

Yong, CY, Sudirman, R, Mahmood, NH, & Chew, KM (2013a). Comparison of human jogging and walking patterns using statistical tabular, scatter distribution and artificial classifier. In *Advanced materials research* (vol. 646, pp. 126–133). Trans Tech Publications Ltd.

Yong, CY, Sudirman, R, Mahmood, NH, & Chew, KM (2013b). Human body and body part movement analysis using gyroscope, accelerometer and compass. In *Applied mechanics and materials* (vol. 284, pp. 3120–3125). Trans Tech Publications Ltd.

Yong, CY, Sudirman, R, Mahmood, NH, & Chew, KM (2013c). Human hand movement analysis using principle component analysis classifier. In *Applied mechanics and materials* (vol. 284, pp. 3126–3130). Trans Tech Publications Ltd.

Smart Urban Traffic Management for an Efficient Smart City

11

M. El Khaili, L. Terrada, A. Daaif, and H. Ouajji

ENSET Mohammedia, Hassan II University of Casablanca, Casablanca, Morocco

Contents

DOI: 10.1201/9781003172772-11

11.1 INTRODUCTION

In recent years, the world has experienced an impressive development of the multimedia world. This is due to the technical and technological progress and major innovations that have revolutionized the world of telecommunication, IT cloud (Cloud Computing), social media: the Internet of Things…

This development is shown as the extension and the invasion of the IoT in the physical world. While the Internet usually does not extend beyond the electronic world, the IoT represents the exchange of information and data from devices present in the real world to the Internet. Therefore, the IoT is regarded as the third evolution of the Internet, known as Web 3.0. The objects constituting the Internet of objects' is called connected', 'communicating', or smart' (Vineela & Sudha, 2015). Currently, Connected or intelligent objects are everywhere; they invaded the world and impact our personal and professional lives. They generate billions of information that must be processed and analyzed to make them usable.

Urban Traffic Management is a very complex problem. Millions of people every day use urban road networks and suffer from congestion. Researchers are led to solve traffic flow management problem in order to reduce the congestion and to improve the air quality. Apart from long-term structural solutions, it is possible in principle to control the urban traffic flow either acting on the individual cars or acting on the traffic signalization.

The aim of our project is to demonstrate that connected objects can consist of a participatory information collection (Crowdsourcing) for the control, supervision, or conduction of a system. We treat the case of index measurement of air quality in the city of Casablanca.

In this chapter, we give an overview of the IoT and urban traffic control. We also present our project of crowdsourcing based on Connected Objects for measuring air pollution and urban road network information. We propose our approach to control urban traffic based on DL methods. A pilot experiment will be done in Casablanca (Morocco) which has more than 4 million inhabitants.

11.2 SMART TRANSPORT IN SMART CITY ENVIRONMENT

Today, Smart cities are switching from internal Information Technology (IT) Infrastructure to External Information Technology Infrastructure via the Internet, in order to handle a huge volume of data generated by several city systems (Smith, 2011). Hence, real-time processing and fast data analysis are required by Smart City Applications to react promptly to the various events occurring in a city according to Explainable Artificial Intelligence. Thus, the edge and fog computing paradigms promise to address Big Data storage and analysis in the field of smart cities (Farahzadi et al., 2017).

In this respect, we review existing service delivery models on adopting these two emerging paradigms. We also describe Intelligent Urban Traffic System as a key performance indicator for

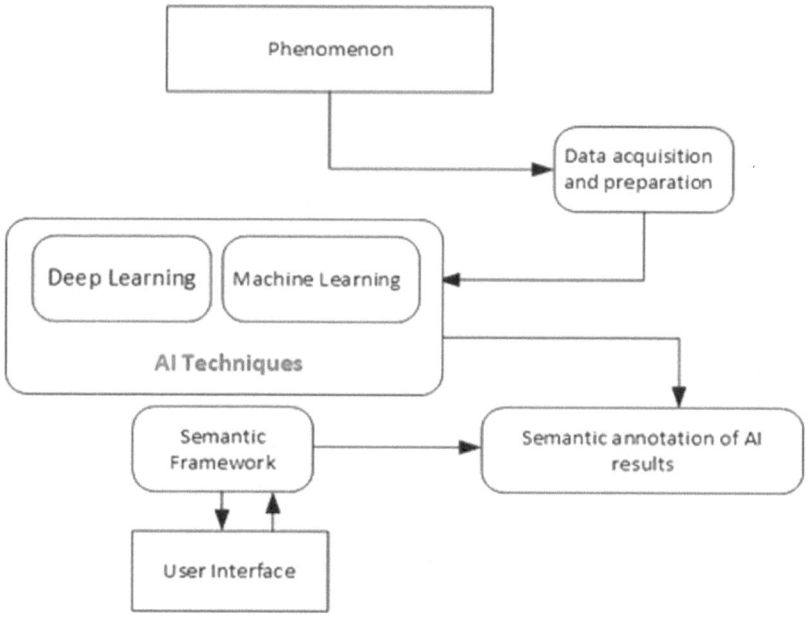

FIGURE 11.1 Architecture of XAI.

smart urban supply chain (Badidi et al., 2020; Chakir 2020; El Khaili et al., 2019; Wang et al., 2020).

11.2.1 Explainable Artificial Intelligence (XAI)

Explainable Artificial Intelligence (XAI) is an emerging concept of the field of artificial intelligence and an extension to the existing techniques of AI. It aims to allow human users to interact and cooperate with systems operated by AI. These systems must be capable of explaining their abilities, then showing them to the users when the explanations are built (Petrović & Tošić, 2020).

The human user takes into account the given results and provides feedback to autonomous systems to confirm the results; the user can also correct, modify the results by giving additional information or instructions, or use an interface to take a specific action as shown in Figure 11.1. Technically, the XAI has appeared in order to reduce the gap between the business sectors and research community, which prevents the implementation of DL models in sectors that are traditionally delayed in the Digital transformation of their processes (Barredo Arrieta et al., 2020).

11.2.2 Edge and Fog Computing for Smart City

According to Gartner, Edge computing is defined as follows: 'Edge computing is part of a distributed computing topology where information processing is located close to the edge, where things and people produce or consume that information' (Badidi et al., 2020). In Edge Computing, data storage and processing are closer, while the devices generate and collect the data instead of relying on a data centre that can be afield. Proximity is required because applications need real-time data for their operations and very low latency (Dastjerdi & Buyya, 2016; Farahzadi et al., 2017; White & Clarke, 2020). Users consider Cloud solutions for business processes with expensive bandwidth (Figure 11.2).

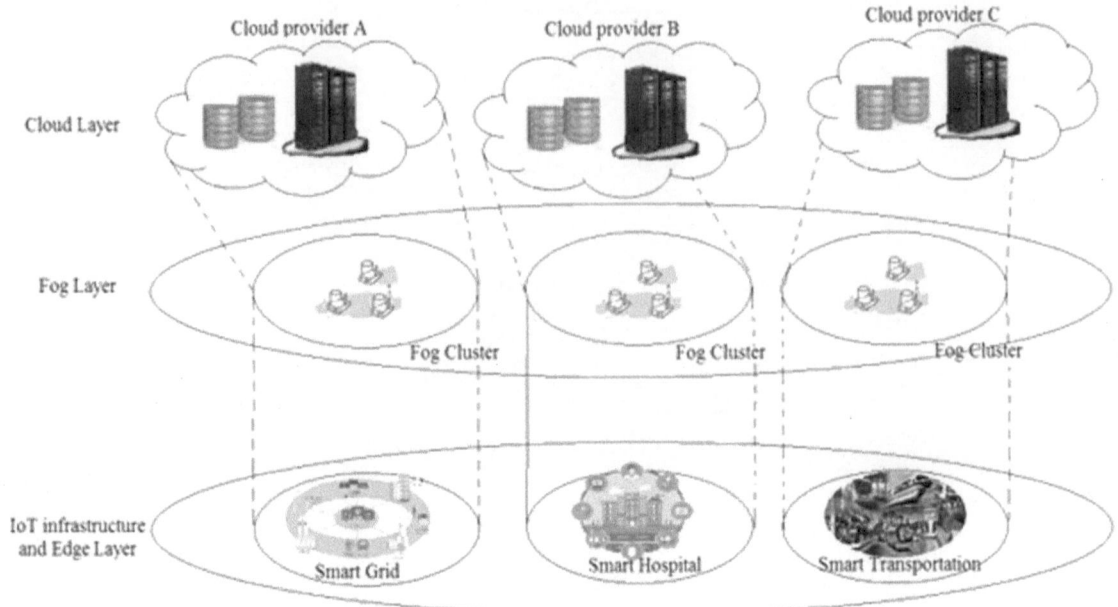

FIGURE 11.2   Smart City Fog and Cloud Infrastructure.

11.2.3 Fog Data as a Service Delivery Model

The Data as a Service (DaaS) is the fourth delivery model in fog-based environments. DaaS has emerged as a data delivery model, which is based on the same concepts as the cloud delivery models such as IaaS, PaaS, and SaaS. Data are delivered in DaaS model to users on demand regardless of their location (Hawas & Mahmassani, 1996; Jiang et al., 2017; Misra et al., 2014; Shahgholian & Gharavian, 2018).

11.2.4 Toward a Fog-Based Real-Time Big Data Pipeline

The implementation of smart city applications is based on a connected data pipeline. Real-time data analysis requires a fog-based data pipeline (Yang & Recker, 2007; Farooq & Djavadian, 2019; Alfaseeh et al., 2018). A real-time big data pipeline for smart cities should have important characteristics to meet city stakeholders' demands, such as Scalable messaging system, Data storage, ML libraries, Backend store, Dashboard, and visualization tools. Badidi et al. propose a process with five phases to build a fog-based data pipeline for a smart city as follows: data ingestion, data preprocessing, data processing and analytics, visualization and reporting, and decision making (Figure 11.3) (Badidi et al., 2020).

11.2.5 Smart Transportation Systems (STS) and Vehicular Fog Computing

Smart Transportation Systems are mainly based on evolving technologies that enable vehicles to communicate with other vehicles V2V or road users and roadside infrastructure V2I (Lee & Park, 2008). STS improve road safety and traffic efficiency in cities and reduce energy consumption and transport emissions (Jing et al., 2018).

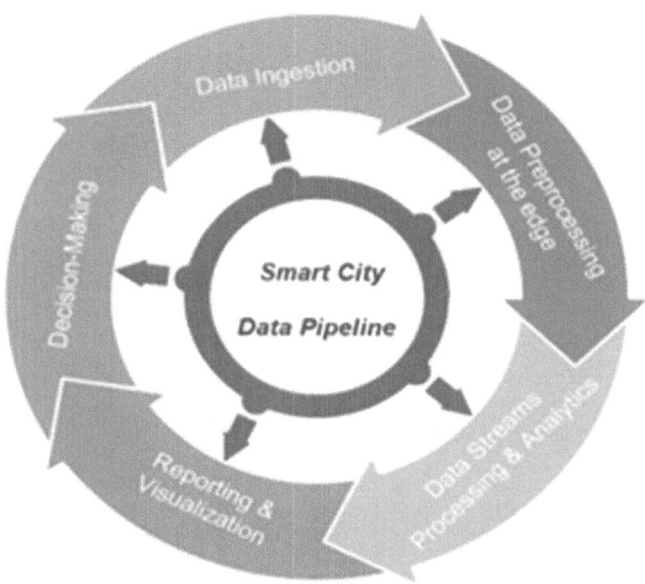

FIGURE 11.3   Phases of a Smart City Data Pipeline (Badidi et al., 2020).

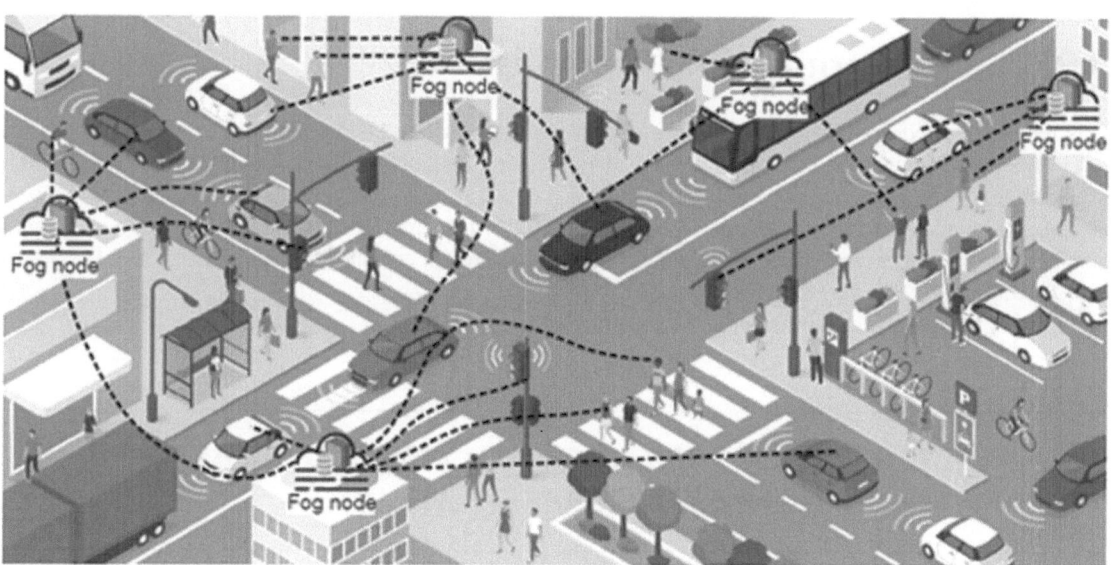

FIGURE 11.4   Fog Computing in Intelligent Transportation System (Badidi et al., 2020).

As illustrated in Figure 11.4, the vehicles receive information about traffic status and congested roads. They are informed of road accidents and emergency evacuation routes, as well as services such as finding appropriate parking spots or nearest gas stations (Badidi et al., 2020). Traffic conditions adjust traffic lights and traffic signs. Emergency vehicles such as ambulances and police cars can be recognized and assigned to a specific lane. The various entities of the roads can communicate with each other in real time via the Internet of Vehicles (IoV).

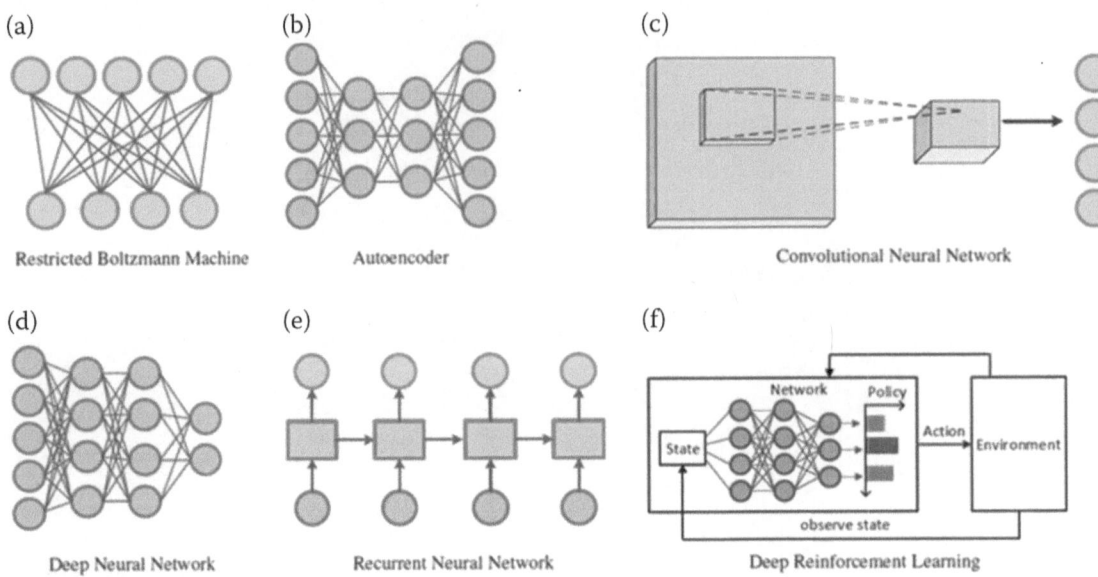

FIGURE 11.5 Structures of Different Deep Learning Models (Wang et al., 2020): (a) Restricted Boltzmann Machine, (b) Autoencoder, (c) Convolutional Neural Network, (d) Deep Neural Network, (e) Recurrent Neural Network, and (f) Deep Reinforcement Learning.

11.2.6 Deep Learning (DL) Methods

DL is widely applied in many fields and has shown relevant results. DL is different to ML methods. DL can ensure powerful information extraction and processing capabilities based on massive computation resources. DL expands, greatly, the edge computing applications in several scenarios, improving performance, efficiency, and management. (Wang et al., 2020). IoV and DL advances will ensure intelligent transportation management, such as autonomous driving, traffic prediction, and traffic signal control (Badidi et al., 2020; Wang et al., 2020).

Figure 11.5 introduces some typical DL models.

11.3 RELATED WORKS ON URBAN TRAFFIC MANAGEMENT UTM

11.3.1 Urban Traffic Management

The aim of Traffic control systems is to tackle road congestion and minimize traffic-related environmental effects (Garcia & Lopez-Carmona, 2019; Gallotti et al., 2012). Currently, software systems in the urban traffic management area tend to be based on a specific data integration, which shares data externally at a relational database level (Abdelaziz et al., 2017; Tlig et al., 2014; Duch & Arenas, 2006).

Transportation data availability is required to solve traffic issues. Berrouk et al. propose a Big Data solution for road congestion problem that aims to reduce and optimize the traffic flow in urban areas, using real-time traffic data to compute the congestion index for each road in the network and then generates recommendations to reassign the traffic flow (Berrouk et al. 2019).

The context of the new application of planning that we describe in this chapter is developing semantic technology in order to better capture and exploit real-time and historical urban data sources, while pursuing a higher level of data integration. We aim to make Urban Traffic Management Control UTMC systems more adaptable by raising the level of traffic control software integration via semantic component interoperability. In this regard, we have the longer-time aim of utilizing an autonomic approach to UTMC in particular, and road transport support in general.

11.3.2 Urban Traffic Management Approaches

Transport operators are not able to get detailed local strategies in real time that will handle unexpected events such as road closures or traffic signal timings changes. These events are the main cause of delays and increased air pollution because of excessive congestion and stationary traffic (Guli'c et al., 2016; Krajzewicz et al., 2012; Luo et al., 2016; McCluskey Thomas & Mauro, 2017Lacuna 2015).

This research study is based on a control theory approach that can be used to get a solution that can offer continuous responses to changing situations. Under changing state conditions, researchers have designed MPC algorithms which can continuously adjust the controlled signal timings in real time. While there are several examples of the application of general AI techniques to road traffic monitoring and management, the trial of UTM systems embodying an AI planning engine within a real urban traffic management centre, with an evaluation performed by transport operators and technology developers, is novel (Terrada et al., 2020a; Zegeye et al., 2009; Bandeira et al., 2016; Watling & Hazelton, 2003; Benedek & Rilett, 1998; Silva et al., 2018).

11.3.3 Traffic Lights Management

The traffic lights can be controlled using a method based on parameter estimation. The simplest approach is to use the Traffic microscopic representation. This type of model is characterized by the emphasis on studying the individual behaviour of each vehicle on the road network or on the length of the queues in a discrete time system. Microscopic models contain four levels of representation for the road network load model: x intersections configuration; x links; x lane choice; and x vehicle-following. Based on this microscopic representation, an online adaptive traffic signal control algorithm is developed and the behaviour of traffic lights is studied.

The system is modelled by using a state called traffic status. The current number of vehicles that form a queue at the level of a lane is represented by a vector. Also, a vector corresponding to the traffic light signal is defined. For each moving lane, it is possible to decide whether the colour of the traffic light signal should be maintained or changed based on action vector values. The network loading model and traffic signal control model are generally constructed in discrete-time, as much as possible to the real circumstance.

By comparing different methods, results show that Approximate Dynamic Programming ADP with Adaptive Phase Sequence APS mode has a quite good performance of queue length and traffic delay reductions (Sun & Liu, 2015; Yang & Recker, 2007; Farooq & Djavadian, 2019; Alfaseeh et al., 2018).

11.3.4 DL Approaches for Urban Traffic Management

Integrated with the latest advances in DL, IoV gives us smart solutions for intelligent transportation management, such as traffic prediction, autonomous driving, and traffic signal control.

11.3.4.1 Traffic Analysis and Prediction

Understanding the mobility patterns of vehicles and people is a critical problem for urban traffic management, city planning, and service provisioning. Given the distributed features of mobile edge servers, edge computing is naturally ideal for vehicle traffic analysis and prediction. Traditional approaches mostly used time-series analysis (Gallotti et al., 2012) or probabilistic graph analysis, which may not sufficiently capture the hidden spatiotemporal relationships therein. As a powerful learning tool, DL stands out as an effective method in this direction. They further pointed out the potential of applying different DL approaches in urban traffic prediction. Polson et al. leveraged deep neural networks to mine the short-term characteristics of the traffic situation of a road segment to predict the near-future traffic pattern (Badidi et al., 2020). A Stacked AutoEncoder SAE is leveraged to learn the generic traffic features from the historical data. Many authors further considered the impact of weather on traffic situations and incorporated the weather information into a deep belief network for integrated learning. In Deep Transport (Liang et al., 2013; Mahmassani et al., 2013; Nagatani, 2002; Nagatani, 2000), the authors considered the mobility analysis at a larger scale, that is, the citywide scale. Long–short-term memory (LSTM) model is used for future movement prediction.

11.3.4.2 Autonomous Driving

Intelligent sensing and perception are the most critical issues in autonomous driving (Khiat et al., 2018). The vehicles first collect the information from various sensors such as cameras and radars, and then conduct an intelligent perception and decision. Purely using vehicle-based and cloud-based solutions may not well satisfy the requirement of high computation capacity, real-time feedback, enough redundancy, and security for autonomous driving. Edge computing however provides a promising solution with powerful computation and low latency communication (Khan & Anderson, 2016). With the benefits of V2X communications, part of the learning-based perception can be offloaded to the edge server for processing (He et al., 2019; White et al., 2020).

11.3.4.3 Traffic Signal Control

With the above traffic analysis and prediction, the combination of edge computing and DL actually can do more things towards intelligent transportation management. Among them, intelligent traffic signal control is one of the most representative applications and has also been explored by researchers for years. A good control policy is able to reduce the average waiting time, traffic congestion, and traffic accident. The early traffic signal control methods usually rely on fuzzy logic or genetic algorithm. The key challenge however lies in how to achieve collaborative and intelligent control among multiple or even citywide traffic lights for large-scale traffic scheduling. Towards this goal, Reinforcement Learning (RL) and multi-agent RL turn out to be promising solutions where each agent (can be implemented as an edge) will make control policy for a traffic light considering not only its local traffic situation but also other agents' traffic situations. Tabular Q-Learning was first applied in an isolated intersection for signal control (Badidi et al., 2020; Wang et al., 2020).

To improve the collaboration among traffic lights, a Multi-Agent Q-Learning MAQL method is proposed to consider the queue length for cooperative scheduling. The latest work further integrated the state-of-the-Art Actor-Critic A2C RL algorithm and the multi-agent learning as a Multi-Agent Actor-Critic MA2C approach to comprehensively combine the traffic features for intelligent control.

11.4 NETWORK FOR URBAN DATA COLLECTION

11.4.1 Internet of Things (IoT)

IoT, also called the Internet of Everything or the Industrial Internet, is a new technology paradigm illustrated as a global network of machines and devices able to interact with each other (Vineela &

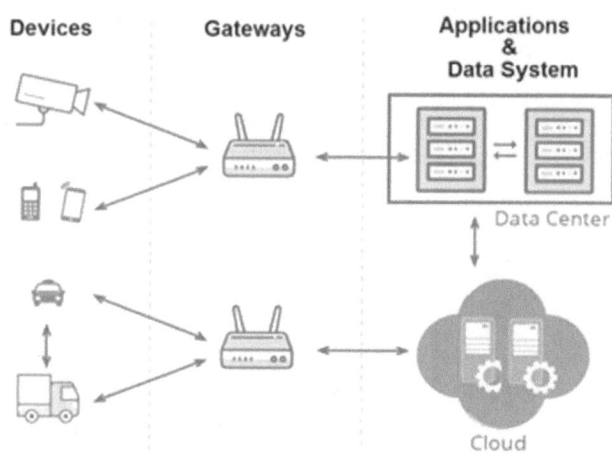

FIGURE 11.6 Architecture Modelling of IoT.

Sudha, 2015). IoT technologies generate billions of information that must be processed and analyzed then stored to make them usable. Today, connected objects begin to take part in our daily lives and are translated into several and different objects in multiple fields of application (Jean, 2011; Rose et al., 2015).

Figure 11.6 illustrates the role of the various processes of the architecture of IoT:

- Sensors to transform a physical quantity analogue to a digital signal
- Connectors allow interfacing a specialized object network to a standard IP network (LAN) or consumer devices
- Store calls made to aggregate raw data produced in real time, Meta tagged, arriving in un-predictable ways
- Present indicates the ability to return the information in a comprehensible way by humans, while providing a means to do it and/or interact

Today, the IoT has simplified processes with Cloud (Terrada et al., 2019), from which the collected data is centralized in a single space. The IoT can be used in plenty and different domains, such as Healthcare, Industry, sport, Smart cities, Smart grids, and logistics. However, this new innovation is still limited by sensitive security issues.

11.4.2 Air Pollution Measurements

V2X, which stands for 'vehicle to everything', is the umbrella term for the car's communication system, where information from sensors and other sources travel via high-bandwidth, low-latency, high-reliability links, paving the way to fully autonomous driving (Chen et al., 2017).

There are several components of V2X, including vehicle-to-vehicle (V2V), vehicle-to-infrastructure (V2I), vehicle-to-pedestrian (V2P), and vehicle-to-network (V2N) communications. In this multifaceted ecosystem, cars will talk to other cars, to infrastructure such as traffic lights or parking spaces, to smartphone-toting pedestrians, and to data centres via cellular networks. Different use cases will have different sets of requirements, which the communications system must handle efficiently and cost-effectively.

What kind of transportation experience can we expect in a V2X world? Researchers from Huawei's German Research Centre in Munich outlined their vision in He et al. (2019) (Figure 11.7):

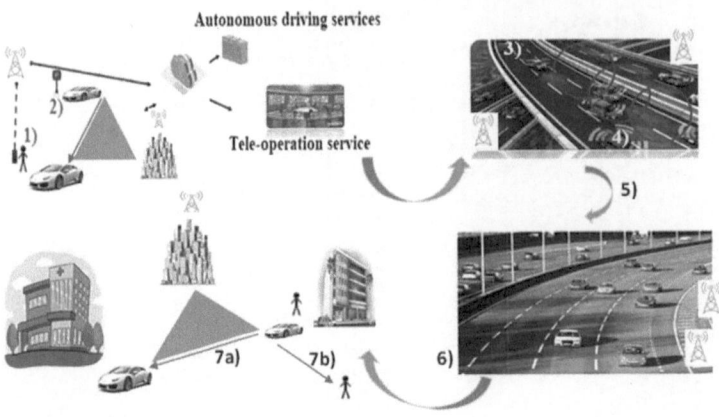

Autonomous driving services

Tele-operation service

(1) User hails an on-demand car via an app; (2) Vehicle self-drives or is teleoperated to the user, whose personal transportation app connected over the mobile radio network adjusts the car to selected presets; (3) Vehicle searches for the platoon best suited to the user's preferences and performs cooperative maneuvering to join the selected platoon, saving energy by allowing car following at very short gaps; (4) Two-way navigation guides the vehicle around a congested stretch of road; (5) User takes control to exit the highway and drive on a particularly scenic road suggested by the connected navigation system; (6) Vehicle drops the user off at the destination; (7) The vehicle self-drives, either to an automated parking lot (7a) or to a different user (7b).

FIGURE 11.7 Vision of Huawei's German Research Centre in Munich.

The Huawei researchers noted that:

While this example might seem far into the future, most of the technology needed to enable it (high precision maps, real time traffic information, sensors inside the vehicle such as radars, cameras, ultrasonic, etc.) are either already available or will be in the near future. The most prominent missing component is a high reliability, low latency communications system.

11.5 OUR PROJECT ARCHITECTURE

The recent years have witnessed the adoption of Blockchain technology in the IoT network. The adaptation has increased due to various security features of Blockchain, such as decentralization, distributed immutable ledger, transparency, tamper-proof, and auditability of data. These features have the potential to address the security and privacy vulnerabilities that occurred with the massive number of smart devices in the IoT environment. The architecture consists of devices, miners, and shared immutable ledgers used to safely communicate among IoT devices. The immutable nature of a shared distributed ledger allows storing the transactions on the Blockchain in a more secure manner.

The measurement of pollution in the air is traditionally based on the deployment by the local authority of fixed sensors in the city. Several initiatives have emerged at the intersection of the citizen sensors and the Open Hardware. The design of low-cost sensors increases coupled smartphones to enable a collaborative data collection (crowdsourcing) and so multiplies the observations and measurements (Figure 11.8).

Our project now aims to design objects connected miniaturized and portable ("wearables") capable of informing users about the quality of the air around them. They would be the equivalent, for the measurement of the environment, "tracer activities" which measure the efforts and physical performance. This information will be collected and processed in a GIS to help authorities in traffic management. Our project calls this new generation of connected objects "enviro-trackers" ("environmental tracers"). A pilot experiment is planned on the urban area of Greater Casablanca, a city of 4.2 million inhabitants (El Khaili et al., 2019; Terrada et al., 2018, 2020b).

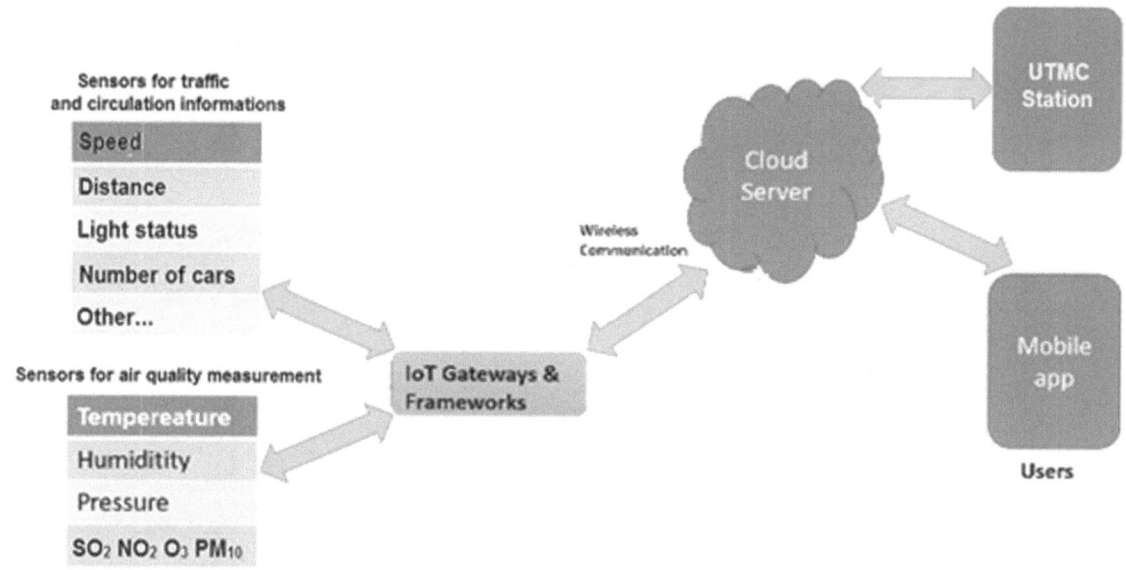

FIGURE 11.8   Deploying Our Project.

Our project takes place in three phases:

- Design of connected objects and embedded applications
- Design of the pilot platform with GIS and web portal
- Operations of results (Decision support based on the availability of information via multi-modal facilities)

11.5.1 Hardware Implementation for Data Collection

The proposed system is based on communication between a Treatment Unit and multiple Tracers. The network shares the same concept of habitual network such as addresses for planning and routing system. Each Tracer can transmit data to the Treatment Unit frequently or after a request of the Treatment Unit. Our architecture is based on:

- Sensors connectivity (Khiat et al., 2016; Khiat et al., 2017)
- Wireless technology
- Wireless data acquisition
- Wireless Sensor Network (WSN) devices

We propose two type of tracers: static and moving tracers. The static tracers are fixed in facilities like buildings but the moving tracers are fixed on buses or used as applications on smartphones of voluntary citizens. The Airparif Air Quality Assessment System is based on four measurements done by SO2 sensor, NO2 sensor, O3 sensor, and PM10 sensor. Each tracer is equipped with at least three sensors.

11.5.2 Urban Traffic Model

An urban circulation network (Figure 11.9) consists of several roads that can be interconnected by cross point. Each road is viewed as a bidirectional graph. It is described in a single direction by a list of elements representing nodes.

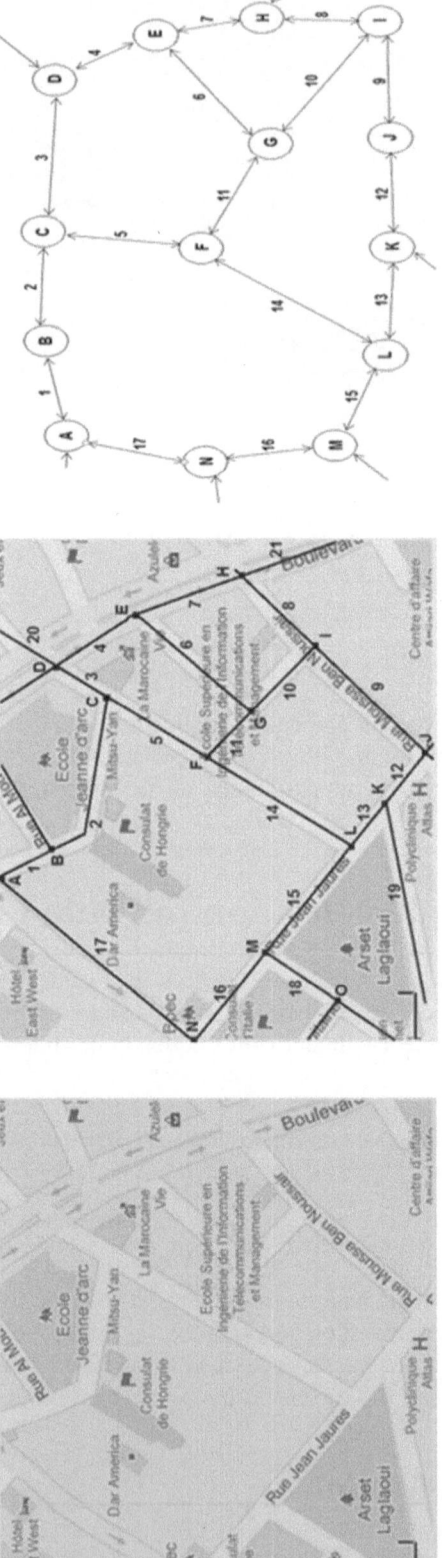

FIGURE 11.9   Road Map: From the Map to an Oriented Graph Representation.

Any road consists of a symmetric set of elements describing it in one direction; it always starts from an entry followed by several intermediate elements and ends in an exit. The second direction is drawn by inverting the input and the output.

In an intelligent road, all the elements must be instrumented by vehicle traffic sensors. The sensors delimit segments for which the number of vehicles can be timely determined. As described in the next section, other sensors can be interposed between the elements at strategic points to increase the number of monitored segments.

When a vehicle passes through a sensor, the latter generates a time-stamped immutable event containing information about the vehicle. By processing the events, it is possible to know at any time the number of vehicles in the corresponding segment.

11.6 OUR APPROACH FOR URBAN TRAFFIC FLOW MANAGEMENT

11.6.1 Concept

Urban traffic prediction aims to exploit sophisticated models to capture hidden traffic characteristics from substantial historical mobility data and then makes use of trained models to predict traffic conditions in the future. Due to the powerful capabilities of representation learning and feature extraction, emerging DL becomes a potent alternative for such traffic modelling. In this section, we present the importance usage of DL in predictions of various traffic indicators (traffic speed, traffic flow, and accident risk).

The Treatment Unit listens to all tracers and collects data. The index of air quality is computed per region which corresponds to its tracer. The indices can be displayed on a map of the city or used for regulating traffics and other ways of management for the city's facilities (urban logistics).

The initial phase of our collaborative project concentrated on the semantic enrichment of the data. The raw data was taken from transport and environment sources and integrated into a Data Hub. The method was to take real-time feeds and process them until they produced logical facts about a traffic scenario, which could serve as part of an initial state of an AI planner. The abstract system architecture is used to test the generation and operation of the control strategies. To work towards that, however, the data enrichment and strategy generation had to be tested in a real scenario, hence rather than taking in real-time current data, we adjusted the system so that what would be translated into the current state would be from historical data. This would allow checking the performance of the system against the observed performance from historical data, in order to evaluate it offline.

As a basis for exploring exceptional or emergency traffic conditions, we chose to use historically averaged traffic data from a time/day when the road links were mostly congested: morning rush hour on a non-holiday weekday. The main data source was the probabilistic processes. From this and other transport engineer documentation records our partners extracted, for the selected region within the urban area, the following:

- The topology of the road links (direction of each link and intersections between roads)
- The vehicle capacity of all the road links taking into account the differing size of vehicles
- The average traffic flows between links in the number of vehicles per second: this number represents the number of vehicles existing in a particular intersection at a fixed time of day, when the corresponding traffic signal phase is green

- The traffic signal features: position, phases of signals, minimum and maximum time that a signal phase can be set for
- Inter-green timings between each of the phases of the signals; inter-green intervals are used between two traffic light phases for clearing the intersection from vehicles and allowing pedestrian crossings

These data items made up the initial state of a data file in planning terms. The goal language of the planner is what the actions in the domain model can do. In this case the goals are made up of numerical constraints denoting predicates on the occupancy levels of each road link.

11.6.2 Mathematical Models of Route Flow and Crossroads

From the description lists of the roads, an Initial Oriented Graph IOG is constructed. The elements of the lists are represented by IOGV vertices or nodes. The edges IOGE of the graph represent the succession of these elements. The representation of the IOG graph is given in an XML file.

In order to determine all the possible paths from all the entrances of the network to all the possible exits, the IOG must be transformed into a new oriented graph TOG describing the entire network in both directions. The TOG nodes will be represented by TOGV and the edges by TOGE. From TOG, the Dijkstra Shortest Path Algorithm (DSPA) is performed to determine the list of all possible paths (Figure 11.10).

As mentioned, the Initial Oriented Graph IOG description is given a collection of vertices and a collection of edges.

A vertex is represented by an XML Element "vertex" having the following attributes:

- name: Sensor identifier (ID)
- type: Element type (Enumeration)
- label: Name of the highway (string)
- locality: The locality name of the sensor position (string)
- long: Longitude (double)
- lat: Latitude (double)
- factor: Attendance factor

The edges are represented by the XML Element "edge" having the following attributes:

- source: Source node (IDREF)
- target: Destination node (IDREF)

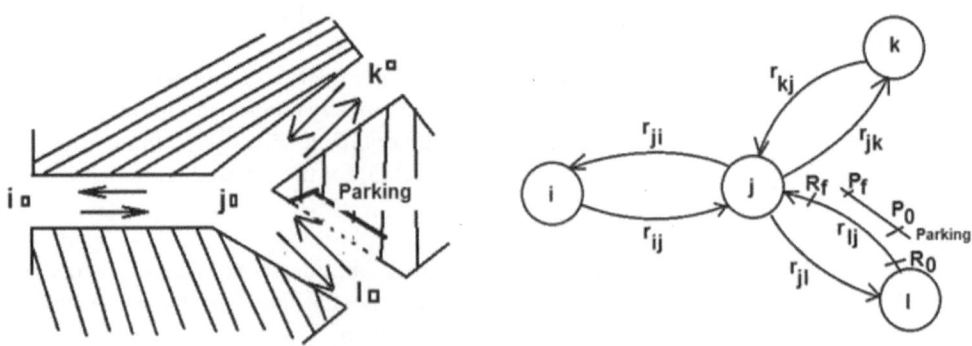

FIGURE 11.10   Convert the Map to an Oriented Graph Representation.

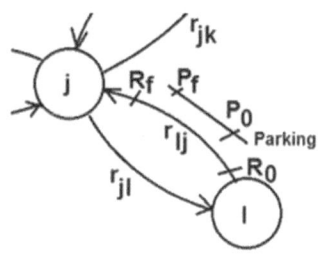

FIGURE 11.11 Sensors Data of a Way between Two Crossroads and Its Parking.

- speed: Segment limit speed (double)
- distance: Distance between the two nodes (double)
- lanes: Number of lanes (int)

The "type" attribute of the "vertex" Element can take one of the following values: {I (Entrance), IO (Entrance/Exit), X (Exchange), R (Service Area), T), S (Sensor), O (Exit)} (Figure 11.11).

In general, the flow of vehicle arrivals from all the arteries is modelled by a Poissonian flow: if the number of vehicles is a few hundred, the relevance of such modelling is not questionable, at least in terms of first approximation.

The possible improvements of this approximation are considered by some authors. Note also that most of the developed algorithms adapt immediately to the case where we replace this Poissonian arrival hypothesis by another arrival law insofar as there is independence of the number of arrivals between t and t + 1 compared to events prior to time t. Now, in a large number of works, it is assumed that the arrivals are of the ON/OFF type. We sometimes consider "geometric": between t and t + 1, there is an arrival of a vehicle with the probability α ($0 < \alpha < 1$) and no arrival with the probability $(1 - \alpha)$. The models already developed are verified with such a hypothesis and lead to much faster computer processing than those associated with a Poissonian arrivals law.

The question which therefore arises is whether the assumption of ON/OFF arrivals, in the sense above, is a reasonable simplification in order to justify our choice of the type of sources in the remainder of the chapter or not. There is obviously no universal answer to such a question. To advance the reflection, we will study, in two revealing particular cases, what happens if we replace the Poissonian hypothesis (therefore implicitly assuming that we are in a case where it is indisputable) by the hypothesis of ON/OFF arrivals.

11.6.2.1 Basic Process Model

Poisson's law is to traffic models what Bernoulli's law is to models of information sources. Poisson's Law is the most natural and simplest way to describe the process of generating vehicle streams from many independent sources. In general, Poisson's law is considered to be the law of low probabilities.

Let us return to the description of the Poisson distribution in urban transport networks. Suppose the vehicle generation rate of a source is λ vehicles per unit time, and consider an arbitrary time interval of length Δt. Poisson's law establishes that the probability that k vehicles (k integer) are generated during the time interval in question is:

$$P_K = \frac{(\lambda. \Delta t)^K}{K!} e^{-\lambda.\Delta t} \tag{11.1}$$

Another way of describing Poisson's law is to characterize the random duration I separating two consecutive events. The probability, Prob $\{I \le t\}$, for this duration to be greater than a positive real t given according to the expression:

$$Prob\{I \ge t\} = e^{-\lambda t} \tag{11.2}$$

Or, on an infinitesimal scale: $Prob\{I \in [t, t + dt]\} = e^{-\lambda t}\lambda dt \tag{11.3}$

There are several traffic models based on Poisson's law. There is the model of the n Poissonian sources of respective rates λ_i vehicles per channel. The cumulative rate is:

$$\lambda = \sum \lambda_i \tag{11.4}$$

We have also seen the very important model of the infinite population, at cumulative intensity λ.

We will start with the simplest case of the basic model proposed by Pellaumail (1992). This model is a Markovian process X which evolves in discrete time and which is integer. Its law of evolution is defined by the equation as follows:

$$X(k + 1) = \sup(0, X(k) - 1) + A(k) \tag{11.5}$$

where A (k) is an integer random variable independent of X (k) and of fixed distribution.

This X process models the delays of vehicles accessing a fixed flow lane on which there are no other vehicles, the law of A(k) determining the law of the flow of arrivals. At the start, we will study the stationary regime of X in the case where A(k) follows a Poisson law with parameter α. We show, for example, that in a steady state, the expectation of X(k) is:

$$\beta = \alpha + \frac{\alpha^2}{2(1 - \alpha)} \tag{11.6}$$

Let us now consider the case where we suppose that A (k) follows the following ON/OFF law:

$$\begin{cases} Prob[A(k) = 1] = \alpha \\ Prob[A(k) = 0] = 1 - \alpha \end{cases} \tag{11.7}$$

Elementary calculations show that in a steady state, the associated law of X is:

$$\begin{cases} Prob[X(k) = 1] = \alpha \\ Prob[X(k) = 0] = 1 - \alpha \end{cases} \tag{11.8}$$

Pellaumail et al. propose an algorithm to simulate the flow of traffic on telecommunications networks (Pellaumail al., 1994; Pellaumail al., 1992). They find that convergence towards this stationary regime is very rapid: for applications to ATM networks, it is therefore obviously this stationary regime that plays the most important role. However, we have just noted that the stationary law associated with ON/OFF arrivals is profoundly different from the stationary law associated with Poissonian arrivals. In particular, in the ON/OFF case, the expectation of X(k) tends towards α while in the Poissonian case, this same expectation tends towards β which tends towards infinity when α tends towards the value 1 while the mean of the law of arrivals is the same in both cases.

11.6.2.2 Fluid Modelling of Road Network Flows

Recall that in telecommunications, the discrete event simulation of ATM/SDH networks has often been based on models at the burst level. Under this approach the ATM cell forms the elementary level of abstraction 'the molecule' as well as the simulation events corresponding to the transit of individual cells through network components. This offers some advantages, but on the other hand the computation time is often too high: for example, simulating traffic on a communication link at 155 MBit/s for three seconds involves processing about 106 events (Pellaumail et al., 1994; Da Costa et al., 2009; Jabin, 2002).

Under the fluid approach, the sources of traffic flow are represented by models at the level of 'burst-level models'; which makes it possible to use a timescale that is much larger than the passage time of a vehicle. We then consider that the sources do not emit individual vehicles but a continuous flow similar to a fluid that can be described by its instantaneous flow (expressed in vehicles/s).

Although there are many analytical studies of fluid models, there is very little dedicated work on simulations based on these models, and more specifically on traffic simulation of urban road networks.

In this section, we describe the equations that govern the operation of the fluid model followed by the principles used to build the complete model to simulate urban road networks (Dixit, 2013; Jones & Farhat, 2004).

11.6.2.3 Dynamics of a Fluid Reservoir

Consider a fluid reservoir characterized by a finite capacity B and a limited output flow rate c (constant positive integer). This reservoir is supplied by one or more sources of fluid, the total flow rate of which may possibly exceed c. In the following, we will develop the equations describing the dynamic behaviour of such a reservoir, in particular with regard to the occupation of the reservoir and the "outflow" process, and that for a particular class of arrival process. This development is carried out in order to apply these equations to the discrete event simulation of a network of fluid reservoirs (Figure 11.12).

Let $\Lambda(t)$ be the total flow rate of the fluid entering the reservoir at time t; then, the volume of fluid $A(t_a, t_b))$ which arrives at the reservoir during the time interval $[t_a, t_b]$ is given by:

$$A(t_a, t_b) = \int_{t_a}^{t_b} \Lambda(t)dt \tag{11.9}$$

By convention, we denote by A(t) the quantity A(0, t). In general, we will only be interested in arrival processes such that, for each trajectory of Λ, there is a sequence:

$$\tau_A = \{T_n \in IN | T_n < T_{n+1}, n \in IN; T_0 = 0\} \tag{11.10}$$

Such as: $\Lambda(t) = \Lambda(T_n), \forall t \in [T_n, T_{n+1}]$

$$\Lambda(T_n) \neq \Lambda(T_{n+1})$$

In other words, the trajectories $\Lambda(t)$ must be "stepped" functions, and therefore the trajectories A(t) are continuous and piecewise linear; the sequence of T_n corresponds to the dates when $\Lambda(t)$ changes its value (Figure 11.13).

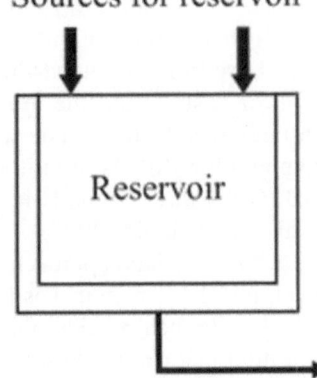

FIGURE 11.12   Presentation of a Reservoir.

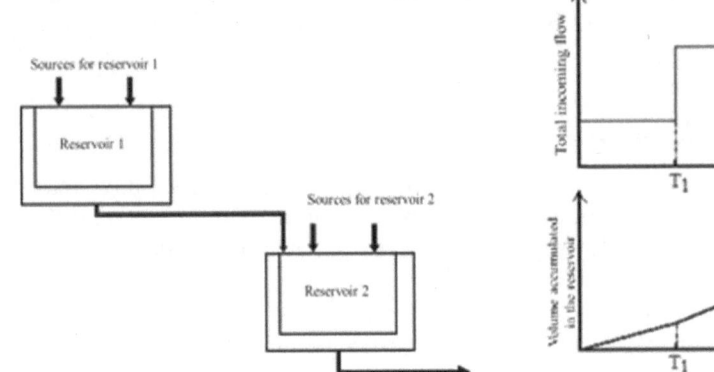

FIGURE 11.13   Trajectories of Q(t) and Flow Λ(t).

Let Q(t) be the quantity or 'level' of fluid contained in the reservoir at time t. The evolution of Q(t) is described by:

$$Q(t) = Q(0) + \int_0^t (\Lambda(s) - c)ds, \ t \geq 0 \tag{11.11}$$

with c the maximum outlet flow from the tank.

We define the set q, for a reservoir of infinite capacity, as follows:

$$q = \{s \geq 0 | \Lambda(s) > c \ \ ou \ \ Q(s) > 0\} \tag{11.12}$$

For a finite capacity tank, it is possible that Q (t) reaches B at a finite instant t. So, we must modify the previous expression of Q to satisfy this constraint, which gives:

$$q = \{s \geq 0 | (\Lambda(s) > c \ \ et \ \ Q(s) < B) o\grave{u}(\Lambda(s) \leq c \ \ et \ \ Q(s) > 0)\} \tag{11.13}$$

In this case, the integral in Equation (11.11) can be interpreted as follows:

- The fluid accumulates at the rate of Λ(t) > c units of volume per unit of time, and this during the intervals when the inlet flow exceeds the maximum outlet flow and when the tank is not completely filled (condition 1: Λ(s) > c and Q(s) < B);

- The tank empties at the rate of $c > \Lambda(t)$ units of volume per unit of time, as long as the inlet flow rate is less than that of the outlet and there is still fluid remaining (condition 2: $\Lambda(s) < c$ and $Q(s) > 0$).
- The condition $\Lambda(s) = c$ and $Q(s) > 0$, corresponding to the case where the fluid level remains constant and strictly positive, could be eliminated without changing the value of the integral in (11.11); however, this condition is included for convenience, since q thus represents the set of time intervals where the reservoir is neither empty nor overflowing.

The resulting trajectories Q (t) are piecewise linear, with slope:

$$Q'(t) = (\Lambda(t) - c) \ \{t \in q\} \tag{11.14}$$

The slope changes occur either at the times when the reservoir becomes empty or saturated, or at the dates T_n of change in value of $\Lambda(t)$ such as $T_n \in q$. By developing the Equation (11.11) for $t \in [T_n, T_{n+1}]$, we obtain:

$$
\begin{aligned}
Q(t) &= Q(0) + \int_0^{T_n} (\Lambda(s) - c)_{\{s \in q\}} \, ds + \int_{T_n}^t (\Lambda(s) - c)_{\{s \in q\}} \, ds \\
&= Q(T_n) + \int_{T_n}^t (\Lambda(s) - c)_{\{s \in q\}} \, ds \\
&= \min \{B, (Q(T_n) + (\Lambda(T_n) - c)(t - T_n))^+\}
\end{aligned}
\tag{11.15}
$$

From which we deduce the recurring equation:

$$
\begin{aligned}
Q(T_{n+1}) &= \min \{B, (Q(T_n) + (\Lambda(T_n) - c)(T_{n+1} - T_n))^+\} \\
&\text{with } (x)^+ \equiv \max(0, x)
\end{aligned}
\tag{11.16}
$$

Parking is one of the essential functions of urban networks. To simulate this function, we made the necessary modifications so that the surpluses will be directed to a finite capacity buffer without forgetting the development of an objective selection criterion 'Which vehicles to store (park)?'. The answers to this question constitute the strategies for using the car park, since the support elements are offered during the phase of movement on a lane.

Single-lane parking will be represented by a reservoir with entry and exit access as shown in Figure 11.14. It suffices to mention the inlet flow according to the state of the tank since the inlet flow must be cancelled once the tank is saturated.

11.6.3 Traffic Lights Management

Within traffic systems, it is desirable to obtain optimal solutions for the fluidization of traffic. The trend is to develop the most efficient algorithms of adaptive traffic control using concepts such as parameter estimation, fuzzy logic, or artificial intelligence. The crossroad is an intersection characterized by the fact that the traffic lights alternate at one moment only one of the green traffic lights. In this chapter, it will be

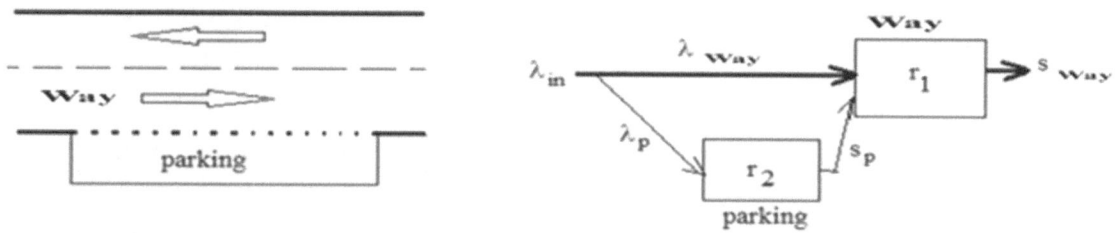

FIGURE 11.14   Parking Management for a Single Lane Way.

presented just a concept of traffic lights configuration mode and how these configuration modes can influence the vehicle movement.

Traffic lights are alternately synchronized, taking into account the direction of movement of the cars (Figure 11.15). Directions that are given green at the same time are chosen in such a way as to avoid collisions between cars. In this way, cars coming from the East and moving towards the West and those coming from West with the direction of going to East will have green at the same time and will bypass the imaginary centre of the intersection in order to follow the route.

11.6.4 Implementation Example

The implementation of our Urban Traffic Management and Control (UTMC) is based on the road intersections of four directions, each containing two lanes (Figure 11.16). Each lane is equipped with two sensors: one is located near the traffic light to count the number of vehicle departures and the other is installed at a variable distance, which depends on the maximum time allowed for the green light, from the first sensor to detect the arrival of vehicles (Terrada et al., 2020c). Each traffic light controller defines a time called cycle which is a sequence of phases and each of the phases has a green light time necessary for two movements which occur simultaneously.

All the sensors communicate and transfer the traffic information to the UTMC station which calculates the length of the queues for each direction and its average waiting time in order to control the traffic flow. Depending on the air quality in the area, the UTMC station may lengthen or shorten the light

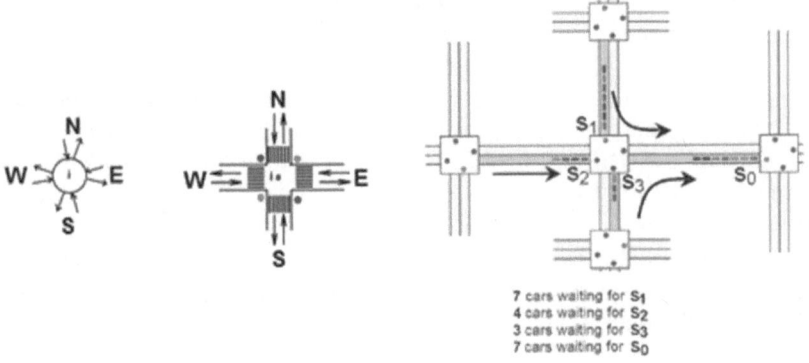

7 cars waiting for S_1
4 cars waiting for S_2
3 cars waiting for S_3
7 cars waiting for S_0

FIGURE 11.15 Example of Crossroad with Traffic Lights.

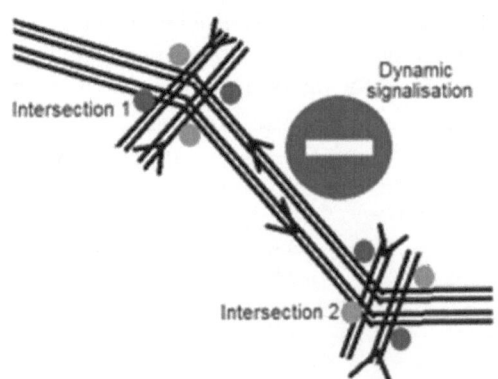

FIGURE 11.16 Road Intersections Relating to UTMC.

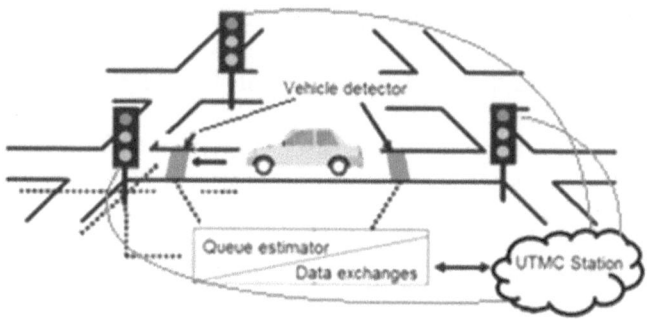

FIGURE 11.17   Exchange of Information.

cycles and control the dynamic signalling. The temporary ban on a lane, for example, will reduce traffic to improve air quality and users will be informed of alternative routes to arrive at their destinations as soon as possible.

In addition to the information provided on the desired route via a radio broadcast or road signs, a speed warning can be offered to the driver as a driving assistance as illustrated in Figure 11.17. Thus, a certain speed is suggested to drivers in order to have a change from light to green at the next intersection in sight. The panel is connected to the lights above and, knowing how many seconds will pass before they turn green, the electronic controls calculate and display an appropriate speed to maintain. The UTMC station records information on the routes and feedback from drivers to build a knowledge base to improve the predictability of the system.

11.7 IMPACT OF UTM ON URBAN SUPPLY CHAIN

According to the Council of Supply Chain Management Professionals; Supply Chain Management includes coordination and collaboration between all stockholders throughout the Supply Chain (Vitasek, 2013; Boudoin et al., 2014). Basically the SCM combines all the activities related to the movement of resources and flows (material, financial and informational) throughout the Chain as shown in Figure 11.18.

The Supply chain is technically subject to strong mutations due to its structure (Terrada et al., 2019). Collaborative solutions can only tackle the risk by reviving large urban structures environmentally, socially, and economically. Therefore, urban planners should be imperatively involved in urban planning issues.

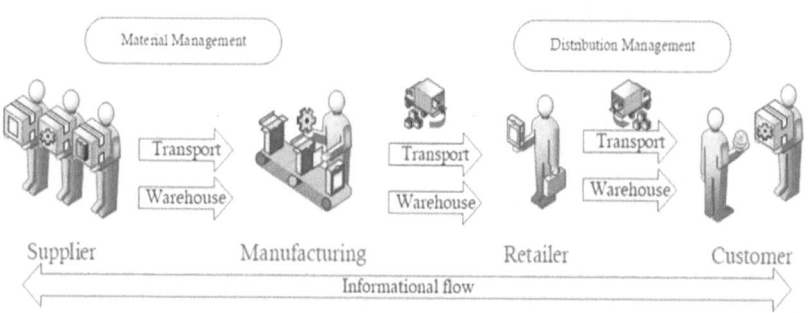

FIGURE 11.18   Supply Chain Management Flows.

11.7.1 Urban SC towards GrSC in the Literature

Green Supply Chain Management (CrSCM) is the conciliation of environmental management and Supply Chain Management (SCM).

Industrial organizations aim to implement green practices in their SCM in order to minimize the effect of environmental pollution, hence, many research studies have been conducted in various activities of the supply chain processes in several manufacturing industries in SMEs and large scale manufacturing companies and to study the impact of the ecofriendly practices in the six main activities, i.e. green sourcing, green manufacturing, green warehousing, green distribution, green packaging, green transportation (Tomar & Oza, 2016; Srivastava, 2007).

11.7.2 Urban Air Pollution and the GrSCM

According to literature transport is one of the major causes of air pollution and emission of Greehouse Gas in an urban zone because of congestion. Boudoin et al. estimate that urban goods movements could be the origin of 70% of particulate matter, 35% of Nitrogen oxides and 25% of GHG emissions (Boudoin et al., 2014). Thus, GrSCM should be handled by urban logistics whose strategy is to efficiently and effectively manage urban freight transport and other traffic flows, in order to reduce environmental impacts (Lacuna et al., 2015).

Actually, the supply chain in the context of air pollution is associated mostly with the emission of carbon dioxide or carbon, especially in cities, which requires the implementation of a measurement system of pollution in the air in the context of urban traffic control, the proposed solution may help transport operators to reduce traffic concentrations in a targeted urban zone to improve effects of predicted road traffic pollution, or optimize the flow of saturated road links due to an emergency road closure, or produce a strategy to deal with a forthcoming complex situation such as optimizing the light timings in order to deal deliver goods on time to the customers or retailers.

11.7.3 Improvement of Key Indicators of Urban Supply Chain

Transport and delivery must meet sustainable constraints, both for private companies and for public actors. We summarize the key indicators of the urban Supply Chain as shown in Figure 11.19 connecting both to the economic, ecological, and operational in alignment with the expectations of private and public actors (lacuna et al., 2015; Issaoui et al., 2019; Issaoui et al., 2020a; Issaoui et al., 2020b).

11.8 CONCLUSION

In this chapter, we present an approach for Urban Traffic Control Management based on IoT and connected objects, in order to tackle a worldwide issue which is air pollution in cities which means hazards for health and worse quality of life. In the context of Urban Supply Chain, we proposed Mathematical models of route flows and crossroads, to minimize congestion, optimize traffic flow management, and thus, reduce the effect of environmental and air pollution. Traffic signalization control has been presented meanwhile a mechanism to control cars is also needed. While the reported tests are encouraging, a more thorough evaluation is required to better assess real-world potential.

Smart cities aim to exploit resources and space efficiently; however, they have to ensure a clean place to live by making the urban transportation more environment friendly, which emerges the concept

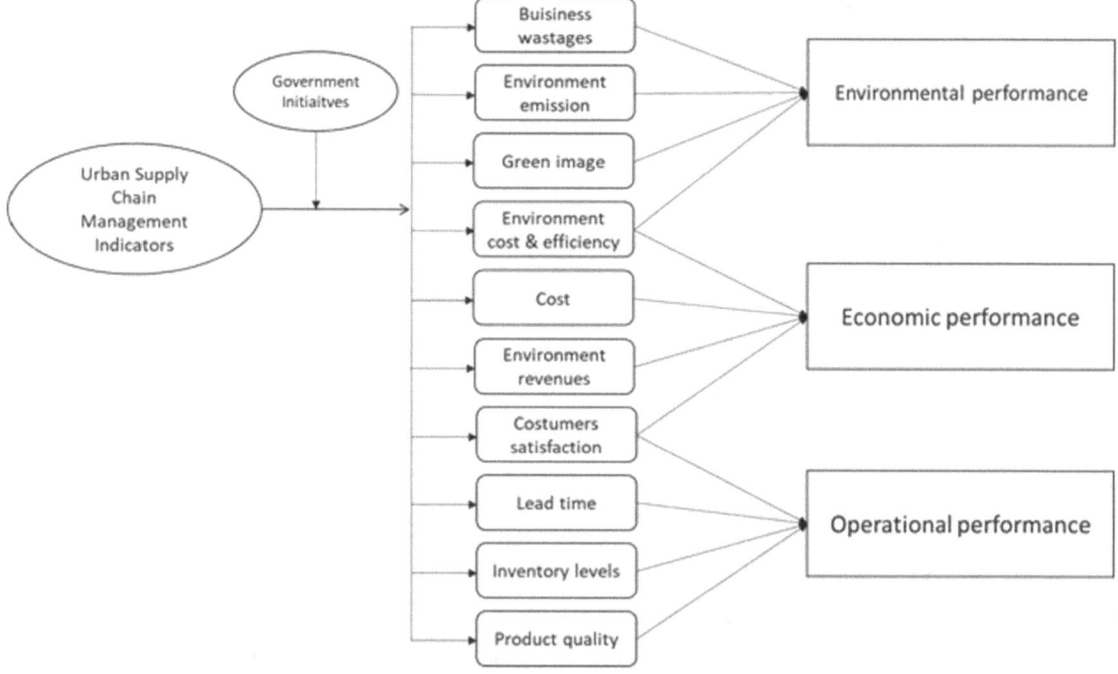

FIGURE 11.19   Urban Supply Chain indicators.

of Ecological Smart City that combines intelligent approaches and green practices. Our study shows that Urban Traffic Control Management is related to the concept of Green Supply Chain, whose strategy is to manage urban freight transport and other traffic flows by integrating green practices for a Smart and Ecological City. As a perspective of our research, we aim to use LabVIEW Graphical programming as decision support for the Smart City.

REFERENCES

Abdelaziz, D, Omar, B, Mohamed, Y, & Abra, OEK (2017). An efficient distributed traffic events generator for smart highways. *International Journal of Advanced Computer Science and Applications*, 8(7), 117–127, 10.14569/IJACSA.2017.080717

Alfaseeh, L, Djavadian, S, & Farooq, B (2018). Impact of distributed routing of intelligent vehicles on urban traffic. In 2018 IEEE International Smart Cities Conference (ISC2), Kansas City, MO, USA, pp. 1–7.

Badidi, E, Mahrez, Z, & Sabir, E (2020). Fog computing for smart cities' Big Data management and analytics: A review. *Future Internet*, 2020, *12*, 190, 10.3390/fi12110190

Bandeira JM, Carvalho, DO, Khattak, AJ, et al. (2016). Empirical assessment of route choice impact on emissions over different road types, traffic demands, and driving scenarios. *International Journal of Sustainable Transportation*, *10*(3), 271–283.

Barredo Arrieta, A, Díaz-Rodríguez, N, Del Ser, J, Bennetot, A, Tabik, S, Barbado, A, Garcia, S, Gil-Lopez, S, Molina, D, Benjamins, R, Chatila, R, & Herrera, F (2020). Explainable Artificial Intelligence (XAI): Concepts, taxonomies, opportunities and challenges toward responsible AI. *Information Fusion*, *58*, 82–115, 10.1016/j.inffus.2019.12.012

Benedek, CM, & Rilett, LR (1998). Equitable traffic assignment with environmental cost functions. *Journal of Transportation Engineering*, *124*(1), 16–22.

Boudoin, D, Morel, C, & Gardat, M (2014). *Supply chains and urban logistics platforms, sustainable urban logistics: Concepts, methods and information systems*, pp. 1–20. Springer.

Chakir, I, El Khaili, M, & Mestari, M (2020). Logistics flow optimization for advanced management of the crisis situation. *Procedia Computer Science*, *175*, 419–426, 10.1016/j.procs.2020.07.059.

Chen, S, Hu, J, Shi, Y, Peng, Y, Fang, J, Zhao, R, et al. (2017). Vehicle-to-Everything (v2x) services supported by LTE-based systems and 5G. *IEEE Communications Standards Magazine*, *1*(2), 70–76.

Da Costa, G, Dufour, G, & Sanchez, D (2009). Fluid modelling for energy saving in grid computing. In *International Conference RENPAR'19*.

Dastjerdi, AV, & Buyya, R (2016). Fog computing: Helping the Internet of Things realize its potential. *Computer*, *49*(8), 112–116.

Dixit, VV (2013). Behavioural foundations of two-fluid model for urban traffic. *Transportation Research Part C: Emerging Technologies*, *35*, 115–126, 10.1016/j.trc.2013.06.009.

Duch, J, & Arenas, A (2006). Scaling of fluctuations in traffic on complex networks. *Physical Review Letters*, *96*, 218702.

El Khaili, M, Alloubane, A, Terrada, L, & Khiat, A (2019). Urban traffic flow management based on air quality measurement by IoT using LabVIEW. In M Ben Ahmed, A Boudhir, & A Younes (Eds.), *Innovations in Smart Cities Applications Edition 2. SCA 2018. Lecture Notes in Intelligent Transportation and Infrastructure*. Springer.

El Khaili, M, Bakkoury, J, Khiat, A, & Alloubane, A (2018). Crowdsourcing by IoT using LabVIEW for measuring the air quality, ACM ISBN: 978-1-4503-6562-8, New York, NY, USA.

Farahzadi, A, Shams, P, Rezazadeh, J, & Farahbakhsh, R (2017). Middleware technologies for cloud of things - A survey. *Digital Communications and Networks*, 10.1016/j.dcan.2017.04.005.

Farooq, B, & Djavadian, S (2019). *Distributed traffic management system with dynamic end-to-end routing, U.S. provisional patent service no. 62/865,725*, The United States Patent and Trademark Office (USPTO).

Gallotti, R, Bazzani, A, & Rambaldi, S (2012). Towards a statistical physics of human mobility. *International Journal of Modern Physics C*, *23*, 1250061.

Garcia, A, & Lopez-Carmona, M (2019). Multimap routing for road traffic management. In Y Demazeau, E Matson, J Corchado, & F De la Prieta (Eds.), *Advances in practical applications of survivable agents and multi-agent systems: The PAAMS Collection. PAAMS 2019. Lecture Notes in Computer Science*, vol. 11523. Springer. 10.1007/978-3-030-24209-1_16

Guli'c, M, Olivares, R, & Borrajo, D (2016). Using automated planning for traffic signals control. *PROMET Traffic & Transportation*, *28*(4), 383–391.

Hawas, YE, & Mahmassani, HS (1996). Comparative analysis of robustness of centralized and distributed network route control systems in incident situations. *Transportation Research Record*, *1537*(1), 83–90.

Issaoui, Y, Khiat, A, Bahnasse, A, & Hassan, O (2019). Smart logistics: Study of the application of blockchain technology. *Procedia Computer Science*, *160*, 266–271, 10.1016/j.procs.2019.09.467.

Issaoui, Y, Khiat, A, Bahnasse, A, et al. (2020a). Toward smart logistics: Engineering insights and emerging trends. *Archives of Computational Methods in Engineering*, *28*, 3183–3210. 10.1007/s11831-020-09494-2

Issaoui, Y, Khiat, A, Bahnasse, A, & Ouajji, H (2020b). Smart logistics: Blockchain trends and applications. *Journal of Ubiquitous Systems & Pervasive Networks*, *12*(2), 9–15.

Jabin, P (2002). Various levels of models for aerosols. *Mathematical Models and Methods in Applied Sciences*, *12*, 903–920.

Jean, C (2011). Objets Connectes. http://www-clips.imag.fr/geod/User/jean.caelen/Publis_fichiers/ObjetsConnectes.pptx. Last visited July 29, 2018.

Jiang, Z, Chen, X, & Ouyang, Y (2017). Traffic state and emission estimation for urban expressways based on heterogeneous data. *Transportation Research Part D: Transport and Environment*, *53*, 440–453.

Jing, H, Xiaochuan, P, Xinbiao, G, & Guoxing, L (2018). Impacts of air pollution wave on years of life lost: A crucial way to communicate the health risks of air pollution to the public. *Environment International*, *113*, 42–49.

Jones, E, & Farhat, W (2004). Validation of two-fluid model of urban traffic for arterial streets. *Transportation Research Record*, *1876*(1), 132–141, 10.3141/1876-14

Khan, T, & Anderson, M (2016). Accurately estimating origin/destination matrices in situations with limited traffic counts: Case study Huntsville, AL. *Journal of Traffic and Transportation Engineering*, *5*(3), 64–72.

Khiat, A, Bahnasse, A, Bakkoury, J, El Khaili, M, & Louhab, F (2018). New approach based internet of things for a clean atmosphere. *International Journal of Information Technology*, *11*, 89–95, 10.1007/s41870-018-0253-6

Khiat, A, Bahnasse, A, El Khaili, M, & Bakkoury, J (2016). Study and evaluation of vertical and horizontal handover's scalability using OPNET modeler. *International Journal of Computer Science and Information Security (USA)*, *14*(11), 1–7.

Khiat, A, Bahnasse, A, El Khaili, M, & Bakkoury, J (2017). Study, Evaluation and Measurement of IEEE 802.16e Secured by Dynamic and Multipoint VPN IPsec. *International Journal of Computer Science and Information Security (USA)*, *15*(1), 1–6.

Krajzewicz, D, Erdmann, J, Behrisch, M, & Bieker, L (2012). Recent development and applications of SUMO - Simulation of Urban MObility. *International Journal on Advances in Systems and Measurements*, *5*(3&4), 128–138.

Lacuna, M, Colomer-Llinàs, M, & Meléndez-Frigola, J (2015). Lessons inurban monitoring taken from sustainable and livable cities to better address the Smart Cities initiative. *Technological Forecasting & Social Change*, *90*, 611–622.

Lee, J, & Park, B (2008). Evaluation of route guidance strategies based on vehicle-infrastructure integration under incident conditions. *Transportation Research Record*, *2086*(1), 107–114.

Liang, X, Zhao, J, Dong, L, & Xu, K (2013). Unraveling the origin of exponential law in intra-urban human mobility. *Scientific Reports*, *3*, 2983.

Luo, L, Ge, YE, Zhang, F, & Ban, X (2016). Real-time route diversion control in a model predictive control framework with multiple objectives: Traffic efficiency, emission reduction and fuel economy. *Transportation Research Part D: Transport and Environment*, *48*, 332–356.

Mahmassani, HS, Saberi, M, & Zockaie, AK (2013). Urban network gridlock: Theory, characteristics, and dynamics. *Procedia - Social and Behavioral Sciences*, *80*, 79–98.

McCluskey Thomas, L, & Mauro, V (2017). Embedding automated planning within urban traffic management operations. Proceedings of the Twenty-Seventh International Conference on Automated Planning and Scheduling (ICAPS 2017), Pittsburgh, USA.

Misra, A, Gooze, A, Watkins, K, Asad, M, & le Dantec, CA (2014). Crowdsourcing and its application to transportation data collection and management. *Transportation Research Record*, *2414*(1), 1–8.

Nagatani, T (2000). Traffic jams induced by fluctuation of a leading car. *Physical Review E*, *61*, 3534.

Nagatani, T (2002). The physics of traffic jams. *Reports on Progress in Physics*, *65*, 1331.

Pellaumail, J (1992). Graphes, Simulation, L-matrices, Hermès, Paris.

Pellaumail, J, Boyer, P, & Leguesdron, P (1994). *Réseaux ATM et P-simulation*, Hermes, Paris.

Petrović, N, & Tošić, M (2020). *Explainable artificial intelligence and reasoning in Smart Cities*. University of Niš, Faculty of Electronic Engineering, Conference Paper March 2020.

Rose, K, Eldridge, S, & Chapin, L (2015). The Internet of Things: An overview - understanding the issues and challenges of a more connected world.

Shahgholian, M, & Gharavian, D (2018). Advanced traffic management systems: An overview and a development strategy. https://arxiv.org/abs/1810.02530.

Silva, DRC, Oliveira, GMB, Silva, I, Ferrari, P, & Sisinni, E (2018). Latency evaluation for MQTT and WebSocket Protocols: An Industry 4.0 perspective. 2018 IEEE Symposium on Computers and Communications (ISCC), pp. 01 233–01 238.

Smith, RD (2011). The dynamics of internet traffic: Self-similarity, self-organization, and complex phenomena. *Advances in Complex Systems*, *14*, 905.

Srivastava, S (2007). Green supply-chain management: A state-of-the-art literature review. *International Journal of Management Reviews*, *9*(1), 53–80.

Sun, J, & Liu, HX (2015). Stochastic eco-routing in a signalized traffic network. *Transportation Research Procedia*, *7*, 110–128.

Terrada, L, Alloubane, A, Bakkoury, J, & El Khaili, M (2018). IoT contribution in supply chain management for enhancing performance indicators. 2018 International Conference on Electronics, Control, Optimization and Computer Science (ICECOCS), Kenitra, pp. 1–5, 10.1109/ICECOCS.2018.8610517.

Terrada, L, Bakkoury, J, El Khaili, M, & Khiat, A (2019). Collaborative and communicative logistics flows management using Internet of Things. In J Mizera-Pietraszko, P Pichappan, & L Mohamed (Eds.), *Lecture Notes in Real-Time Intelligent Systems. RTIS 2017. Advances in Intelligent Systems and Computing*, vol. 756, pp. 216–224. Springer.

Terrada, L, El Khaïli, M, Ouajji, H, & Daaif, A (2020). Smart urban traffic for green supply chain management. 2020 IEEE 2nd International Conference on Electronics, Control, Optimization and Computer Science (ICECOCS), Kenitra, Morocco, pp. 1–6, 10.1109/ICECOCS50124.2020.9314569.

Tlig, M, Buffet, O, & Simonin, O (2014). Decentralized traffic management: A synchronization-based intersection control, 2014 International Conference on Advanced Logistics and Transport.

Tomar, A, & Oza, H (2016). Review of GSCM studies relating to green supply chain management practices and performance. *Abhinav*, 5, 49–55.

Vineela, A, & Sudha, L (2015). Internet of Things – Overview. *International Journal of Research in Science & Technology*, 2(4). https://ia800501.us.archive.org/7/items/2.IJRST020402812/2.IJRST020402(8-12).pdf.

Vitasek, K (2013). CSCMP supply chain management definitions and glossary. *Council of Supply Chain Management Professionals*, August 2013.

Wang, F, Zhang, M, Wang, X, Ma, X & Liu, J (2020). Deep learning for edge computing applications: A state-of-the-art survey. *IEEE Access*, 8, 58322–58336, 10.1109/ACCESS.2020.2982411.

Watling, D, & Hazelton, ML (2003). The dynamics and equilibria of day-to-day assignment models. *Networks and Spatial Economics*, 3(3), 349–370.

White, G, & Clarke, S (2020). Urban intelligence with deep edges. *IEEE Access*, 8, 7518–7530, 10.1109/ACCESS.2020.2963912.

Yang, X, & Recker, W (2007). Modeling dynamic vehicle navigation in a self-organizing, peer-to-peer, distributed traffic information system. *Journal of Intelligent Transportation Systems*, 10(4), 185–204.

Zegeye, SK, De Schutter, B, Hellendoorn, H, & Breunesse, E (2009). Model-based traffic control for balanced reduction of fuel consumption, emissions, and travel time. In Proceedings of the 12th IFAC Symposium on Transportation Systems, Redondo Beach, California, pp. 149–154.

Systematic Comparison of Feature Selection Methods for Solar Energy Forecasting

12

S. El Motaki[1] and A. El-Fengour[2,3]

[1]*University Sidi Mohamed Ben Abdellah, Fez, Morocco*
[2]*University Ibn Tofail, Kenitra, Morocco*
[3]*University of Castilla-La Mancha, Ciudad Real, Spain*

Contents

12.1 INTRODUCTION

Feature selection (FS) has been a highly active research topic in various explainable artificial intelligence fields including machine learning, data mining, image processing, and data analysis, in these last decades (Liu & Yu, 2005). For example, the real-time monitoring of facility energy systems and building-grid innovation is driven by an intelligent structural energy prediction model. Despite having a reduced manufacturing expense

DOI: 10.1201/9781003172772-12

throughout their entire lifecycle, data-driven models usually experience inadequate generalization of the underlying pattern due to the high dimensionality of the data (Zhang & Wen, 2019). FS enables to address highly dimensional characteristics, to improve the interpretability of an intelligent system, and to better generalize it.

FS involves selecting from a large set of features a subset of features that are relevant to a given problem. FS is a critical milestone for discovering and processing knowledge that renders data mining and ML tools more accurately and efficiently. In fact, various irrelevant, misleading, or redundant features may occur in high-dimensional data. Moreover, a certain number of learning algorithms do not perform properly with a large range of attributes. Consequently, scientists and practitioners have introduced FS techniques to preprocess the data before applying any mining or learning model; resulting in speeding up the model train and test mechanism by greatly reducing computation time on the one hand, and on the other hand, improving the results comprehensibility and prediction accuracy and avoiding over-fitting cases.

In general, FS techniques are broadly grouped into three main categories, which are described as follows:

- Filter-based FS: aims at selecting the relevant features, regardless of the data mining algorithm used, by applying statistical measures. These approaches are either univariate, that is, each feature is taken separately and coupled with a score; the selection is elaborated then based on features ranking (Zhang & Wen, 2019; Hall, 2000), or multivariate where different possible attribute subsets are examined and assessed to determine their suitability for differentiation (Manoranjan et al., 2002).
- Wrapper FS: in this category, the FS method relies on the result of a pre-determined data mining or machine learning algorithm to identify the quality of a given subset of features. The required algorithm is used to select features based on its performance (Saeys et al., 2007).
- Embedded FS: in this category of FS techniques, both filter and wrapper approaches are combined. The methods are supported by learning algorithms that have their own integrated function selection methods. Both independent testing and performance evaluation function of feature subsets are employed (Kohavi & John, 1997).

The choice of FS methodology to be used depends on how the selected attributes' usefulness is assessed. The three strategies mentioned above may vary in terms of accuracy, computational time, as well as in tendency to over-fitting. As a matter of fact, filter approaches easily scale to a large volume of data and have a low computational cost with a low risk of over-fitting. Still, their accuracy is relatively limited compared to wrapper methods, which have high accuracy but tend to be more expensive in terms of computational time. Embedded FS is introduced to overcome all the before-mentioned problems by striking the appropriate balance between wrapper approach accuracy and filter-based selection ease-of-use (Manoranjan et al., 2002).

In this chapter, a comparative study has been made between three different FS methods. The stepwise algorithm, which is a wrapper approach, is compared to both least absolute shrinkage and selection operator (Lasso) and random forests (RF) FS, which follow an embedded strategy. Next, we use a simple correlation coefficient score as the filter method to improve the results previously reached. The three algorithms are applied to solar energy forecasting. It is an interesting subject in the renewable energy sector, and data mining and machine learning algorithms have a crucial role in this area.

The remainder of this chapter is planned as follows: Section 12.2 reviews some recent works on FS methods applied to solar energy forecasting. Section 12.3 describes the three methods used in this chapter. In Section 12.4, we discuss the application results through a real-world application. Section 12.5 concludes with a brief discussion.

12.2 RELATED WORK

FS has been a productive scope of research and development since the seventies. It has played an important role in various fields such as data mining, machine learning, bioinformatics, and natural disaster management (Blum & Pat, 1997; Awada et al., 2012; Drotar et al., 2015; Khalid et al., 2014; Liu & Yu, 2005; Rangarajan & Veerabhadrappa, 2010). Solar energy forecasting is one of the recent and dynamic sectors, which is considered an important field of application for FS techniques. A wide range of alternative approaches has been proposed in this sense. Namely, Martin et al. (2016) have proposed a model based on FS algorithms such as linear correlation, ReliefF, and logical information analysis to improve the prediction process of solar energy production in different grid stations by selecting the most relevant meteorological attributes. Linear correlation and ReliefF methods are also addressed in (Goswami et al., 2018); authors have used the methods to select only the significant features to enhance the performance of a numerical weather prediction model. Besides, they have introduced a new FS technique that relies on local information analysis. A work elaborated in (O'Leary & Kubby, 2017) consists of using correlation-based FS to improve the artificial neural network prediction accuracy; they have shown the importance of removing noisy and complex weather features in improving the solar energy forecasting results.

Similarly, a forecasting method that involves a neural network coupled with an improved version of shark smell optimization algorithm has been proposed as a hybrid forecasting system (Abedinia et al., 2018). Specifically, the metaheuristic algorithm is designed to tune neural network parameters. Furthermore, the model consists of a two-phased FS algorithm, using the mutual information and interaction profit theoretical criteria, which eliminates irrelevant input features.

In Zhang and Wen (2019), the authors propose a systematic FS approach for developing a predictive model for building energy. They proposed a model that aims at combining statistical data analysis, the physics of buildings and engineering experiments. The system includes a preliminary processing step using domain knowledge coupled with a statistical FS method to remove irrelevant and duplicate features, namely a filtering method. Then, the wrapping method is used to determine the optimal set of features.

At the same token, a novel FS algorithm is presented in (García-Hinde et al., 2016); it relies on using a bootstrapping of support vector machine classifiers to evaluate each feature relevance. The proposed method selects the most appropriate features for the best solar radiation prediction. In the same interest of using FS techniques for solar forecasting improvement, authors in (Hossain & Ali, 2013) have demonstrated that FS methods can significantly increase the performance of machine learning models applied to real-life historical meteorological data.

In the same context, we propose a work that seeks to conduct a comparative study between three methods that have different selection policies that were not previously adopted to predict solar energy. Mainly, the contributions of this work are as follows:

- We introduce a systematic review of different FS approaches – stepwise, Lasso, and RF – for solar energy prediction, based on geographical features. To our knowledge, no previous research has investigated these algorithms for selecting solar energy predictors. The results are evaluated by linear regression and support vector regression (SVR) prediction methods.
- The comparative study is made based under realistic conditions. The dataset used contains various target metrics involving meteorological and geographical information (more details are presented in Section 12.4), which makes the study more exhaustive.

12.3 FEATURE SELECTION ALGORITHMS

12.3.1 Least Absolute Shrinkage and Selection Operator

Lasso is a method of withdrawal and variable selection for linear regression models introduced by (Tibshirani, 1996). It belongs to the family of embedded methods. The purpose of Lasso regression is to generate a subset of predictors so that the prediction error for a quantitative response variable is minimized by imposing a constraint on the model that forces some coefficients to shrink to zero. Subsequently, the variables with non-zero coefficients are more highly associated with the response variable; and variables with zero coefficients are excluded. Mathematically, Lasso is expressed as an estimation of a multiple regression function. It can be defined by a penalization function K as follows:

$$K(\beta) = argmin\left\{\frac{1}{2}\right\} \sum_{i=1}^{n}(y_i - \beta_0 - \sum_{j=1}^{P} x_{ij}\beta_j)^2 + \lambda \sum_{j=1}^{P}|\beta_j| \tag{12.1}$$

where $y = (y_1, y_2, ..., y_n)$ is a $1 \times n$-vector that denotes the predicted variable, $X = (1, x_1, x_2, ..., x_n)$ is $n \times (n + 1)$-matrix of predictors, $\beta = (\beta_0, \beta_1, ..., \beta_n)$ is a column vector of $p + 1$ size representing the associated impacts of each predictor in X, and λ is the penalty setting that controls the trade-off between the loss of precision and the model complexity. Lasso limits the variability of the estimations by setting some of the coefficients precisely to zero, which makes it feasible to produce models that can be easily interpreted.

12.3.2 Random Forests FS

Random forests is an embedded method that is very popular and accurate for high and improperly constructed classification and regression tasks based on an idea of model aggregation (Breiman, 2001). The key concept behind the RF framework is to produce a large number of impartial decision trees from random samples of training data with substitution, where each tree decides on a class and the forest selects the classification with the highest number of ratings among all the trees in the forest.

Basically, for a given input X and K trees, we assume that $\tilde{y}_k$ is the prediction of the tree t_k:

$$\tilde{y} = majorityVote\{\tilde{y}_k\}_1^K \tag{12.2}$$

The basic technique for measuring the significance of features in the prediction is introduced by "Breiman (2001)" and denoted by out-of-bag (OOB) importance score. It seeks to calculate the margin between the original mean error and the randomly swapped mean error in OOB sampling. The method reorganizes all measurements of a selected feature in OOB for all trees stochastically and uses the RF model to predict this permuted feature and determine the mean error. This permutation aims to discard the current correlation between the given feature and the y values and then examine the impact of this overwriting on the RF model. Features with greatly decreasing mean error are supposed to be the most relevant.

12.3.3 Stepwise Regression

The stepwise regression, which is a wrapper method, determines the best predictive features to include in a model from a more extensive set of potentially predictive features for linear regression, logistic, and other traditional regression models (Mundry & Nunn, 2009). Two approaches are possible when implementing stepwise regression. The first (known as descending FS) involves using a model that includes all the features expected to influence the target variable potentially and then sequentially removes the least essential features from the original model based on an adjustment-quality measure which adjusts the number of features included in the model. The process continues, and other attributes are dropped in successive descendant steps until the adjustment measure can no longer be improved. The second basic approach (called ascending FS) starts with a model that includes only a constant and then expands this model to include features from a set of potential features that provides the most significant improvement in the adjusted fitting measure. The process is replicated to add further features using a set of incremental ascending steps and stops when there is no possible improvement in the adjusted fitting measure.

In this work, we use a combination of ascending and descending FS. First, we create a subset of features that may be relevant using an ascending approach. Then, we use the descending FS tool to determine which features should be excluded from the subset previously built. The adjusted measure used is the R-squared metric.

12.4 REAL-WORLD APPLICATION

12.4.1 Dataset

For the current study, we use a dataset provided by the Open Power System Data (OPSD) platform for open and free-of-charge data for power system modelling (Open Power System Data – A platform for open data of the European power system 2020). This database covers conventional power plant data for Germany and other European countries, including the technical characteristics of each individual power plant, for example, geographical information, the main source of energy, etc. Then, the dataset is processed and coupled with a weather dataset provided by the same platform. For further details, the reader is referred to Predicting wind and solar generation from weather data using Machine Learning (2020).

We assume that the predicted variable is the actual solar energy production in *MW*. Table 12.1 presents a brief overview of some geographical and meteorological features used for this analysis.

12.4.2 Feature Selection Results and Discussion

Having applied the FS methods studied over the dataset mentioned above, we get a list of the most relevant features for solar energy forecasting. We use the Scikit-learn package of Python, which includes a wide range of machine learning libraries, including RF and Lasso models; however, we implement the stepwise algorithm manually since, for our knowledge, no built-in function method for ascending and descending stepwise exists in *Python statsmodels* package.

The Figure 12.1(a, b, and c) exhibits, respectively, the features selected by Lasso, RF, and stepwise algorithms. We observe that by employing Lasso classier, SWTDN, SWGDN, the temperature T, the cumulated hours, and other attributes exhibited in Figure 12.1(b) are the most relevant features among all existing features with a score lying between 0.1 and 0.8. The same attributes are selected using the RF technique (as shown in Figure 12.1(c)) with slightly higher score values. This result may be explained by the fact that both methods follow the same selection strategy since they belong to the same FS category (embedded). From Figure 12.1(c), we can notice that features with high score values SWTDN, SWGDN, and the temperature T are identified as relevant attributes by the stepwise algorithm also. Other features,

TABLE 12.1　Description of some ground, wind, temperature, and air features presented in the dataset used for predicting solar energy generation

FEATURE	DESCRIPTION
Lat	The latitude of each geographical chunk examined (m)
Lon	The longitude of each geographical chunk examined (m)
h1	Height above ground level (2 m above height of displacement)
h2	Height above ground level (10 m above height of displacement)
v1	Velocity at height h1 (m/s)
v2	Velocity at height h2 (m/s)
v_50	Velocity at height 50 m above ground level (m/s)
z0	Roughness length (m)
SWTDN	Full horizontal radiation at the top of the atmosphere (W/m²)
SWGDN	Full ground horizontal radiation (W/m²)
T	Temperature at h1 level (K)
Rho	Surface air density (kg/m³)
p	Surface air density (Pa)

which have not been marked as significant features by the first two algorithms, appear, such as the latitude of each geographical chunk and the surface air density (Rho).

To evaluate the accuracy of the selected features, we train a linear regression and SVR for each selected feature subset. From the analysis of Figure 12.1(d), it follows that the set of selected features by the stepwise method provides an accuracy that exceeds 96% for linear regression, with a slight difference in comparison with the Lasso and RF algorithm, which gives an accuracy that varies between 90% and 95%. Generally, the linear regression accuracy obtained is satisfactory for all the selection methods studied. However, the regression with SVR is not sufficiently accurate since it does not exceed 70% for all the selected set of features, especially those obtained using the Lasso method. This is likely due to the existence of a correlation between the selected features.

To further illustrate the low scores achieved by SVR, a fundamental analysis was conducted to examine the correlation between the features. Figure 12.1(e) exhibits the correlations between the different features selected. Dark colours represent a high correlation between features. We notice that different selected features are significantly correlated. For example, the latitude parameter is correlated with velocity (v1, v2, v_50) parameters. The temperature attribute T is correlated with the height above ground-level features (h1, h2). Another important aspect is the correlation between the predicted variable (actual solar energy generation) and both full horizontal radiation at the top of the atmosphere SWTDN and full horizontal radiation at full ground horizontal radiation SWGDN. Consequently, we intend to maintain the features that are highly correlated with the predicted variable and to exclude the features that correlate with the maintained features. By performing this process, the accuracy of SVR increases to about 80%.

12.5 CONCLUSION

FS is a critical part of the regression and classification problems when processing large volumes of datasets with noisy, misleading, or redundant data. The use of FS with ML models can be significant in improving performance in terms of accuracy. In this chapter, three different methods have been used for

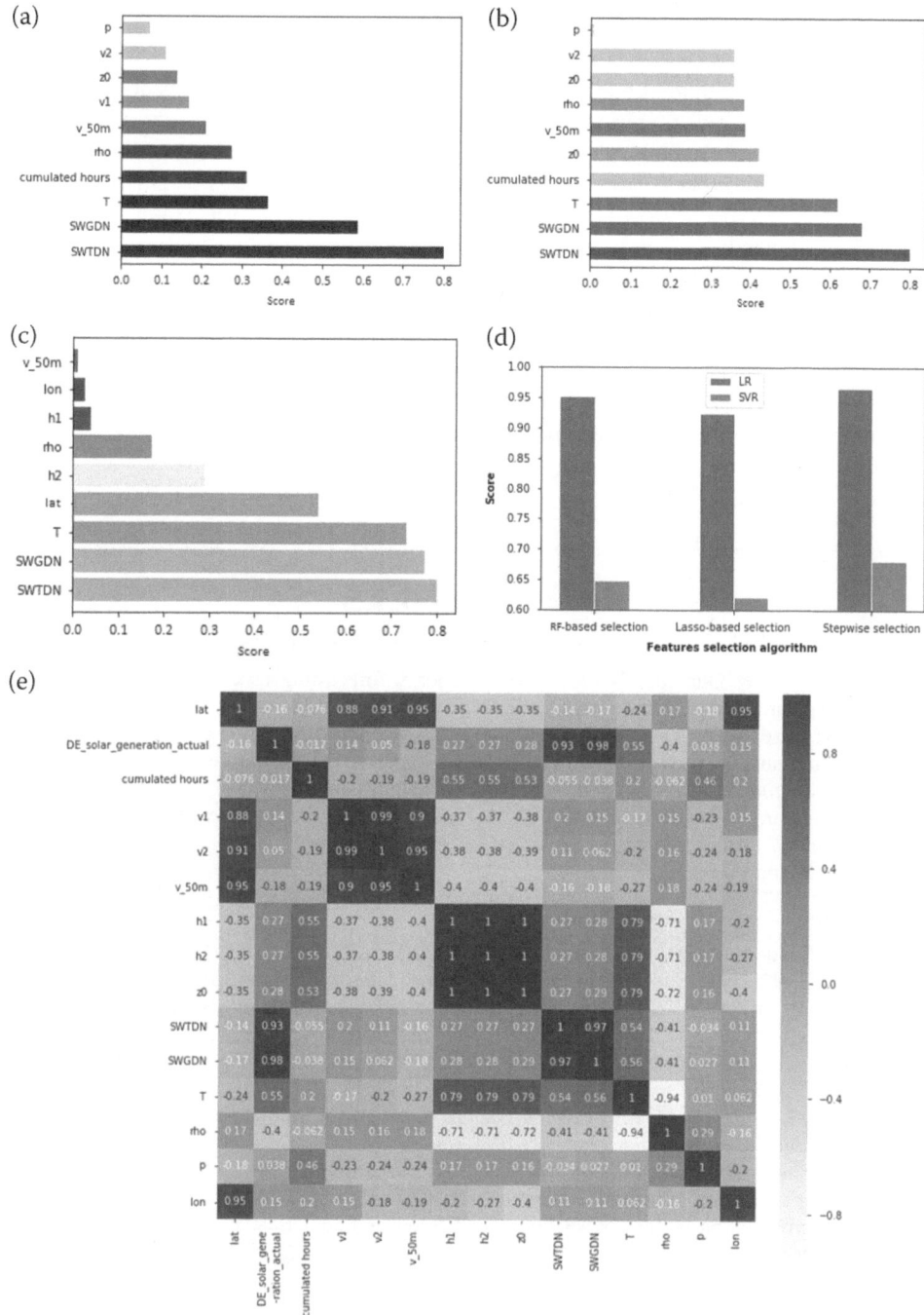

FIGURE 12.1 (a) Topmost important features selected by lasso algorithm. (b) topmost important features selected by rf algorithm. (c) topmost important features selected by rf algorithm. (d) bar chart exhibiting the r-squared score values achieved by linear regression and svr models applied the most relevant features selected. (e) correlation map representing the correlation value between the selected features.

selecting the most relevant features to predict solar energy production based on various meteorological and geographical parameters.

We have addressed methods of different categories (filter-based and embedded FS). The results obtained show that the stepwise approach, which belongs to the wrapper category, outperforms both embedded Lasso and RF algorithms. However, a further process was needed to enhance the prediction accuracy using SVR. The process consists of excluding correlated features even if they have been selected as relevant. The type of data being processed can explain this; for example, the geographical aspect features are dependent. It is then useful to keep only the one that is most correlated with the predicted variable and eliminate the others. It should be noted that regression algorithms are applied with the default learning hyper-parameters.

Eventually, this chapter's findings demonstrate FS's ability to enhance solar energy prediction quality. Moreover, the presented comparative study provides a sound background for new insights on this subject. It highlights the advantages and the limitations of the methods addressed. Thus, we fix a promising future research direction: developing and applying other ML techniques for selecting the most relevant features to improve the solar energy prediction for a smart environment.

REFERENCES

Abedinia, O, Amjady, N, & Ghadimi, N (2018). Solar energy forecasting based on hybrid neural network and improved metaheuristic algorithm. *Computational Intelligence, 34*, 241–260.

Aler, R, Ricardo, M, José, M, & Inés, M (2015). A study of Machine Learning techniques for daily solar energy forecasting using numerical weather models. In C David, B Lars, V Salvatore, & B Costin (Eds.), *Intelligent distributed computing VIII*. Springer International Publishing, pp. 269–278.

Awada, W, Khoshgoftaar, T, Dittman, D, Wald, R, & Napolitano, A (2012). A review of the stability of feature selection techniques for bioinformatics data. *2012 IEEE 13th International Conference on Information Reuse Integration (IRI)*, pp. 356–363.

Blum, AL, & Pat, L (1997). Selection of relevant features and examples in Machine Learning. *Artificial Intelligence, 97*(12), 245–271.

Breiman, L (2001). Random forests. *Machine Learning, 45*(10), 5–32.

Dash, M, & Liu, H (1997). Feature selection for classification. *Intelligent Data Analysis, 1*, 131–156.

Drotar, P, Gazda, J, & Smekal, Z (2015). An experimental comparison of feature selection methods on two-class biomedical datasets. *Computers in Biology and Medicine, 66*, 1–10.

García-Hinde, O, et al. (2016). Feature selection in solar radiation prediction using bootstrapped SVRs. *2016 IEEE Congress on Evolutionary Computation (CEC)*, pp. 3638–3645.

Georgiev, G, Valova, I, & Gueorguieva, N (2011). Feature selection for multiclass problems based on information weights. *Procedia Computer Science, 6*, 189–194.

Goswami, S, Chakraborty, S, Ghosh, S, Chakrabarti, A, & Chakraborty, B (2018). A review on application of Data Mining techniques to combat natural disasters.

Hall, MA (2000). Correlation-based feature selection for discrete and numeric class Machine Learning. Proceedings of the Seventeenth International Conference on Machine Learning. Morgan Kaufmann Publishers Inc., pp. 359–366.

Hossain, M, & Ali, A (2013). The effectiveness of feature selection method in solar power prediction. *Journal of Renewable Energy, 2013*, 1–9.

John, G, Kohavi, R, & Pfleger, K (1994). Irrelevant features and the subset selection problem. *Machine Learning: Proceedings of the Eleventh International*. Morgan Kaufmann, pp. 121–129.

Khalid, S, Khalil, T, & Nasreen, S (2014). A survey of feature selection and feature extraction techniques in machine learning. *2014 Science and Information Conference*, pp. 372–378.

Kira, K, & Rendell, LA (1992). The feature selection problem: Traditional methods and a new algorithm. *Proceedings of the Tenth National Conference on Artificial Intelligence*. AAAI Press, 129–134.

Kohavi, R, & John, G (1997). Wrappers for feature subset selection. *Artificial Intelligence*, *97*, 273–324.

Liu, H, & Yu, L (2005). Toward integrating feature selection algorithms for classification and clustering. *IEEE Transactions on Knowledge and Data Engineering*, *17*(4), 491–502.

Manoranjan, D, Kiseok, C, Peter, S, & Huan, L (2002). Feature selection for clustering - A filter solution. *Proceedings of the 2002 IEEE International Conference on Data Mining*. Washington, DC: IEEE Computer Society. 115–.

Martin, R, Aler, R, Valls, J, & Galvan, I (2016). Machine learning techniques for daily solar energy prediction and interpolation using numerical weather models. *Concurrency and Computation: Practice and Experience*, *28*, 1261–1274.

Mundry, R, & Nunn, CL (2009). Stepwise model fitting and statistical inference: Turning noise into signal pollution. *The American Naturalist*, *173*, 119–123.

O'Leary, D, & Kubby, J (2017). Feature selection and ANN solar power prediction. *Journal of Renewable Energy*, *2017*, Article ID 2437387, 7 pages. https://doi.org/10.1155/2017/2437387

Open Power System Data - A platform for open data of the European power system (2020). *Open Power System Data - A platform for open data of the European power system*.

Predicting wind and solar generation from weather data using Machine Learning (2020). *Predicting wind and solar generation from weather data using Machine Learning*.

Rangarajan, L, & Veerabhadrappa (2010). Article: Bi-level dimensionality reduction methods using feature selection and feature extraction. *International Journal of Computer Applications*, *4*(7), 33–38.

Saeys, Y, Inza, I, & Larranaga, P. (2007). A review of feature selection techniques in bioinformatics. *Bioinformatics*, *23*(8), 2507–2517.

Tibshirani, R (1996). Regression shrinkage and selection via the Lasso. *Journal of the Royal Statistical Society. Series B (Methodological)*, *58*, 267–288.

Zhang, L, & Wen, J (2019). A systematic feature selection procedure for short-term data-driven building energy forecasting model development. *Energy and Buildings*, *183*, 428–442.

Indoor Environment Assistance Navigation System Using Deep Convolutional Neural Networks

13

M. Afif[1], R. Ayachi[1], and M. Atri[2]

[1]*Laboratory of Electronics and Microelectronics (EµE), Faculty of Sciences of Monastir, University of Monastir, Monastir, Tunisia*
[2]*College of Computer Science, King Khalid University, Abha, Saudi Arabia*

Contents

13.1 INTRODUCTION

Accurate indoor objects identification in order to identify a set of indoor landmark objects is highly recommended to help blind and visually impaired people (VIP) during their daily activities. It also belongs to a wide range of artificial intelligence and smart environments. Applications for indoor objects detection and recognition and building new components with accurate knowledge of the surrounding environments is necessary to ensure the autonomy for blind and VIPs highly contributes to building new smart cities components. Object detection and classification present key components and primary

DOI: 10.1201/9781003172772-13

functionalities required to contribute to a smart cities application. The blind and the VIPs are unable to get information about their surrounding environments. According to the latest statistics of the World Health Organization (WHO) 188.5 million persons suffer from visual impairments, 217 million persons present moderate to severe visual impairments, and 36 are blind (http://www.who.int/mediacentre/factsheets/fs282/en/).

Deep convolutional neural networks (DCNN) present the best solution for solving a wide variety of tasks belonging to the computer vision area. These deep learning models have shown good performances with promising gains for many artificial intelligence tasks.

Decreasing the number of parameters and increasing computational efficiency are major factors to ensure best contribution for many applications as indoor assistance navigation for VIPs and robotic navigation. However, people with visual impairments are able to see visual information. In this chapter, we introduce a deep convolutional neural network DCNN to recognize indoor landmark objects

Deep neural networks are on the top of the majority computer vision solution for variable applications and tasks. The outstanding success proved by deep learning models. Many researchers focused on approving algorithms' performances by using DCNN models.

The quality of deep CNN models is getting higher and significantly improved by using deeper layers and networks.

Our main goal in this work is to classify and recognize landmark indoor objects as doors, signs, stairs, etc.

Wayfinding and moving around indoor environments assistance navigation can conribute for visually impaired people is a challenging problem as it depends on many factors such as object shapes, size, colour, and position. For this fact, an increasing interest is attributed to building deep CNN models to solve these issues. In this chapter, a vision system based on deep CNN models is used to recognize landmark indoor objects in real-world scenarios. This application can be implemented in embedded systems and used for mobile vision.

Our aim from the proposed work is to develop an indoor assistance navigation system based on deep convolutional neural network especially dedicated for blind and VIPs to deeply explore their surroundings. The proposed indoor object detection system contributes to present a smart environment system in which the proposed work can identify a set of indoor landmark objects highly recommended for mobility of blind and VIPs. In this work, we evaluate various types of benchmark neural networks used to develop an indoor objects classification system to facilitate life and to contribute to a smart cities application. The proposed work has been evaluated using the McIndoor dataset and the experimental results have shown good performances of the Pre-trained deep CNN models used for our applications.

The remainder of this chapter is set as follows: in Section 13.2 we provide an overview of the state-of-the-art works regarding the indoor objects detection and recognition task. In Section 13.3 we detail the different architectures used to develop the indoor object classification and recognition systems. In Section 13.4, all the experiments conducted in this chapter are presented and Section 13.5 concludes the chapter.

13.2 RELATED WORK

Various classical methods used for indoor objects detection and classification using machine learning (ML) were elaborated (Loussaief & Abdelkrim, 2016; Guerfala et al., 2016). Building appropriate applications and representations of the indoor environments still presents a challenging problem for the robotic community. For this fact, indoor mapping is very useful for localization and navigation (Booij et al., 2007), path-finding (Bhattacharya & Gavrilova, 2008).

As indoor objects detection and recognition presents a very important component in computer vision and artificial intelligence and thanks to the huge development of deep learning-based models, various works have been proposed to address the problem of object detection including pedestrian detection (Ayachi et al., 2020a; Ayachi et al., 2020b), road sign detection (Ayachi et al., 2021; Ayachi et al., 2020), indoor objects detection (Afif et al., 2020; Afif et al., 2020), and indoor objects recognition and classification (Afif et al., 2020). Several tasks associated with robot navigation depend basically on acquiring knowledge for indoor environments (Collet et al., 2009; Gupta et al., 2014). As RGB-D sensors provide not only image colour but also depth, these types of cameras were highly used in indoor robot navigation guidance (Asif et al., 2017). SLAM techniques were also elaborated for indoor navigation. Chae et al. (Jiang et al., 2016) proposed a system for mapping and recognizing indoor objects using depth sensors. In Ding et al. (2017) authors employed performances of convolutional neural networks (CNN) where they used them for an object recognition task.

Recently, artificial neural networks gained increasing attention and were used in the computer vision field. DL methods were employed for recognition tasks (Zhang et al., 2016), detection tasks (Bianco et al., 2016), and for segmentation (Gupta et al., 2014).

Ding et al. (2018) take advantage of deep convolutional neural networks in order to develop an application for indoor object recognition. To improve detection, the authors propose a prior-knowledge stage based on deep learning methods.

Indoor object recognition presents a key component for much of an autonomous indoor navigation assistance. Ding et al. (2017) proposed a pipeline used for indoor object recognition based on deep CNN models. Georgakis et al. (2017) proposed a new method to explore the ability of using synthetic images to train and test state-of-the-art models. This work is used especially for object instance detection.

13.3 RECOGNIZING INDOOR OBJECTS: APPROACH ADOPTED BASED ON DEEP CNN

Among the five senses, vision is the most important. Human beings present a great ability to recognize and identify the surroundings.

During the last few years, deep convolutional neural networks have achieved great performances to achieve good detection and recognition accuracies in a huge set of applications. In order to address the problem of blind and VIPs' mobility, we employed the use of different CNN models in order to build new indoor objects recognition systems to contribute to new smart environments applications.

13.3.1 Indoor Objects Recognition Using EfficientNet

As a first step, we choose to take advantage of EfficientNet (Tan & Le, 2020) neural network. This deep learning model presents a new technique of compound scaling. It consists of scaling the model among three network dimensions: width, depth, and network resolution. This compound scaling contributes to better neural networks performances.

The compound scaling consists of, at the first time to perform, a grid search to find the best scaling compound under fixed resources constraints. EfficientNet presents a very powerful way to obtain better performance improvements of the classification of the neural networks. EfficientNet scales up the network among multiple dimensions (depth, width, and resolution). After various experiments and grid searches, compound scaling is under the following coefficients:

Depth = 1.20

Width = 1.10

Resolution = 1.15

Figure 13.1 presents an idea about the complex compound scaling performed by EfficientNet architecture.

Deep convolutional neural networks are on the top of artificial intelligence solutions for a wide range of applications. These deep learning models are crucial and mainstream to yielding good results compared to the state-of-the-art models.

In this chapter, we explore deep CNN models in order to build systems for indoor objects recognition. Indoor navigation and wayfinding still present a very challenging compute vision problem. It presents a key component for autonomous robotic navigation systems and for VIPs' indoor navigation. All our experiments were performed on real-world images. Figure 13.2 presents the details of EfficientNet architecture.

We deployed various Pre-trained deep CNN models on our work to perform a system which aims to indoor assistance navigation. For our indoor images classification task, we used transfer learning technique with uses of a Pre-trained model trained for one task and we fine tune it on our data. EfficientNet architecture introduces the use of the following powerful block named mobile bottleneck convolution (MBConv) (Figure 13.3).

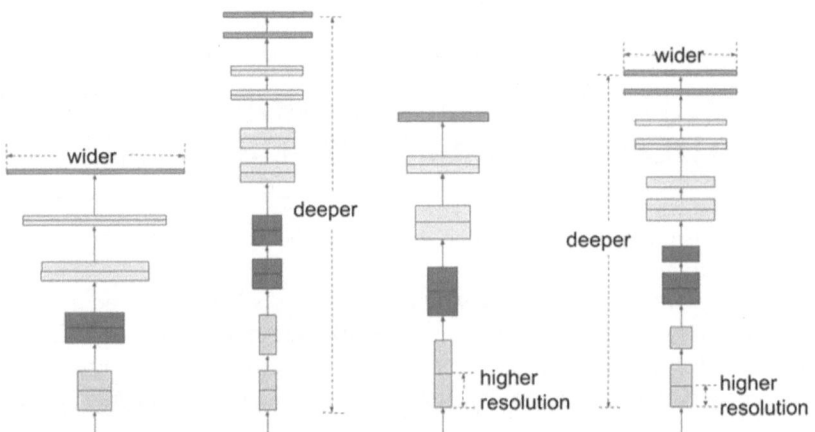

FIGURE 13.1   Architecture of Compound Scaling Used in EfficientNet.

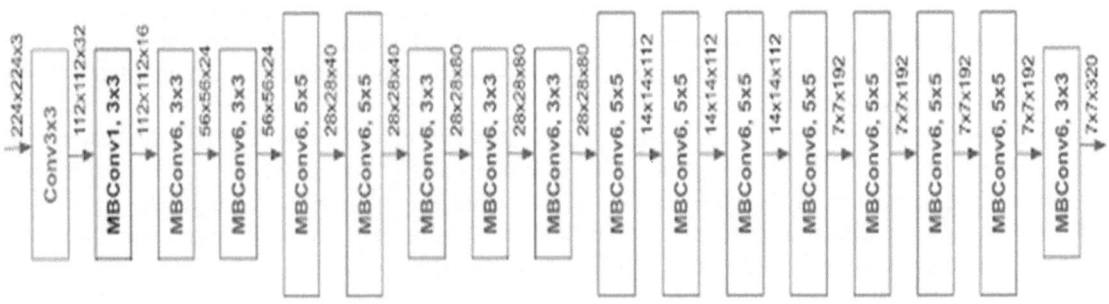

FIGURE 13.2   EfficientNet Architecture.

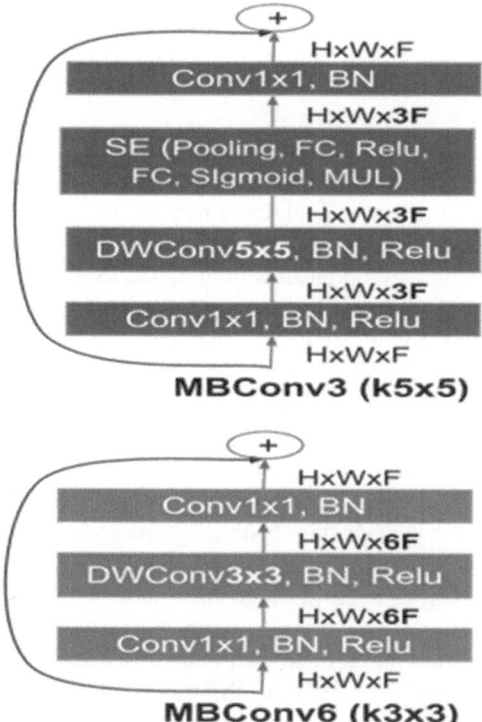

FIGURE 13.3 MB Conv Block Details.

It is very important to build robust and efficient indoor objects classification systems. We aim from this part of the work to help blind and VIPs in their daily mobility to more explore their surrounding environments. The proposed indoor object recognition system consists of two main parts:

1. Train: it aims to extract a set of rules in order to contribute to predict class objects. Training is performed on the training subset.
2. Test: it aims to perform final predictions about the objects' class names.

In Figure 13.4, we present the proposed pipeline used for indoor objects classification adopted during our experiments.

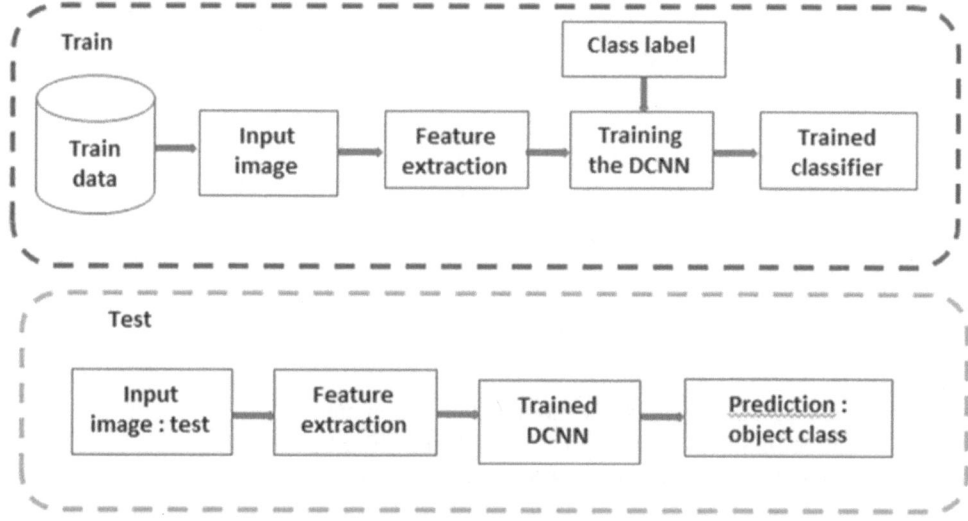

FIGURE 13.4 Proposed Pipeline Used in Our Work.

Generally, a deep CNN model consists of two main parts:

- Features extraction background based on convolutional layers
- Classification part consisting of fully connected and softmax layers

The Pre-trained model extracts general features from input data and classifies them based on those extracted from the second part. Transfer learning technique is the fact of building a new model working on your dataset based on a Pre-trained model. We then reuse the feature extraction part and just retrain the classification part on our dataset. Using the transfer learning method ensures training the model with less computational resources and faster training time.

Among all skills and senses used to fully interact with real-world environments, perception is the most important. To ensure a best and autonomous indoor navigation for VIPs, we have to prepare a vision system providing a set of capabilities allowing persons with visual impairments to move and to interact with the real-world scenery.

Training DCNN-based object recognition requires a large number of annotated data due to the large number of parameters to be learned during the training process. To ensure a best recognition of indoor object is learned we have to train the deep CNN on a challenging annotated data. Including challenging situations such as lighting condition variations, viewpoint variation, and intra-class and inter-class variation.

The best approach to perform a good system with best performances to recognize indoor objects is exploring deep convolutional neural networks.

Training such category of applications requires a large amount of annotated data which is highly time- and resource-consuming. We demonstrate the effectiveness of multiple Pre-trained deep CNN models to build a system dealing with landmark indoor object detection.

To perform our vision system for indoor object assistance navigation we resort to three preferred deep learning models.

13.3.2 Indoor Objects Recognition Using Inception Family

In our implementations we used the three versions of inception classifier, but the more robust with the best performances of classification was Inception v3 (Szegedy et al., 2015). It presents 42 layers; this inception version provides more computational efficiency with fewer parameters than its previous versions. This inception version makes lower error rate for image classification in ILSVRC (http://image-net.org/challenges/LSVRC/). Inception v3 is the enhanced inception architecture. It presents less number of parameters with more computational efficiencies. By reducing the number of parameters required for the deep CNN model we ensure that the network can go deeper. Inception v3 architecture presents a very promising inception category with different types of parameter optimization. In this pre-trained model a 5×5 convolution will be replaced by two 3×3 convolutions. When using a convolution layer with a convolution filter of 5×5, the number of parameters is set to $5 \times 5 = 25$, while by using $2 * 3 \times 3$ filter we have $3 \times 3 + 3 \times 3 = 18$. This technique reduces considerably the model's complexity. This convolution operation reduces complexity and contributes for on block composing inception v3 architecture named 'inception module A'. Figure 13.5 presents the inception module A architecture.

One convolution of 3×3 is replaced by two convolutions of 1×3. This modification participates in the 'inception module B'. In Figure 13.6 the inception module B is presented.

Another module named 'inception module C' is proposed in inception v3 model. All these modules contributes for reducing the number of parameters of the whole network and minimize the risk of overfitting.

Another block is introduced with some modifications compared to those introduced in inception V1. One auxiliary classifier on the top of the last 17×17 layers. The batch normalization is also used in the auxiliary classifier. The feature map downsampling is done using max pooling. This technique is either

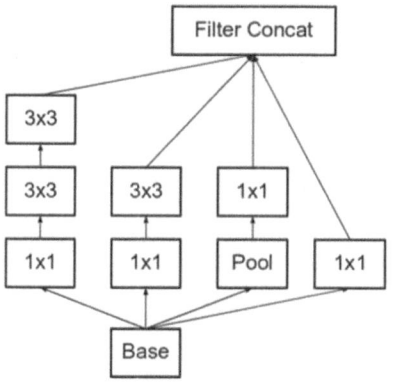

FIGURE 13.5 Inception Module A Architecture.

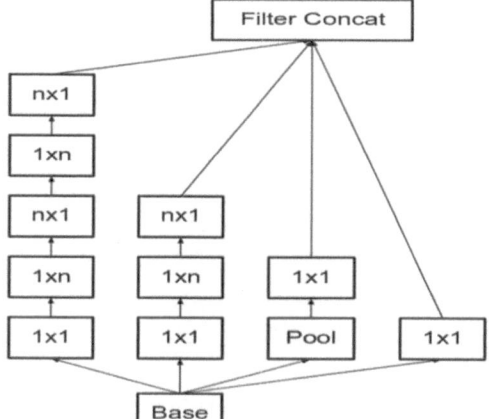

FIGURE 13.6 Inception Module B Architecture.

too greedy if max pooling is followed by convolution or too expensive if the convolution is followed by max pooling. For this fact a grid sized reduction is proposed in the inception v3 architecture (Figures 13.7 and 13.8).

13.3.3 Indoor Objects Recognition Using ResNet

ResNet (He et al., 2016) model was trained on the ImageNet (http://image-net.org/challenges/LSVRC/) dataset using 152 layers ($\approx 8 \times$ deeper than VGG (Simonyan & Zisserman, 2014), and still having less complexity).

When training a deep CNN, the degradation problem appear which indicates that not all the network can be similar to optimize. To address the degradation problem, ResNet architecture introduces 'the deep residual learning framework'. In Figure 13.9, we present the residual block architecture.

The underlying mapping is noted as H(x). Stacked nonlinear layers fitting on another mapping are (Figure 13.10):

$$F(x) = H(x)–x \tag{13.1}$$

The original mapping is:

$$F(x) + x \tag{13.2}$$

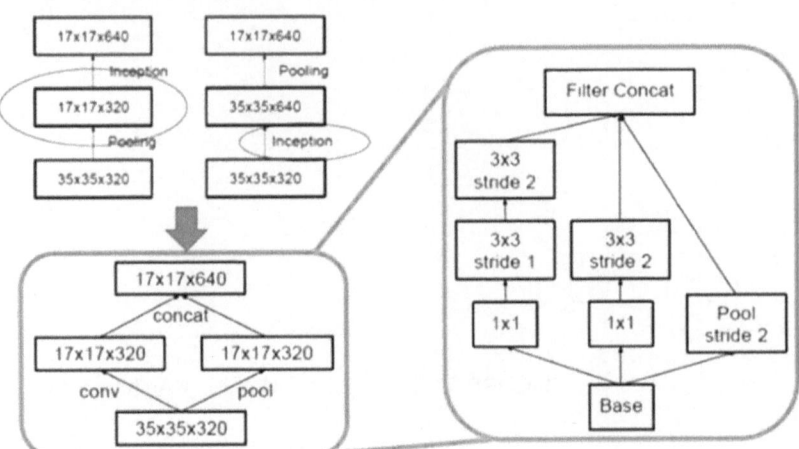

FIGURE 13.7   Grid Size Reduction.

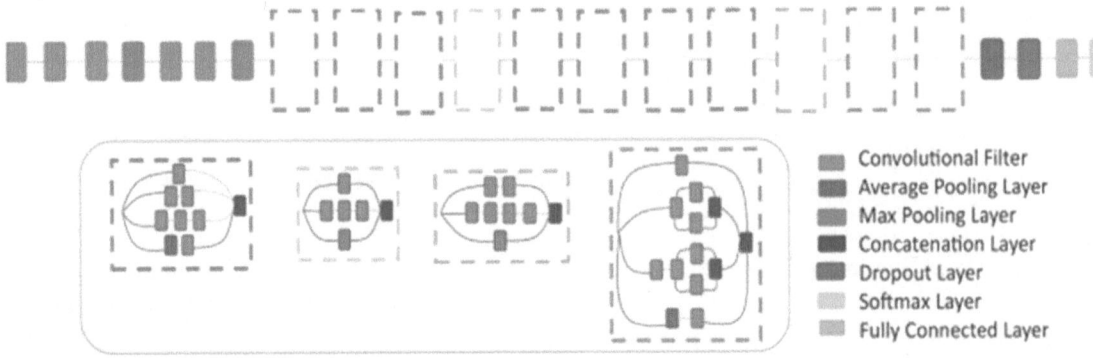

FIGURE 13.8   Inception V3 Architecture.

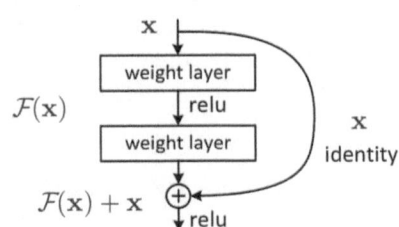

FIGURE 13.9   Residual Block Framework.

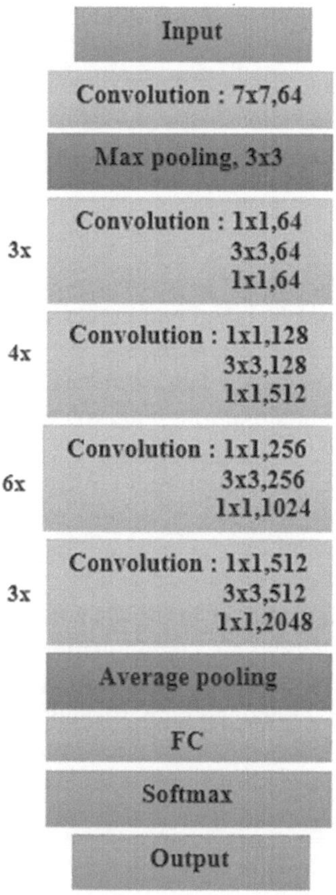

FIGURE 13.10   ResNet 50 Architecture.

In the residual block, the shortcut connections are used to perform identity mapping. All the outputs of the shortcut connections are added to those of the stacked layers. The network will be trained end to end and can be implemented using various deep learning libraries and frameworks.

13.4 EXPERIMENTS AND RESULTS

The proposed method has been evaluated using the EfficientNet (B0, B1, B2) family, the inception model family (inception v1, v2, and v3), and ResNet (34 layers and 50 layers). Experiments were performed on MCIndoor 20000 indoor objects dataset which provides 20000 images divided into natural and processed images. This dataset provides three indoor landmark objects (doors, stairs, and signs).

Objects provided in this dataset are isolated from their surrounding environments which make them very suitable for classification and recognition tasks. Images presented undergo processing as applying salt and pepper filter, Gaussian noise, rotation, Poisson function, and so on. To perform implementation of these deep CNN models, we used the TensorFlow framework (https://www.tensorflow.org), NVIDIA libraries and tools (https://developer.nvidia.com/digits), and Python. During the training process, we used the SGDW optimizer. We used this method as it presents the best error rate minimization with the minimum vibration. Pre-trained models were trained using the transfer learning technique. This technology allows knowledge transfer acquired for one task to be used to treat better a second task. Generally, transfer learning is in other words taking a model trained on large datasets to transfer its knowledge to a smaller dataset. In our work, we freeze the earlier convolution layers of the network and we only train the fully connected layers to make predictions on our task.

Table 13.1 reports all the implementation details elaborated when training and testing the pre-trained models.

During the proposed experiments, the dataset has been divided into 3 main parts: train, validation, and test. We used the validation subset in order to evaluate the ability of the network used on learning the main features from the dataset MCIndoor 20000. In order to evaluate the efficiency and the robustness of the proposed work, we used the two versions of the MCIndoor 20000 dataset (original and processed). We note that the original version of MCIndoor 20000 dataset contains 1700 original images and the rest are processed images. The second part of the dataset has been obtained by applying data augmentation technique to the original dataset. The main techniques used for data augmentation are flipping, rotation, salt, and pepper filter, and so on.

Table 13.2 reports all accuracies obtained using EfficientNet B0, B1, and B2, inception V1, V2, and V3, and ResNet 34 and 50. We compare our results achieved with those presented in Bashiri et al. (2018). The proposed work achieves better results than Bashiri et al.'s method for the original images and more efficient results for the processed images when using the proposed experiments.

Based on the obtained results in Table 13.2, we note that our work achieves better results than those obtained in Bashiri et al. (2018); the obtained results are almost higher than 93% as a recognition rate for all the networks evaluated. We also note that based on the achieved results, the better result has been achieved using EfficientNet family (B0, B1, and B2) in our case. We almost obtained high recognition

TABLE 13.1   Experiment settings

Training steps	8000
Learning rate	0.01
Testing percentage	30%
Validation percentage	20%
Train batch size	100
Test batch size	1
Validation batch size	100

TABLE 13.2   Indoor object classification accuracies obtained

MODEL NAME	MCINDOOR 20000 (ORIGINAL IMAGES)	MCINDOOR 20000 (PROCESSED IMAGES)
EfficientNet B0	98.6	99.2
EfficientNet B1	98.6	99.6
EfficientNet B2	98.7	99.9
Inception v1	93.9	96.3
Inception v2	94.7	96.9
Inception v3	95.7	97.1
ResNet 34	98.7	99.3
ResNet 50	98.8	99.9
Bashiri et al. (2018)	90.2	99.8

accuracy of 99.2%, 99.6%, and 99.9% for EfficientNet B0, B1, and B2, respectively. The obtained results are more competitive than those obtained in the state-of-the-art works.

13.5 CONCLUSION

We have employed various state-of-the-art classifications as EfficientNet, Inception family and ResNet 34 and 50 layers models and we have increased its performances by modifying the optimizer using stochastic Gradient descent with warm SGDW. We take advantage of deep learning techniques that were proven in order to highly contribute to our recognition task. Results achieved are very encouraging and outperform the state-of-the-art. We trained and tested the Pre-trained models on MCIndoor 20000 dataset. The proposed indoor object recognition model can be used for mobile applications, robot navigation, and for VIP's assistance navigation. We also note that the proposed indoor objects recognition system can highly contribute to a smart environment.

Potential future work will be dedicated to implementing a deep CNN model used for real-time indoor object detection on embedded systems.

REFERENCES

Afif, M, Ayachi, R, Pissaloux, E, et al. (2020). Indoor objects detection and recognition for an ICT mobility assistance of visually impaired people. *Multimedia Tools and Applications*, *79*(41), 31645–31662.

Afif, M, Ayachi, R, Said, Y, et al. (2020). An evaluation of retinanet on indoor object detection for blind and visually impaired persons assistance navigation. *Neural Processing Letters*, 51, 2265–2279.

Afif, M, Said, Y, Pissaloux, E, et al. (2020). Recognizing signs and doors for indoor way finding for blind and visually impaired persons. In *2020 5th International Conference on Advanced Technologies for Signal and Image Processing (ATSIP)*. IEEE, pp. 1–4.

Asif, U, Bennamoun, M, & Sohel, F (2017). RGB-D object recognition and grasp detection using hierarchical cascaded forests. *IEEE Transactions on Robotics*, *33*(3), 547–564.

Ayachi, R, Afif, M, Said, Y, et al. (2020a). Pedestrian detection for advanced driving assisting system: A transfer learning approach. In *2020 5th International Conference on Advanced Technologies for Signal and Image Processing (ATSIP)*. IEEE, pp. 1–5.

Ayachi, R, Afif, M, Said, Y, et al. (2020b). Traffic signs detection for real-world application of an advanced driving assisting system using deep learning. *Neural Processing Letters*, *51*(1), 837–851.

Ayachi, R, Afif, M, Said, Y, et al. (2021). Real-time implementation of traffic signs detection and identification application on graphics processing units. *International Journal of Pattern Recognition and Artificial Intelligence*, *35*(7), 2150024.

Ayachi, R, Said, Y, & Abdelaali, AB (2020). Pedestrian detection based on light-weighted separable convolution for advanced driver assistance systems. *Neural Processing Letters*, *52*(3), 2655–2668.

Bashiri, F, LaRose, E, Peissig, P, & Tafti, A (2018). MCIndoor20000: A fully-labeled image dataset to advance indoor objects detection. *Data in Brief*, *17*, 71–75.

Bhattacharya, P, & Gavrilova, M (2008). Roadmap-based path planning - Using the voronoi diagram for a clearance-based shortest path. *IEEE Robotics and Automation Magazine*, *15*(2), 58–66.

Bianco, S, Celona, L, & Schettini, R (2016). Robust smile detection using convolutional neural networks. *Journal of Electronic Imaging*, *25*(6), 063002.

Booij, O, Terwijn, B, Zivkovic, Z, & Kröse, B (2007). Navigation using an appearance based topological map. In *International Conference on Robotics and Automation*. IEEE, pp. 3927–3932.

Collet, A, Berenson, D, Srinivasa, S, & Ferguson, D (2009). Object recognition and full pose registration from a single image for robotic manipulation. *IEEE International Conference on Robotics and Automation*, Kobe, Japan, 12–17 May 2009, pp. 48–55.

Ding, X, Luo, Y, Li, Q, Cheng, Y, Cai, G, Munnoch, R,... Wang, B (2018). Prior knowledge-based deep learning method for indoor object recognition and application. *Systems Science & Control Engineering*, 6(1), 249–257.

Ding, X, Luo, Y, Yu, Q, Li, Q, Cheng, Y, Munnoch, R, ... Cai, G (2017, September). Indoor object recognition using pre-trained convolutional neural network. In *2017 23rd International Conference on Automation and Computing (ICAC)*, Huddersfield, UK, IEEE, pp. 1–6.

Georgakis, G, Mousavian, A, Berg, A, & Kosecka, J (2017). Synthesizing training data for object detection in indoor scenes. arXiv preprint arXiv:1702.07836.

Guerfala, M, Sifaoui, A, & Abdelkrim, A (2016). Data classification using logarithmic spiral method based on RBF classifiers. In *7th International Conference on Sciences of Electronics, Technologies of Information and Telecommunications (SETIT)*. IEEE, pp. 416–421.

Gupta, S, Girshick, R, Arbeláez, P, & Malik, J (2014). Learning rich features from RGB-D images for object detection and segmentation. In *Computer vision - ECCV 2014* (8695, pp. 345–360). Springer.

He, K, Zhang, X, Ren, S, & Sun, J (2016). Deep residual learning for image recognition. In *Proceedings of the IEEE Conference on Computer Vision and Pattern Recognition* (pp. 770–778). http://image-net.org/challenges/LSVRC/

Jiang, L, Koch, A, & Zell, A (2016). Object recognition and tracking for indoor robots using an RGB-D sensor. In Intelligent autonomous systems 13. *Advances in intelligent systems and computing* (302, pp. 859–871). Springer.

Loussaief, S, & Abdelkrim, A (2016). Machine learning framework for image classification. In *7th International Conference on Sciences of Electronics, Technologies of Information and Telecommunications (SETIT)* . IEEE, pp. 58–61.

Nvidia digits system for deep learning model implementation on Nvidia GPU. https://developer.nvidia.com/digits. Last accessed 2018/9/21.

Simonyan, K, & Zisserman, A (2014). Very deep convolutional networks for large-scale image recognition. arXiv preprint arXiv:1409.1556.

Szegedy, C, Liu, W, Jia, Y, Sermanet, P, Reed, S, Anguelov, D,... Rabinovich, A (2015). Going deeper with convolutions. In *Proceedings of the IEEE Conference on Computer Vision and Pattern Recognition* (pp. 1–9).

Tan, M, & Le, Q (2020). Efficientnet: Rethinking model scaling for convolutional neural networks. arXiv 2019. arXiv preprint arXiv:1905.11946.

Tensorflow a deep learning framework. https://www.tensorflow.org. Last accessed 2018/9/21.

WHO: Vision impairment and blindness. http://www.who.int/mediacentre/factsheets/fs282/en/. Last accessed 2021/1/8.

Zhang, F, Duarte, F, Ma, R, Milioris, D, Lin, H, & Ratti, C (2016). Indoor space recognition using deep convolutional neural network: A case study at MIT campus. arXiv:1610.02414.

Pixel-Based Classification of Land Use/Land Cover Built-Up and Non-Built-Up Areas Using Google Earth Engine in an Urban Region (Delhi, India)

14

A. Kumar[1], A. Jain[1], B. Agarwal[1], M. Jain[1], P. Harjule[2], and R.A. Verma[1]

[1]Indian Institute of Information Technology, Kota, India
[2]Department to Mathematics, MNIT Jaipur, Jaipur, India

Contents

DOI: 10.1201/9781003172772-14

14.1 INTRODUCTION

14.1.1 Background

For many years, the world has experienced a rapid rise in the Urbanization of areas with changes in LU/LC cover patterns. Urbanization takes place when cities assimilate the near-by rural regions, mostly in terms of sprawl diffusions appearing out from the middle of the city or directly across major corridors of transportation (Baum-Snow, 2007; Sudhira et al., 2004).

But these human settlements have proved to be very uneven and undistributed urbanization. Between 1950 and 2014, the urban area share in the population count increased from 30% to 54% and by 2050, this may increase to an enormous amount of 2500 million in Afro-Asia (Goldblatt et al., 2016). In order to achieve sustainable development, it is very necessary to look towards the physical, geographic, social, economic, and cultural impact of this urbanization on the human settlements.

Recent times have witnessed rapid development in rural and urban areas of India as well. This is mainly due to the Industrial Revolution and the migration to urban areas in the hope of access to better facilities and income. Due to an uneven distribution of population, this urbanization has shown un-distributed patterns which can lead to misinformation in the Population data and census and other geographical information covering a particular area. With the availability of satellite data and powerful computational platforms like GEE, it has actually been possible to analyse various regions and study their change in urban sprawl. Although many studies in the past have used GEE for the identification of urban areas and LU/LC urban analysis, very few research has been performed over India.

14.1.2 Methodology

To explore the power of GEE for analysis over the urban regions of India, in this study, we perform the pixel-based classification of areas of Delhi into Built-Up and Non Built-Up respectively. First of all, the dataset mentioned in Goldblatt et al. (2016) was acquired, and points over the study area of Delhi were obtained and preprocessed using the Landsat 7 Top of Atmosphere (TOA) percentile composite satellite data of the year 2014, with no negative sun elevation. Additionally, NDVI and NDBI bands were added as features to facilitate the task of classification. We use three most prominent classifiers available in GEE, CART, Linear SVM, and Random Forest for the classification task, with 10-fold cross validation for assessment.

14.1.3 Chapter Organization

The rest of the chapter is organized as follows: Section 14.2 presents existing work done by researchers for analysis of remote sensing data using Machine Learning models and Google Earth Engine (GEE). Section 14.3 discusses the aim of this chapter, while Section 14.4 explains Google Earth Engine (GEE), various Machine Learning models and Cross-Validation in detail. Section 14.5 describes the proposed method and Section 14.6 describes the various steps of the experiment. Section 14.7 presents with the results obtained. Section 14.8 discusses the conclusion of the chapter, and Section 14.9 puts lights on the future scope in this field.

14.2 RELATED WORK

Previously, the literature measured the extent of the urbanized areas by door-to-door household surveys and analysing its data, nightlights and some mobile phone records. But with the advancements in satellite images and band availability, urban research has moved digitally towards the development of the Remote Sensing Data and eventually using them for the classification purposes. The conventional techniques of obtaining demographic data and exploration of environmental specimen are not sufficient enough for multi-complex ecological analysis (Mallupattu & Sreenivasula Reddy, 2013). On the other hand, the use of Remote Sensing Satellite data along with Geographic Information Systems (GIS) has expanded as one of the essential technologies that can be used to obtain precise and suitable information on various LU/LC patterns over large areas.

With the rapid developments in the analysis of highly-accurate Machine Learning models, increasing availability of Satellite Images has enabled the extraction of features of various LU/LC classes using highly accurate pixel-based classification models.

Hua et al. (2017) use many ML techniques including Decision Tree Classification (DTC), CART analysis, SVM etc. and found DTC to be the best method out of these. Jiang et al. (2010) also proposed classification methods based on DTC for Remote Sensing Images. Ha et al. (2020) use the Random Forest Classification model using Landsat satellite data to monitor the LU/LC variation and Urbanization of Rural areas in Vietnam.

Sidhu et al. (2018) use Google Earth Engine (GEE) to evaluate its utility for performance over vector and raster information on GlobCover, Landsat and MODIS imagery. Shetty (2019) conducted LU/LC classification using various Machine Learning Classifiers available in GEE over the region of Dehradun, India and consequently analysed their performances.

Goldblatt et al. (2016) developed a dataset consisting of about 21,000 polygons, classified manually into "Built-Up" and "Non Built-Up" classes, and which can be used for performing pixel-based classification over different areas of India in GEE using Landsat 7 and Landsat 8 satellite data.

Kumar and Mutanga (2018) use Google Earth Engine (GEE) to evaluate its utility and application like vegetation mapping and monitoring, land cover mapping, disaster management and Earth science, agricultural application, etc. Tamiminia et al. (2020) used Google Earth Engine for geo-big data applications. Tobón-Marín and Cañón Barriga (2020) analyse the change in river plan forms using Google Earth Engine.

14.3 PROBLEM STATEMENT

Although, more and more Remote Sensing Satellite data is now being available, researchers often face difficulty in analysing these data with the help of Machine Learning models, because of the requirement of high-computation processing units. But, with the availability of platforms like the Google Earth Engine (GEE), it has become quite easy to perform geo-spatial analysis, over large areas. Therefore, in this chapter, we use the power of GEE to perform a study of the Built-Up patterns of Delhi, India.

The aim of this experiment is to classify the different regions of Delhi, India, into Built-Up and Non Built-Up areas, using Machine Learning models on Landsat 7 TOA composite data of the year 2014 in Google Earth Engine.

14.4 METHODOLOGY

Under this section, we lay an outline of the software and classification models used in this study.

14.4.1 Google Earth Engine (GEE)

Google Earth Engine (GEE) is a data processing platform based on Google Cloud Platform (GCP) (Goldblatt et al., 2018) and Geographic Information System (GIS) tool used to display maps and synthesize the geographical data and to run geospatial analysis on Google's infrastructure. It is a huge platform with petabytes of satellite images and geospatial data. The data and images on GEE are the contributions of various satellites. Thus, it has a timeline of satellite imagery. As the complexity in the technology is increasing with time, Google Earth Engine provides numerous kinds of ready-to-use datasets within a timeline of thirty years of historical images, which are updated and expanded on a daily basis. For example, thermal satellite sensors provide information about emissivity and surface temperature for both land and sea from satellites like MODIS, ASTER, and AVHRR. There are many more other datasets that provide information about Climate, Weather, and Atmosphere. Various other satellites also cover the geophysical information for Terrain, Land Cover, Cropland, and other geophysical data. GEE has an enormous humanitarian effect when it comes to carrying out various scientific analyses using it.

Various case studies carried out by different international institutes/organizations are present on GEE for further Research and Development practices. These include remote sensing research, predicting disease outbreaks, natural resource management, and many more. For example, Global Forest Watch provides a real-time system for online monitoring of forests worldwide, for their conservation. It is a project developed by the World Resource Institute.

GEE comes along with an Integrated Development Environment (IDE) which is available online. The Code Editor features have been designed in such a way that it enables carrying out various geospatial studies using JavaScript and develop maps using Google Cloud easily. The Earth Engine API, available for use as a Python module, also comes along with the power of Google Cloud Platform (GCP).

14.4.2 Classification Models

The task of pixel-based classification is performed using three different Machine Learning models available in GEE.

14.4.2.1 Classification and Regression Tree (CART) or Decision Tree

A Decision Tree is a supervised machine learning model that differentiates the data into different classes by making simple decisions based on the weights or impact of the particular action. It starts with only one node based on the condition; it divides further into branches which add other nodes that in turn give it a tree-like structure. The process is repeated till the division adds value to the output. It also works well with high dimensional data. The leaf nodes in the decision tree represent the different classes. For the selection of the attributes in the decision tree, various techniques are used, such as Gini Index, Information Entropy among others, based on which it is decided that the node should be further divided or not. It allows an individual to weigh different possibilities of actions against one another depending on their consecutive benefits, costs and probabilities.

14.4.2.2 Support Vector Machine (SVM)

A support vector machine (SVM) is a supervised machine learning model that works on the concept of an N-dimensional space containing hyperplane, which does the classification of the data points distinctly. Hyperplane having the maximum distance (maximum margin) between different data points of both classes is chosen to separate the data points into different classes. Maximum distance assures that the next data points will be classified more accurately. Here, the decision factor/boundaries are the hyperplanes which help in classification of the data points. Data points on different sides of hyperplanes fall under different classes. Number of features is directly proportional to the hyperplane dimension, which can be formulated as:

$$\text{Hyperplane Dimension} = \text{Number of Features} - 1 \qquad (14.1)$$

Support vectors are those data points that fall closer to the hyperplane and eventually affect the orientation of the hyperplane, along with its position. These are used for the maximization of the classifier margin. Changes like deletion of the support vectors affect hyperplane position. SVM works really well when there is a clear margin between the classes and is also effective in higher dimensional data. It trains on the labelled data and based on it a hyperplane is drawn with the maximum margin. Based on the data point position with respect to the hyperplane, it is categorized into various classes. One of the drawbacks of SVM is that when the data is large, training time required is higher. Also, it does not perform well when the classes are overlapping.

14.4.2.3 Random Forest

Random Forest is a supervised machine learning model which is an extension of the decision tree model but in place of a single tree, a large number of trees are created to achieve higher accuracy. In this model, a forest of trees is created where for each tree features are selected randomly and the output is calculated by finding out the average or mean of all the trees. The Decision Trees are sensitive on the training data, and as such that a small change may result in a significant change in the data. Random Forest uses this property of decision trees and uses the technique of bagging, in which features are selected randomly which results in different trees each time. Random Forest works efficiently as each tree works individually and the random selection of features helps in avoiding overfitting and provides more accurate results. Also the error of an individual tree does not affect the output, as the final result is the average of the output of all the trees. It is observed that in case of increasing the number of trees in the Random Forest, the accuracy may increase upto a certain level and only computation cost is increased.

14.4.3 Cross-Validation

The probability of a classifier to correctly classify a random set of samples denotes its accuracy (Kohavi, 1995). To ensure that the classifier performs fairly, the data is split up into train and test sets, respectively. One such technique, Cross-Validation divides the dataset into k subsets (ideally 5 or 10), in order to have least biased estimation (Rodriguez et al., 2009) for k trials. This method makes sure that each subset of data is incorporated once in the test set and is not used for training in that trial. The overall performance is calculated as an average performance of these k trials.

In this experiment, we have used 10-fold Cross-Validation by partitioning the data into 10 random stratified folds. In each of the 10 trials, 9 folds are used for training and the remaining fold is used for testing. The aggregate performance is then measured as an average of all the 10 trials.

14.5 PROPOSED METHOD

In this section, we present the proposed method of the experiment. The steps of the proposed method are as shown in Figure 14.1. First, the dataset mentioned in Goldblatt et al. (2016) was obtained and was loaded in the Google Earth Engine (GEE). The dataset was filtered to include the data points over the study area. Also, the Landsat 7 Satellite data of the year 2014, with the least cloud cover and no negative sun elevation was loaded in GEE. This was used to create the Top of Atmosphere (TOA) composite image, which was then used to preprocess the data points from the dataset for extracting the per-pixel band values of Landsat 7 as features. Additionally, two bands were added to facilitate the classification task: Normalized Difference Vegetation Index (NDVI) and Normalized Difference Built-Up Index (NDBI), whose values were calculated using the existing bands of Landsat 7 for each data point. Each data point also included an output binary class, representing "Built-Up" or "Non Built-Up" respectively.

Once the data points were preprocessed and features were obtained, this set of points was given as input to the three Machine Learning classification models (available in GEE) namely, Classification and Regression Tree (CART) or Decision Tree, Linear SVM, and Random Forest. The assessment of the models was done using the technique of Cross Validation with 10 folds. The data was given to Random Forest consisting of 10, 50, and 100 Trees, respectively, and the out of these the results of the best performing model were then compared with the results of that of CART (Decision Tree) and Linear SVM.

14.6 EXPERIMENT

In this section, we give the details of how the experiment was performed and the data used to perform it.

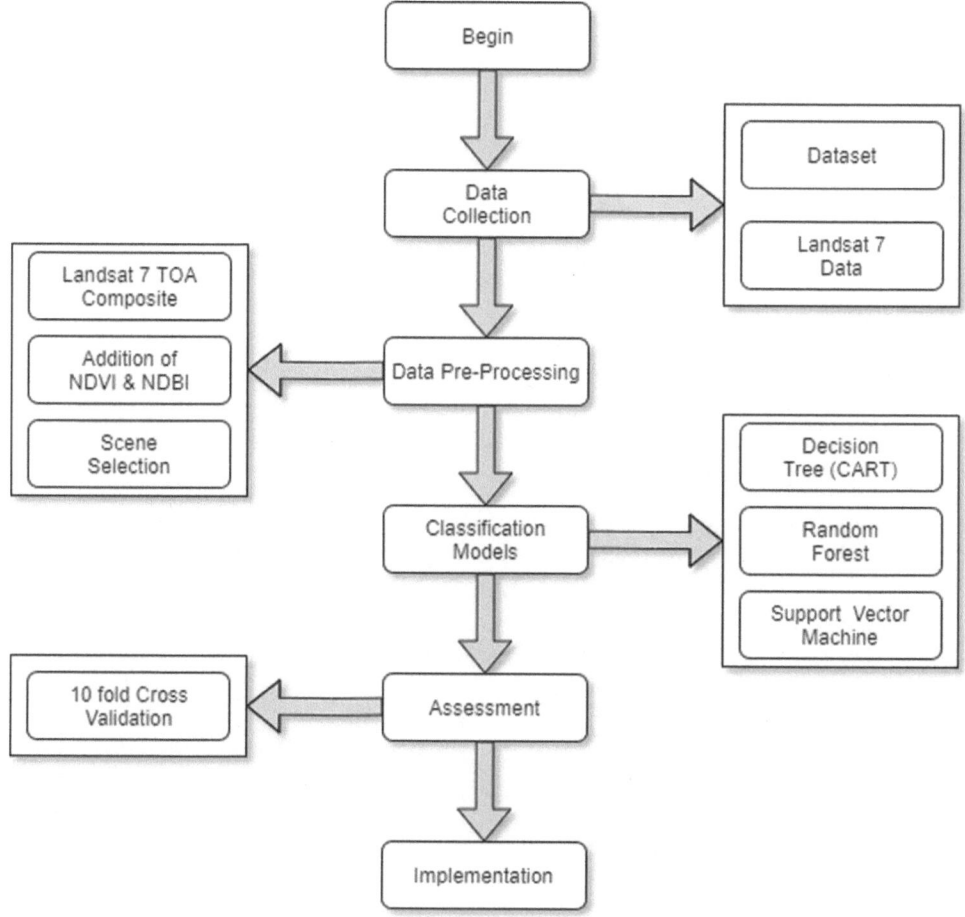

FIGURE 14.1 Experimental Workflow.

14.6.1 Study Area

National Capital Territory (NCT) of Delhi is one of the largest metropolitan areas, which consists of New Delhi, the capital city of India. It is located between 28°-24'-17" N, 76°-50'-24" E and 28°-53'-00" N, 77°-20'-37" E, consisting of an area of about 1484 sq. km. (573 sq. mi.). According to the 2011 census, the total population of NCT of Delhi was about 16.8 million. It is the world's second-largest urban area, as per the United Nations. With the second-highest GDP per capita in India, the NCT of Delhi ranks fifth in the Human Development Index (HDI) among all the states and Union Territories (UTs) of India. It is surrounded by the satellite cities of Gurugram, NOIDA, Ghaziabad, Faridabad among others, which has led to an increase in urban migration to the area in the last few years. By 2021, it is expected that the total population of the NCT of Delhi will be 19.5 million. As of 2012, there are 11 districts in the NCT of Delhi as shown in Figure 14.2.

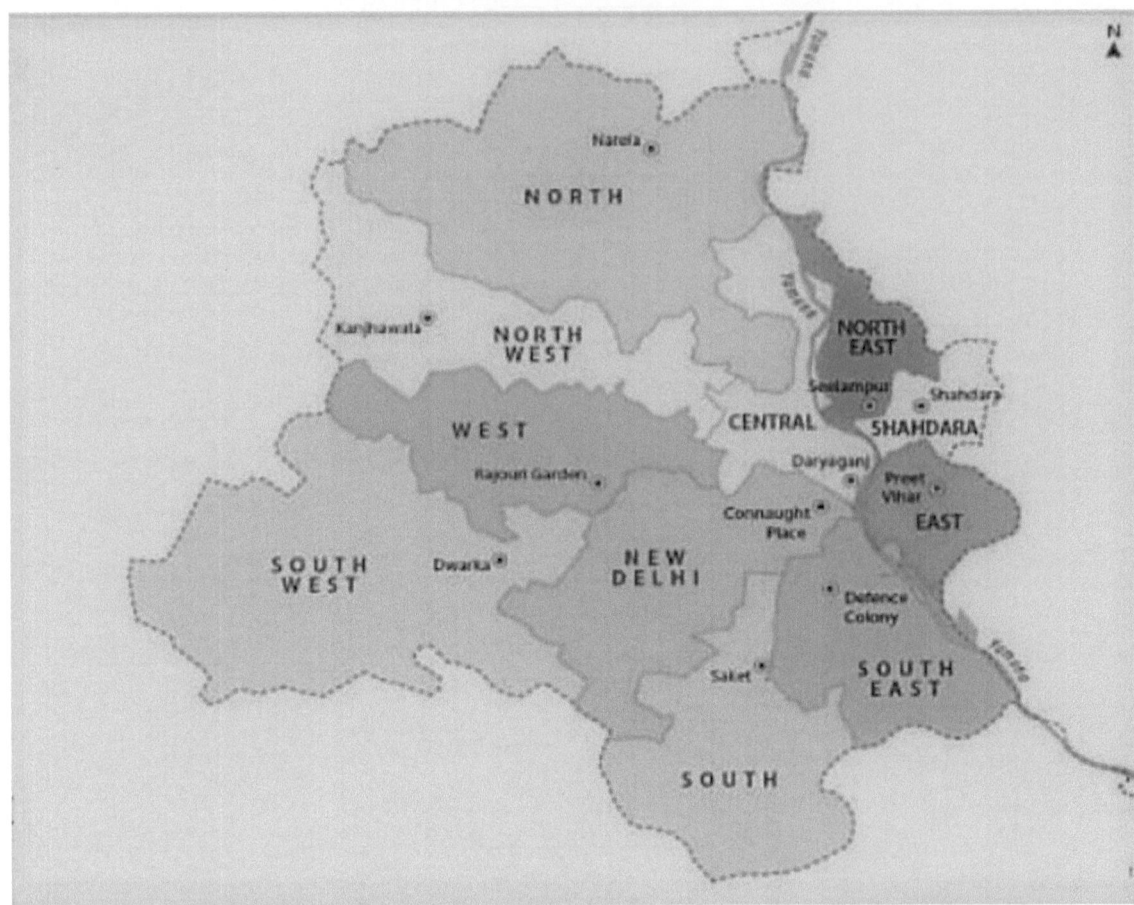

FIGURE 14.2   District-Level Map of the NCT of Delhi.

14.6.2 Dataset

For this study, the dataset used was as mentioned in Goldblatt et al. (2016). The dataset consists of 21,030 polygons (Geographically mapped areas) of 30 x 30 m size, distributed randomly over mainland India. These polygons have been manually classified into Built-Up (BU) and Non Built-Up (NBU) classes, respectively (Goldblatt et al., 2016). The polygons with more than 50% area consisting of man-made structures are marked as Built-Up and the remaining as Non Built-Up. The dataset has been generated using WorldPop, a per-pixel population estimation of 2010–2011 India Census dataset to create the sample polygons. The whole procedure of construction of the dataset as mentioned in Goldblatt et al. (2016) is shown in Figure 14.3. The Satellite View and Map View of the Polygons are shown in Figure 14.4. The dataset consists of 286 polygons over our study of the area as seen in Figure 14.5.

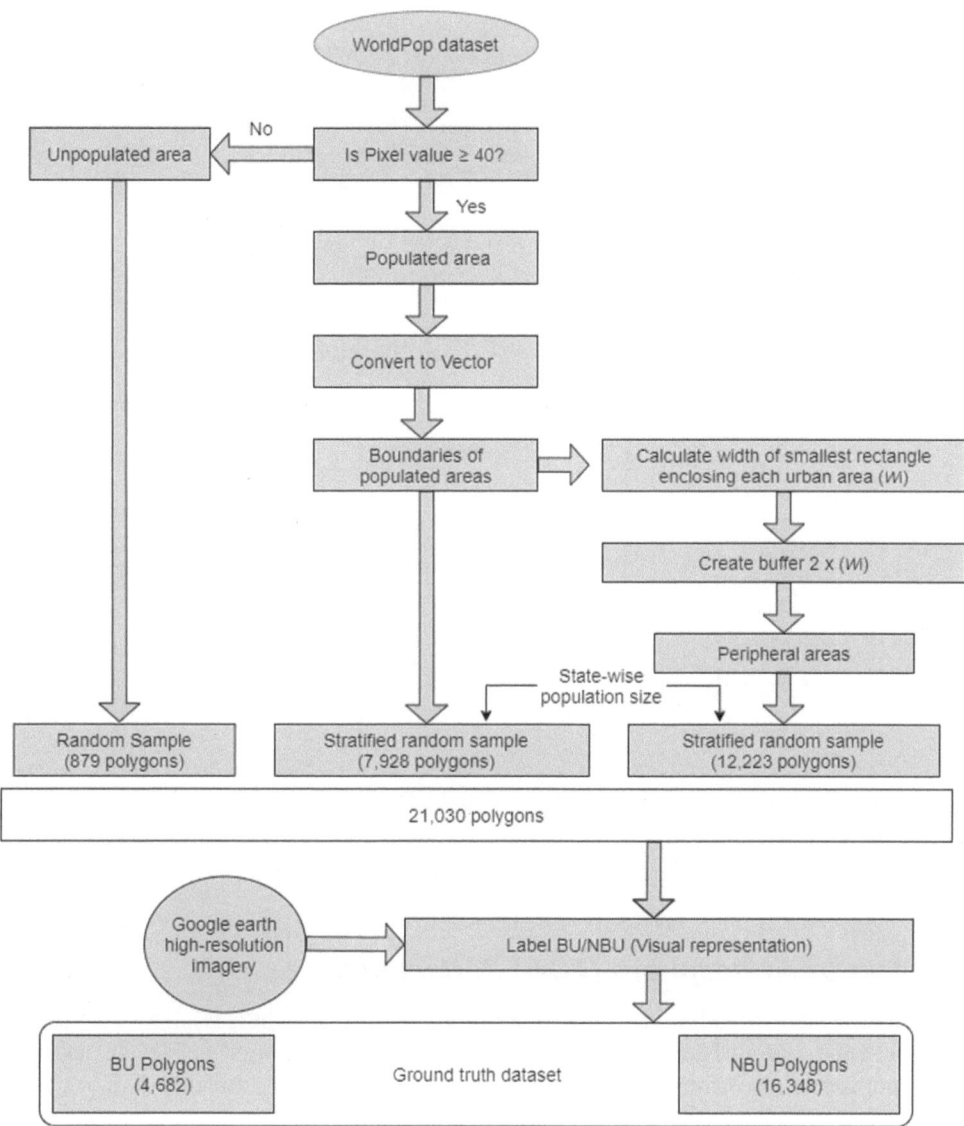

FIGURE 14.3 The Procedure Used to Generate the Dataset (Goldblatt et al., 2016).

14.6.3 Preprocessing and Scene Selection

After retrieving the dataset, for scene selection, Landsat 7 Surface Reflectance (TOA-1) images of year 2014 (available in GEE) were used as input for image classification which consists of 8 different bands of different wavelengths (micrometres), as shown in Table 14.1. Landsat 7 annual TOA percentile composite (2014), which is a composite of preprocessed scenes, filtered with lowest cloud cover and no negative sun elevation, was given as input feed. The composite consists of pixels, evaluated as percentile estimates of each band scaled to 8 bits (for bands 1 to 5 and band 7 to 8) and in the units of Kelvin-100 (for band 6). Two additional indices are added as bands to the Landsat 7 data, so that it can improve the classification: Normalized Difference Vegetation Index (NDVI) (Pettorelli et al., 2005) and Normalized Difference Built-Up Index (NDBI) (Zha et al., 2003).

(a)

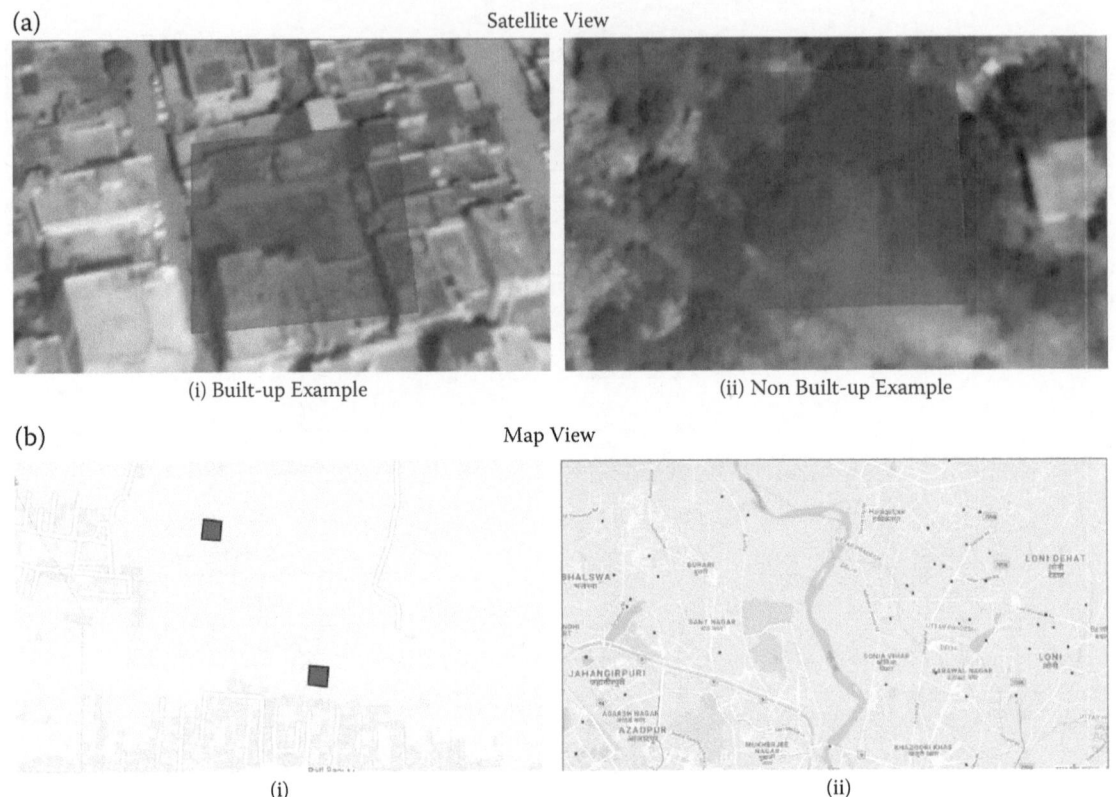

FIGURE 14.4 Polygon Representation as Viewed in Google Earth Engine.

14.6.3.1 *Top of Atmosphere (TOA) Reflectance*

Reflectance is basically defined as the proportion of the measure of light leaving a target to the measure of light striking the target. TOA Reflectance is a unit-less estimation which evaluates the reflected radiation proportion to the incoming solar radiation from a surface. For Landsat 7, it is calculated as:

$$\rho_\lambda = \frac{M_\rho Q_{cal} + A_\rho}{cos(\theta_{SZ})} \tag{14.2}$$

where, ρ_λ = TOA Reflectance

M_ρ = Band-specific multiplicative rescaling factor

A_ρ = Band-specific additive rescaling factor

Q_{cal} = Quantized and calibrated standard product pixel values

θ_{SZ} = Local solar zenith angle

BU example
NBU example

FIGURE 14.5   Satellite View of Built-Up and Non Built-Up Polygons Mapped over an Area of Delhi (as Viewed in Google Earth Engine).

TABLE 14.1   Landsat 7 bands used as features for classification

SPECTRAL BAND NAME	RESOLUTION PIXEL SIZE (M)	WAVELENGTH (MM)	DESCRIPTION
B1	30	0.45–0.52	Blue
B2	30	0.52–0.60	Green
B3	30	0.63–0.69	Red
B4	30	0.77–0.90	Near Infrared (NIR)
B5	30	1.55–1.75	Medium Infrared (MIR) 1
B6_VCID_1	30	10.40–12.50	Low-gain Thermal Infrared (TIR)
B6_VCID_2	30	10.40–12.50	High-gain Thermal Infrared (TIR)
B7	30	2.08–2.35	Medium Infrared (MIR) 2
B8	15	0.52–0.90	Panchromatic
NDVI	30		(B4 − B3)/(B4 + B3)
NDBI	30		(B5 − B4)/(B5 + B4)

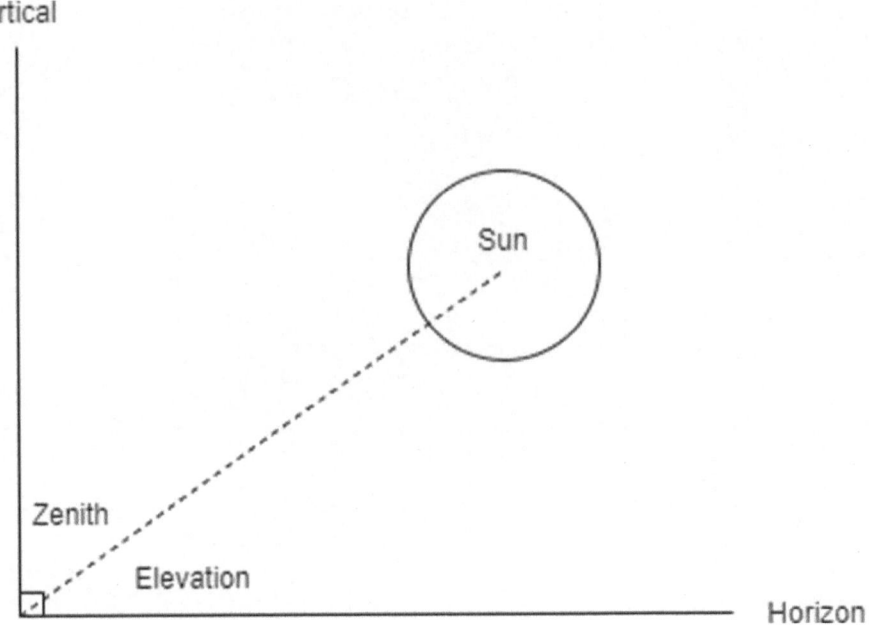

FIGURE 14.6   Solar Zenith Angle and Sun Elevation Angle.

14.6.3.2 Solar Zenith Angle

It is defined as the angle between the geometric centre of the solar disk and the observer's local vertical, as shown in Figure 14.6. It is also formulated as:

$$\theta_{SZ} = 90° - \theta_{SE} \tag{14.3}$$

where, θ_{SZ} = Solar zenith angle
θ_{SE} = Sun elevation angle

14.6.3.3 Sun Elevation Angle

It is defined as the angle between the geometric centre of the solar disk and the observer's local horizon, as shown in Figure 14.6.

14.6.3.4 Normalized Difference Vegetation Index (NDVI)

Normalized Difference Vegetation Index (NDVI) communicates the correlation of the visible red light (which is absorbed by the chlorophyll of plant) and near-infrared wavelength (which is dispersed by the mesophyll of a plant leaf). It is the most-widely used vegetation index to observe greenery globally. Usually, healthy plants have high reflectance in Near Infrared (NIR) in the range of 0.7 to 1.3 μm. High reflectance for NIR and high absorption for the Red spectrum; help to estimate the formula for the NDVI.

$$NDVI = \frac{(NIR - Red)}{(NIR + Red)} \qquad (14.4)$$

Here, for Landsat 7, NIR falls in Band 4 and Red falls in Band 3. Therefore,

$$NDVI = \frac{(B4 - B3)}{(B4 + B3)} \qquad (14.5)$$

14.6.3.5 Normalized Difference Built-Up Index (NDBI)

Normalized Difference Built-up Index (NDBI) communicates the correlation of the medium-infrared (MIR) and near-infrared (NIR) wavelengths. The bare soil and built-up structures reflect more MIR than NIR.

$$NDBI = \frac{(MIR - NIR)}{(MIR + NIR)} \qquad (14.6)$$

Here, for Landsat 7, MIR falls in Band 5 and NIR falls in Band 4. Therefore,

$$NDBI = \frac{(B5 - B4)}{(B5 + B4)} \qquad (14.7)$$

The NDBI value lies in the range of –1 to 1. Higher the value of NDBI, it denotes built-up area and values closer to –1 denote water bodies.

14.6.4 Detection of Built-Up Regions

In order to detect the Built-Up regions using GEE, first, we overlaid the labelled polygons from the dataset over the region of interest, the NCT of Delhi. We collected all Landsat 7 points within the area of these polygons along with per-band reflectance values and index values. These sampled points differ from the original number of polygons used, as they entirely do not overlap with the pixels of the Landsat data points. Each sample point includes an output binary class: Built-Up or Non Built-Up. This set of points, as seen in Figure 14.7, is used as input to the different classification models.

14.7 RESULT AND ANALYSIS

In GEE, various models are available for pixel-based classification. We compare the results obtained by three most prominent models available: CART, Linear SVM, and Random Forest. First, we performed 10-fold cross validation for Random Forest with 10, 50, and 100 trees, respectively, and the average testing accuracy achieved was 64.9%, 65.8%, and 65%, respectively. The highest performance was therefore obtained with 50 Trees, as seen in Figure 14.8 and Table 14.2.

We then performed 10-fold cross validation using Linear SVM and CART (Decision Tree), and compared their results with that of Random Forest consisting of 50 Trees, as shown in Figure 14.9,

(a)
Satellite View

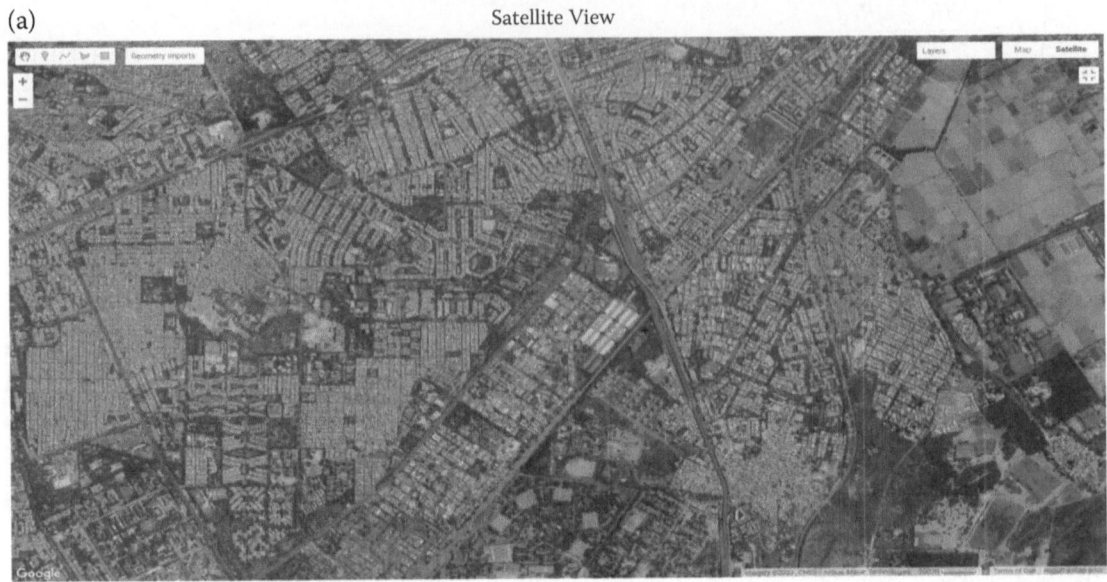

(b)
Map View

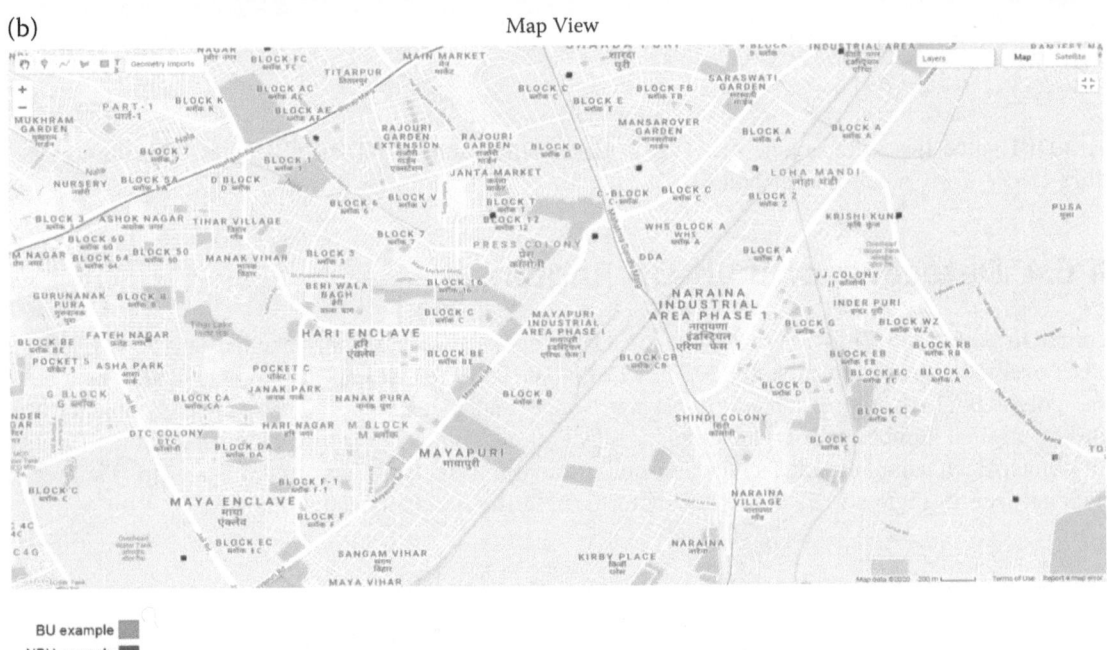

BU example ▨
NBU example ▮

FIGURE 14.7 Training Points After Being Mapped over Points of Landsat 7 Composite (as Viewed in Google Earth Engine): (a) Satellite View and (b) Map View.

Figure 14.10, Table 14.3, and Table 14.4. The average testing accuracy achieved for CART (Decision Tree) was 65.7%, while the Linear SVM performed with an average accuracy of 63.6%. The Ground Truth and Output Points as predicted by all the three models (for one of the folds) for comparison is shown in Figure 14.11, Figure 14.12, Figure 14.13, and Figure 14.14, respectively.

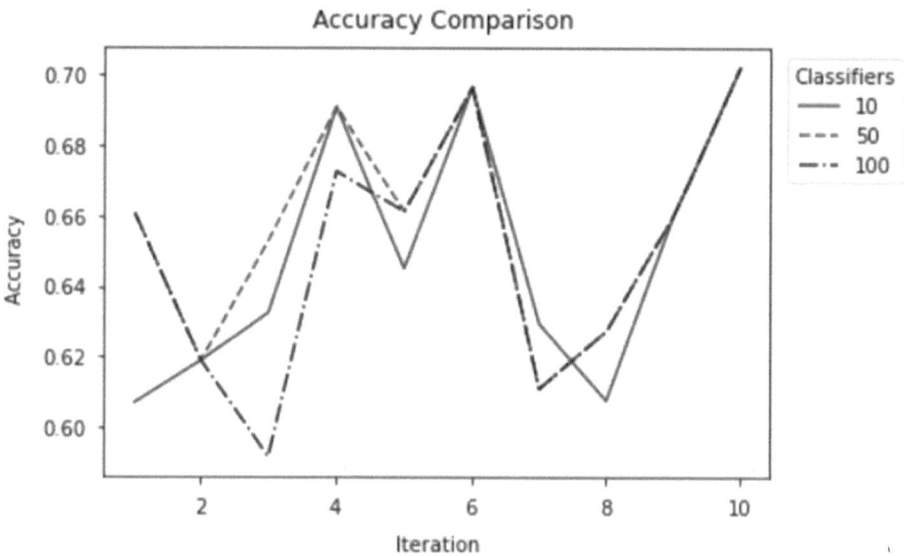

FIGURE 14.8 Accuracy Comparison Graph between Random Forest with 10, 50, and 100 Trees.

TABLE 14.2 Average accuracy comparison between random forest with 10, 50, and 100 trees

NUMBER OF TREES IN RANDOM FOREST	AVERAGE ACCURACY
10	0.6490942916534208
50	0.6582140865247743
100	0.6502734557270007

14.8 DISCUSSION AND CONCLUSION

Various studies have shown that a Random Forest classifier's accuracy is proportional to the number of trees (Rodriguez-Galiano et al., 2012). But, at a certain point, this improvement in performance can decrease as the number of trees is increased if the performance of prediction from learning is less than the computation time for learning with these additional trees (Oshiro et al., 2012). This was the case when we performed 10-fold cross validation on Random Forest consisting of 10, 50, and 100 Trees, and the highest average accuracy was achieved for that with 50 Trees.

Out of all the models used, Linear SVM performs the worst with average testing accuracy of 63.6%. The CART (Decision Tree) and Random Forest (consisting of 50 Trees) gave almost the same performance, with average testing accuracy of 65.7% and 65.8% respectively. This can be because Random Forest generally builds trees, with each tree including some of the features and so it can be possible that some of the trees do not consist of some of the important distinguishing features, and thus it performs similar to a single Decision Tree.

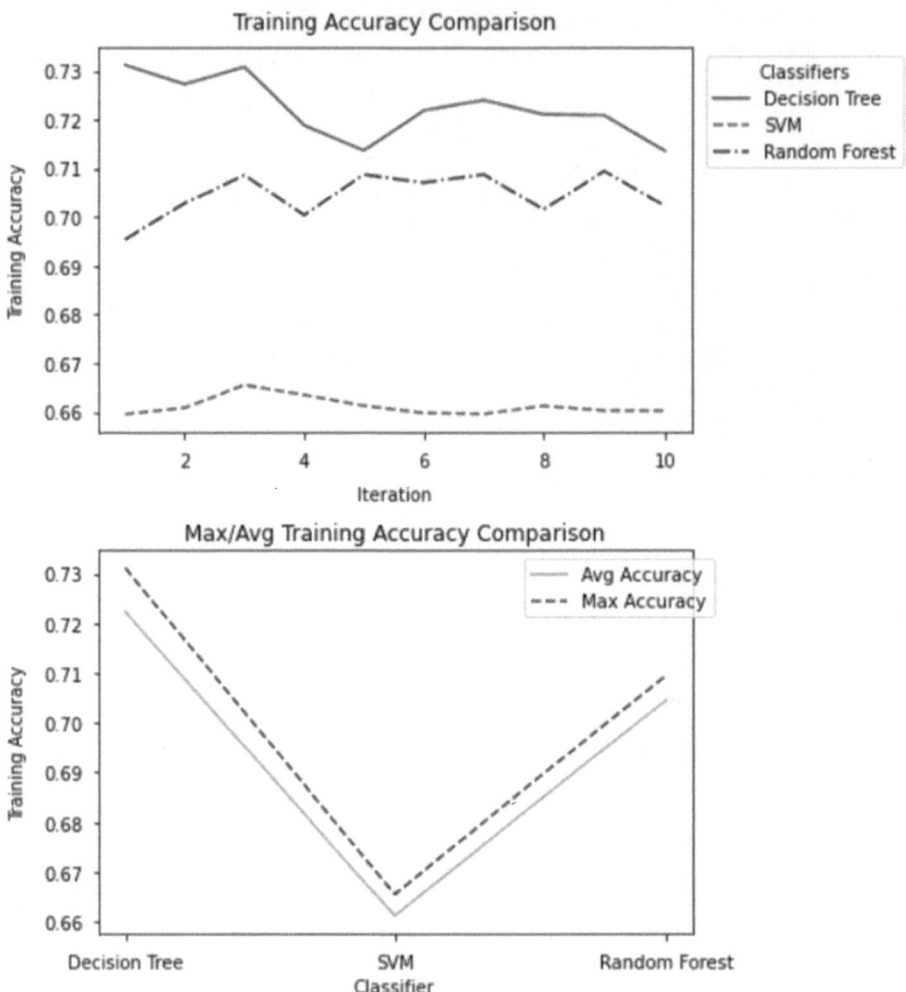

FIGURE 14.9   Training Accuracy Comparison Graphs between Decision Tree, Linear SVM, and Random Forest (with 50 Trees).

14.9 FUTURE WORK RECOMMENDATIONS

The task presented in this study can be extended by incorporating socio-economic and other geographic features over large areas to study the extension of the urban sprawl in a limited timeframe. The performance of the classifiers can also be improved by further tuning their parameters using various techniques, other than cross validation.

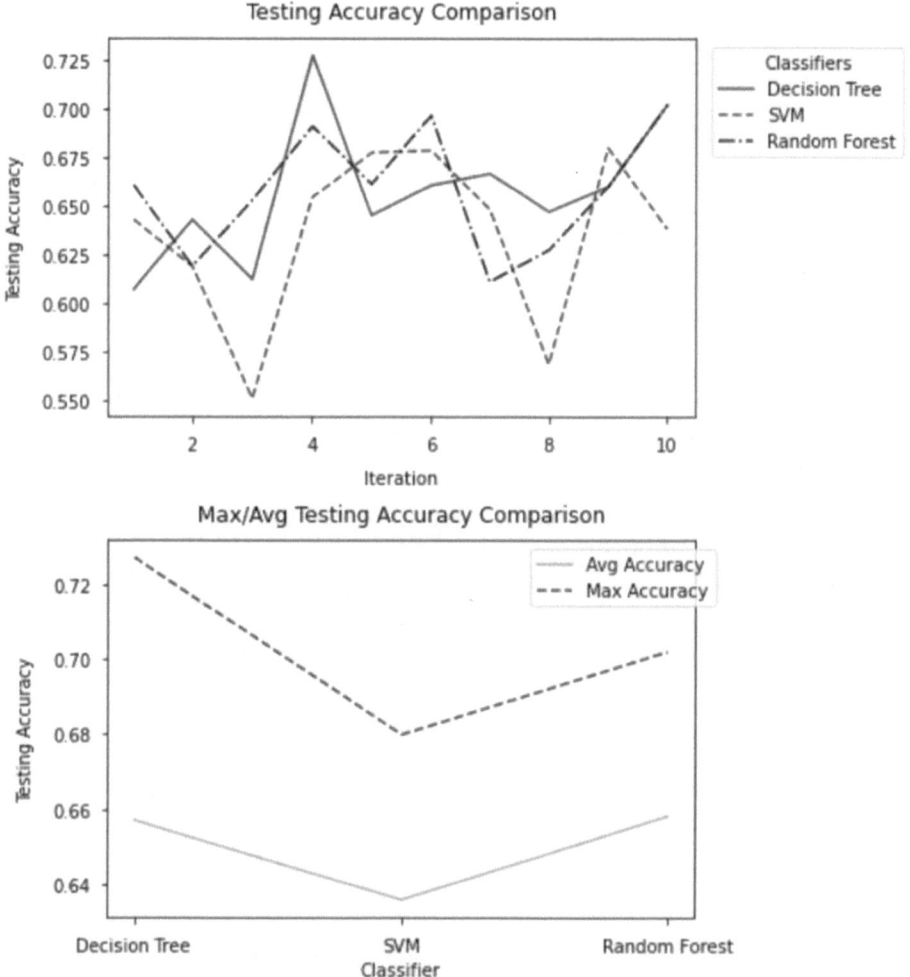

FIGURE 14.10 Testing Accuracy Comparison Graphs between Decision Tree, Linear SVM, and Random Forest (with 50 Trees).

TABLE 14.3 Training accuracy comparison between decision tree, linear SVM, and random forest (with 50 trees)

CLASSIFIER	AVERAGE ACCURACY	MAXIMUM ACCURACY
Decision Tree	0.7223108322725234	0.7311827956989247
SVM	0.661146076403116	0.6655011655011654
Random Forest	0.7044972999807453	0.7094488188976378

TABLE 14.4 Testing accuracy comparison between decision tree, linear SVM, and random forest (with 50 trees)

CLASSIFIER	AVERAGE ACCURACY	MAXIMUM ACCURACY
Decision Tree	0.6571246351039324	0.7272727272727273
SVM	0.6358534879492584	0.68
Random Forest	0.6582140865247743	0.7021276595744681

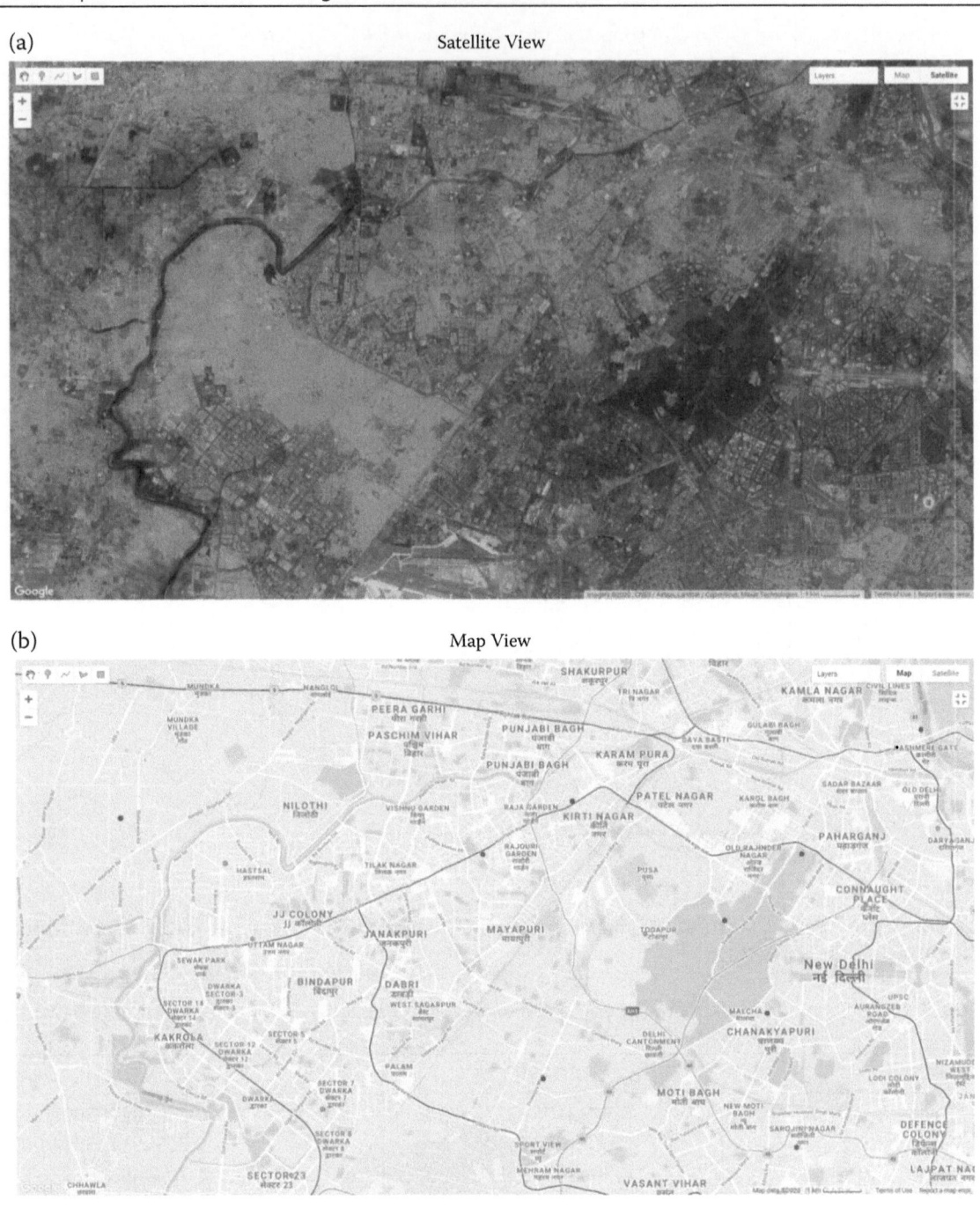

FIGURE 14.11   Ground Truth: (a) Satellite View and (b) Map View.

(a)   Satellite View

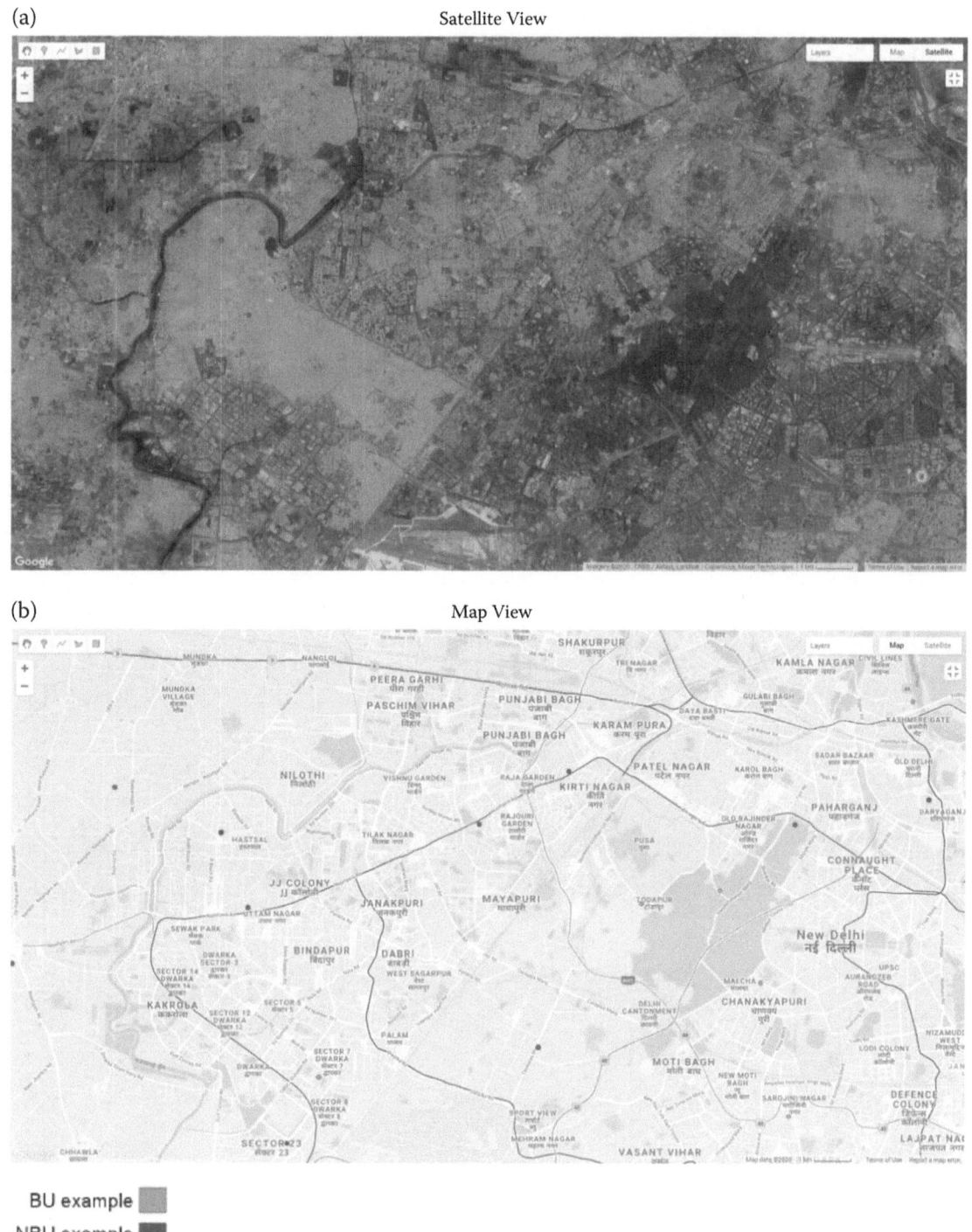

(b)   Map View

BU example ▢
NBU example ▦

FIGURE 14.12   Prediction Output (Decision Tree): (a) Satellite View and (b) Map View.

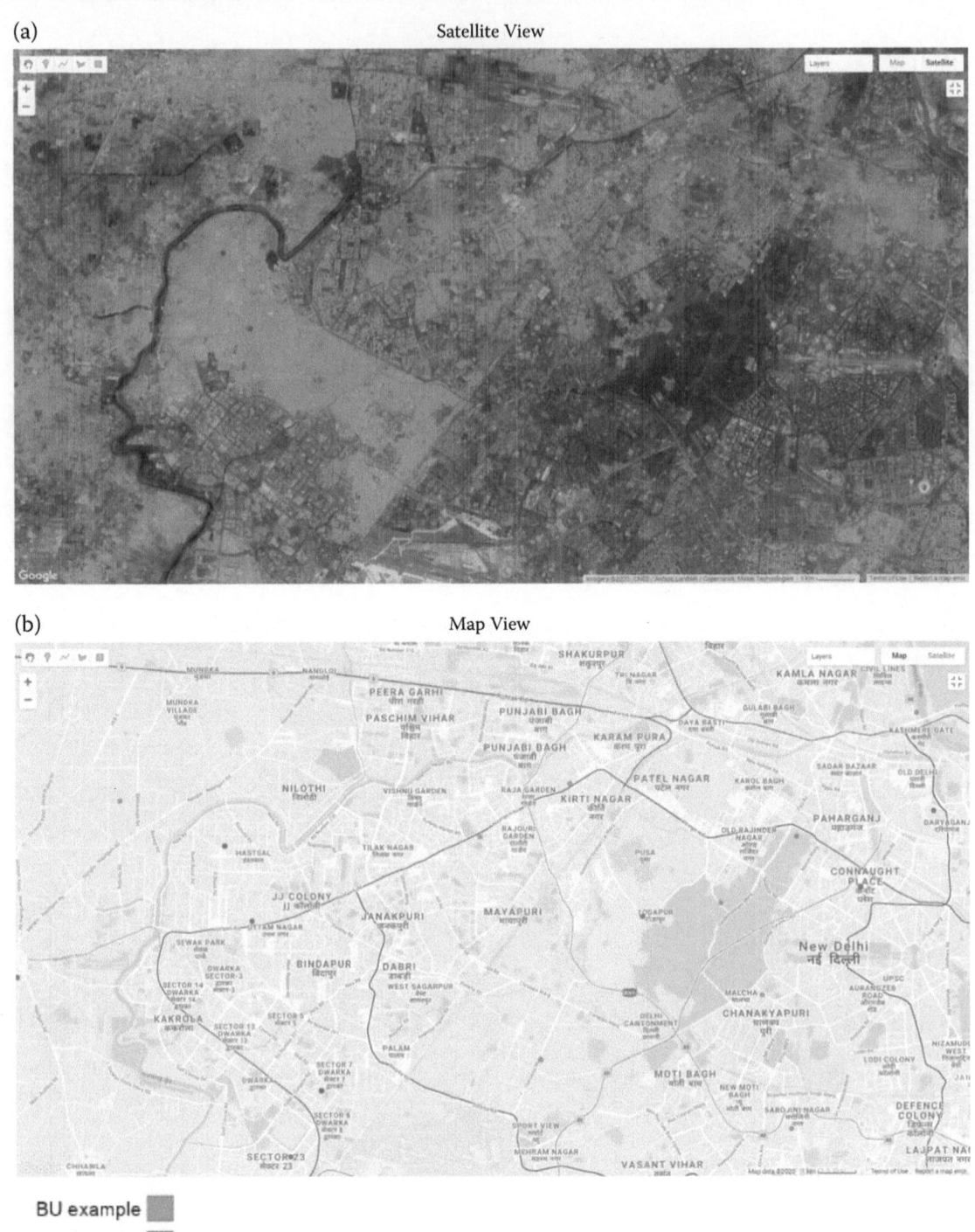

FIGURE 14.13 Prediction Output (Linear SVM): (a) Satellite View and (b) Map View.

(a)   Satellite View

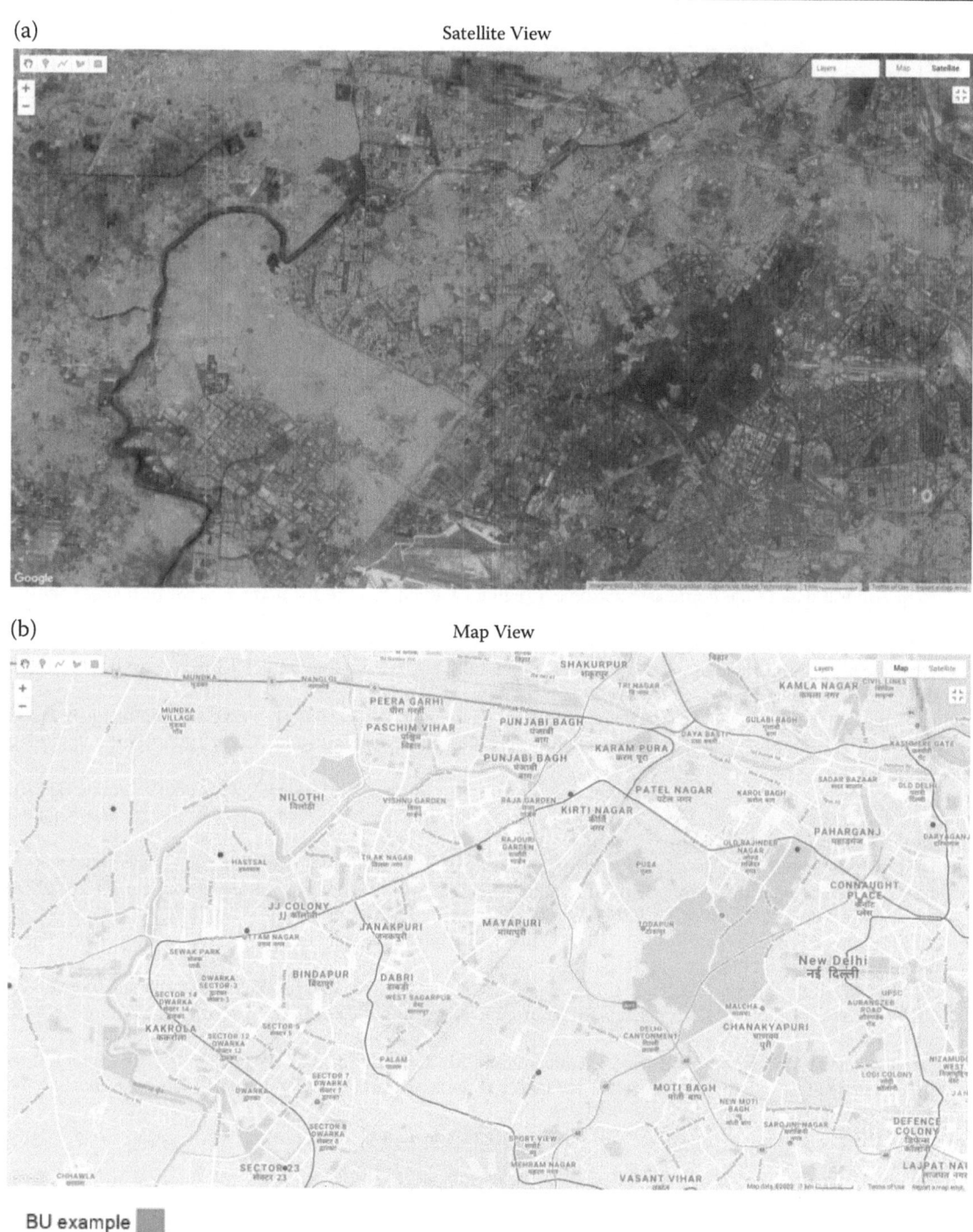

(b)   Map View

BU example
NBU example

FIGURE 14.14   Prediction Output (Random Forest with 50 Trees): (a) Satellite View and (b) Map View.

ACKNOWLEDGEMENTS

The authors thank Dr. Nitesh Patidar (Scientist "B", National Institute of Hydrology, Roorkee, India) for assistance on using Google Earth Engine.

We also thank Dr. Amit K. Khandelwal (Columbia Business School, Columbia University, New York, USA) for providing us the dataset used in this study.

REFERENCES

Baum-Snow, N (2007). Did highways cause suburbanization? *The Quarterly Journal of Economics, 122*(2), 775–805.

Goldblatt, R, Deininger, K, & Hanson, G (2018). Utilizing publicly available satellite data for urban research: Mapping built-up land cover and land use in Ho Chi Minh City, Vietnam. *Development Engineering, 3*, 83–99.

Goldblatt, R, You, W, Hanson, G, & Khandelwal, AK (2016). Detecting the boundaries of urban areas in india: A dataset for pixel-based image classification in google earth engine. *Remote Sensing, 8*(8), 634.

Ha, TV, Tuohy, M, Irwin, M, & Tuan, PV (2020). Monitoring and mapping rural urbanization and land use changes using Landsat data in the northeast subtropical region of Vietnam. *The Egyptian Journal of Remote Sensing and Space Science, 23*(1), 11–19.

Hua, L, Zhang, X, Chen, X, Yin, K, & Tang, L (2017). A feature-based approach of decision tree classification to map time series urban land use and land cover with Landsat 5 TM and Landsat 8 OLI in a Coastal City, China. *ISPRS International Journal of Geo-Information, 6*(11), 331.

Jiang, L, Wang, W, Yang, X, Xie, N, & Cheng, Y (2010). Classification methods of remote sensing image based on decision tree technologies. International Conference on Computer and Computing Technologies in Agriculture.

Kohavi, R (1995). A study of cross-validation and bootstrap for accuracy estimation and model selection. IJCAI.

Kumar, L, & Mutanga, O (2018). Google Earth Engine applications since inception: Usage, trends, and potential. *Remote Sensing, 10*(10), 1509.

Mallupattu, PK, & Sreenivasula Reddy, JR (2013). Analysis of land use/land cover changes using remote sensing data and GIS at an Urban Area, Tirupati, India. *The Scientific World Journal, 2013*, Article ID 268623, 6 pages. https://doi.org/10.1155/2013/268623

Oshiro, TM, Perez, PS, & Baranauskas, JA (2012). How many trees in a random forest? International workshop on machine learning and data mining in pattern recognition.

Pettorelli, N, Vik, JO, Mysterud, A, Gaillard, J, Tucker, CJ, & Stenseth, NC (2005). Using the satellite-derived NDVI to assess ecological responses to environmental change. *Trends in Ecology & Evolution, 20*(9), 503–510.

Rodriguez-Galiano, VF, Ghimire, B, Rogan, J, Chica-Olmo, M, & Rigol-Sanchez, JP (2012). An assessment of the effectiveness of a random forest classifier for land-cover classification. *ISPRS Journal of Photogrammetry and Remote Sensing, 67*, 93–104.

Shetty, S (2019). Analysis of machine learning classifiers for LULC classification on Google Earth Engine. University of Twente.

Sidhu, N, Pebesma, E, & Câmara, G (2018). Using Google Earth Engine to detect land cover change: Singapore as a use case. *European Journal of Remote Sensing, 51*(1), 486–500.

Sudhira, H, Ramachandra, T, & Jagadish, K (2004). Urban sprawl: Metrics, dynamics and modelling using GIS. *International Journal of Applied Earth Observation and Geoinformation, 5*(1), 29–39.

Tamiminia, H, Salehi, B, Mahdianpari, M, Quackenbush, L, Adeli, S, & Brisco, B (2020). Google Earth Engine for geo-big data applications: A meta-analysis and systematic review. *ISPRS Journal of Photogrammetry and Remote Sensing*, *164*, 152–170.

Tobón-Marín, A, & Cañón Barriga, J (2020). Analysis of changes in rivers planforms using google earth engine. *International Journal of Remote Sensing*, *41*(22), 8654–8681.

Zha, Y, Gao, J, & Ni, S (2003). Use of normalized difference built-up index in automatically mapping urban areas from TM imagery. *International Journal of Remote Sensing*, *24*(3), 583–594.

Emergence of Smart Home Systems Using IoT: Challenges and Limitations

15

Nabeel Khan

Department of Information Technology, College of Computer, Qassim University, Buraydah, Saudi Arabia

Contents

DOI: 10.1201/9781003172772-15

15.1 INTRODUCTION

15.1.1 Background

The Smart home system is an application of ubiquitous computing where automated and assistive services are provided in the form of ambient intelligence. Smart home system is also known as Intelligent home or home automation, supporting home with an embedded system that provides plenty of different smart features working in the background without much of user interaction. A Smart home consists of many devices, sensors, actuators, etc., connected via IoT technology and monitors daily-life activities by an intelligent computer system. It also includes user-friendly Graphical User Interface (GUI), wireless sensors, and remote controlling devices (Hasan et al., 2018). AI has been a key enabler for the establishment of Smart cities. Smart cities and automation Smart home systems support remote home control by electronic devices such as smartphones and laptops. A Smart Home System offers an easy and comfortable way to control home environments and improve the quality of life. An automated environment can control and monitor lighting, air conditioners and temperature, home appliances, security systems, and other functions. Smart home systems consider as solution for elderly and disabled people who live independently by providing services and healthcare support in their home environment and connecting remotely to healthcare service providers for immediate medical support in emergency situations. AI became a root technology in creating smart cities due to its characteristic to be able to learn from experience by making self-sustaining systems to assist in various aspects of life. The rapid development of IoT and communication technology causes a lot of challenges and difficulties related to security, privacy, connectivity, and network. Also, it highlights various operational and adoption barriers in the development of smart cities. This chapter will present smart home system functions and analyze the current challenges (Alam et al., 2012).

15.1.2 Methodology

This chapter is focused to determine the current status and future directions for the smart home systems, its challenges, and limitations. This review aims to overcome the current challenges of smart home system and guide future research to propose solutions. The internet and communication technologies transfer the daily life environments into connectivity, smart homes, and smart cities, therefore the motivation behind this chapter is highlighting challenges and difficulties that limit the adoption of the smart home system.

The first phase of this review determines the review sources and materials. Smart home systems research is still an emerging research area and contemporaneous researches are published in a wide variety of journals and conferences. The searching includes both academic journals papers and conferences. We start the literature search with scanning of online academic journal and conference databases. The used databases are as follows:

- IEEE Xplore
- ScienceDirect | Elsevier
- ACM Digital Library
- Google Scholar for academic papers

Search was done by using the keywords 'Smart Home System', 'Home Automation', 'Challenges', 'Difficulties, and 'Limitations' that were found in the title or abstract of the paper. The result of this search leads to the collection of research works which were published from 2011 to 2020. In the second phase of this review, previous studies were analyzed and systematic review was done to identify gaps, challenges, and limitations which are to be covered and addressed by future researches. The result of the analysis phase identified that papers may be specified in security challenges, connectivity and integrating challenges, or represented challenges related to a certain proposed system or explained a variety of challenges from various aspects. This chapter discusses major challenges from different technical, social, and ethical views.

15.1.3 Chapter Organization

The rest of the chapter is organized as follows: Section 15.2 discusses the functions of smart home system. Section 15.3 discusses the challenges and limitations, while Section 15.4 concludes the chapter.

15.2 FUNCTIONS OF SMART HOME SYSTEM

A smart home system consists of many applications built and connected by using IoT technology. These applications have various functions as follows:

15.2.1 Alert and Sensors

Smart home system sensors have the ability to sense, feel, and communicate with its environment. Many sensors such as heat, gas, smoke, temperature, and energy sense collect data remotely and based on the collected data using intelligent decision-making approach respond to the environment and send alerts to the user. Alerts messages can be in the form of text messages, emails, or any other social media (Almusaylim & Zaman, 2019; Malche & Maheshwary, 2017).

15.2.2 Monitor

The Most significant function of Smart home system is monitoring that keeps track of every activity in smart home environments using which any further action can be taken or decision can be made. Monitoring presents data getting from sensor and camera and statistics and suggestions based on data analysis. The user can observe the home environment at any time and location, make decisions, and improve the security (Malche & Maheshwary, 2017).

15.2.3 Control

The user of smart home system can control various activities such as switching on/off lights, air-conditioners, and appliances; lock/unlock doors; open/close windows and doors; and many more. Users can control these appliances remotely (from remote location) or locally (in the same place). The control functions to automate activities such as automatically switching on/off lights when the room is full/empty (Malche & Maheshwary, 2017).

15.2.4 Intelligence and Logic

Smart home or intelligent home behaves in an intelligent way where automated decisions are made. The system appliances are able to learn user behaviour and track user location for example, turning on air-conditioner and preparing coffee as soon as users arrives. Home appliances can schedule tasks at pre-defined times. This function helps to reduce energy consumption and enhances safety (Malche & Maheshwary, 2017).

15.3 CHALLENGES AND LIMITATIONS

Many challenges limit smart systems from wide adoption. These challenges perceived with different technical, social, and ethical directions. These difficulties related to security, privacy, connectivity, and network, operational and adoption. This section discusses some challenges as follows.

15.3.1 Interoperability

In the smart home network, devices, sensors, and systems are made by different manufacturers and connected with different protocols and network interfaces. Every appliance in the home connects to the control system in its unique manner. Challenges in connectivity within the smart home are shown as a consequence of heterogeneity of components in the smart home system. Interoperability is the main challenge because the home system should have easy-to-use and easy-to-connect devices that simply work together in an efficient way in spite of their heterogeneity. Interoperability is defined as the ability of systems, applications, and services to work together reliably in a predictable fashion (Almusaylim & Zaman, 2019; Samuel, 2016).

15.3.2 Self-Management

Self-managed devices mean that smart home devices should have the ability to monitor their operating system, operate and collaborate with other devices, adapt to failures and changes in the environment, and notify the users about potential issues. This challenge for the smart system is to be completely independent of human intervention. Where the system that require frequent administrating and managing will be frustrating for the user (Samuel, 2016; Sultan & Nabil, 2016).

15.3.3 Maintainability

One of the main challenges is designing smart home systems with the ability of easy maintenance and repair of their components and networks at a fast, efficient, and lower cost. Since the changes occur at anytime and anywhere in the environment, the system needs to monitor its progress and make changes in operational parameters to fit the new requirements. For example, providing lower quality with limited energy resources (Samuel, 2016).

15.3.4 Signalling

It is a connectivity issue related to collecting and routing data between devices and servers in an IoT network. Each device in the smart home network can communicate to a server to collect data, or communicate with other devices. This communication from a source device to a destination device requires 100% surety that data stream arrives at its destination quickly and reliably every time (Samuel, 2016).

15.3.5 Usability

Designing the smart system must be simple and easily understandable for a user. A user may be children or those who don't have advanced knowledge of technology or don't want to spend more time trying to understand the system, so the interface must be user-friendly. When an interface is poorly designed it may cause frustration for the user, a higher risk with lots of errors, reduced satisfaction, and least use of the system. Usability is an important issue in domestic use which leads to consistency, personalization, robustness, absence of barriers, adequacy to multiple users, robustness of the input, and interaction logic (Alshammari et al., 2019).

15.3.6 Power Aware/Efficient Consumption

One of the prime issues related to smart home system is power saving and reducing cost. Therefore, the smart system is capable of calculating and controlling the amount of energy consumed by its appliances and operations. Each device connected to the IoT smart system can determine the best time to operate which in turn provides higher efficiency in power consumption. An efficient and qualified IoT system requires minimal battery drain and low power consumption (Almusaylim & Zaman, 2019).

15.3.7 High Cost of Ownership

The first challenge that faces the user is the high cost of Smart home system. Money spent on smart home system varies depending on the brand and amount of installed functionality and capabilities. Cost also spent on hardware, software installation, support and maintenance (Brush et al., 2011).

15.3.8 Security and Privacy

Security and privacy represent the main concerns of users in the smart home system which require authenticity, integrity, reliability, and confidentiality of data through appliances and services. The ratio of security risk rises as the number of devices and appliances connected to the Internet increases and the information turns to remote access and control in new ways. Therefore, the home network needs to secure all the information and data, check the authenticity of every user and the integrity of all data through devices. Table 15.1 presents a summary of some of the security threats that are addressed in the smart home system (Almusaylim & Zaman, 2019).

In smart home environments, sensors and devices monitor the home and collect data which may consist of sensitive and personal information. Hence, it is necessary to protect the private information of the users from threats, leakage, and disclosure. The security and privacy challenges may be considered legal and ethical issues.

15.3.9 Acceptance and Reliability

Installing new smart home systems can constitute a big challenge for some users. The smart system should be designed to operate smoothly and have an easy-to-use and understand interface especially for

TABLE 15.1   Summary of smart home security threats

THREATS	POSSIBLE IMPACTS
Eavesdropping	An attacker eavesdrops the home user traffic illegally such as emails between the internal network of the smart home and the third parties without changing the legitimate communication parties with the goal of violating the communication's confidentiality
Traffic analysis	An attacker observes the traffic pattern between the third party and the owner of the smart home
Denial of service (DoS)	An attacker blocks an authorized user from accessing services or limiting it by making the internal network of the smart home flooded with messages to overload its resources with traffic
Node compromise	An attacker captures and reprograms a legitimate node in the network in order to disrupt the network communication
Sinkhole and wormhole attacks	A malicious node attracts network packets towards it by spreading false routing information to its neighbours in order to make selective forwarding of packets which, in turn, reshape the network's routing behaviour
Physical attack	An attacker gains physical access to sensors which can perform many other attacks such as removing sensors from the network or stealing nodes
Masquerade attack	An attacker pretends a false identity to gain some unauthorized privileges
Replay attack	An attacker gets messages that are sent previously between two parties re-sends them again pretending that it is from an authorized entity
Message modification attack	An attacker alters the content of the message being sent by reordering it or delaying it to produce an illegal effect
Interception attack	An attacker intercepts the data packets sent to the remote user
Season-stealing attack	An attacker waits patiently for authenticated users or nodes to authenticate itself and then fraudulently takes over the session by impersonating the identity of the genuine user or node
Malicious code	Malicious Codes are software threats that cause negative effects on the Smart Home internal network by exploiting its vulnerabilities

elder users. Some of the system reliability problems happened due to unpredictable behaviours. For example, users were waiting several minutes for a system's response. Therefore, smart home devices have to work reliably and have a charming design (Brush et al., 2011).

15.3.10 Calmness and Context-Awareness

It is very important that the used system should be hidden and embedded as a ubiquitous environment to operate in the background and utilize users' behaviour and interactions. The main idea of smart home systems is ease of use; calmness; and not requiring much of user interaction, personalization, and operation (Hasan et al., 2018).

15.3.11 Architectural Readiness

It is one of the obstacles facing the adoption of smart home systems limiting its expansion, whether it's better for a home to be enhanced with smart devices or should be built smart in the first place. Some think that homes should be designed and built as smart from the beginning, while others believe any preexisting building can still take advantage of the smart system and devices which can be added and connected even on phases. Furthermore, these problems require further research which includes eliminating the demand for structural changes for an excellent home automation experience (Brush et al., 2011).

15.4 CONCLUSION

The smart home system is one of the most important applications of IoT. It has a lot of benefits and facilitates our daily life. It provides high efficiency and performance; a safe and secure environment; and saves time, power, and cost. The user can do personalization and customization with easy management and maintenance. However, these smart system faces challenges and difficulties that eliminate the rapid and widespread adoption. This chapter discussed and analyzed the most important challenges and difficulties related to security and privacy, network and connectivity, operational, etc. These challenges require further research to propose the solution system and achieve a good home automation experience. This review research will greatly help other researchers to understand and overcome these barriers which allow smart systems to be easily adopted.

REFERENCES

Alam, MR, Reaz, MBI, & Ali, MAM (Nov. 2012). A review of smart homes—Past, present, and future. *IEEE Transactions on Systems, Man, and Cybernetics, Part C (Applications and Reviews)*, 42(6), 1190–1203.

Almusaylim, ZA, & Zaman, N (2019). A review on smart home present state and challenges: Linked to context-awareness internet of things (IoT). *Wireless Networks*, 25(6), 3193–3204.

Alshammari, A, Alhadeaf, H, Alotaibi, N, Alrasheedi, S, & Prince, M (2019). The Usability of HCI in Smart Home.

Brush, AJ, Lee, B, Mahajan, R, Agarwal, S, Saroiu, S, & Dixon, C (2011, May). Home automation in the wild: Challenges and opportunities. In Proceedings of the SIGCHI Conference on Human Factors in Computing Systems (pp. 2115–2124). ACM.

Guo, X, Shen, Z, Zhang, Y, & Wu, T (2019). Review on the application of artifical intelligence in smart homes. *Smart Cities*, 2, 402–420.

Hasan, M, Biswas, P, Bilash, MTI, & Dipto, MAZ (2018). Smart home systems: Overview and comparative analysis. In 2018 Fourth International Conference on Research in Computational Intelligence and Communication Networks (ICRCICN), Kolkata, India, pp. 264–268.

Malche, T, & Maheshwary, P (2017). Internet of Things (IoT) for building smart home system. In 2017 International Conference on I-SMAC (IoT in Social, Mobile, Analytics and Cloud) (I-SMAC), Palladam, pp. 65–70.

Samuel, SSI (2016). A review of connectivity challenges in IoT-smart home. In 2016 3rd MEC International Conference on Big Data and Smart City (ICBDSC), Muscat, pp. 1–4.

Sultan, M, & Nabil, K (2016). Smart to smarter: Smart home systems history future and challenges. Future of HBI: Human-Building Interaction Workshop, ACM CHI'16.

Acceptance of Blockchain in Smart City Governance from the User Perspective

16

Emre Erturk, Dobrila Lopez, and Weiyang Yu

Eastern Institute of Technology, Napier, New Zealand

Contents

DOI: 10.1201/9781003172772-16

16.1 INTRODUCTION

16.1.1 Objectives

As blockchain continues to rise; it has potential within smart city applications. It is important to analyze this from an end-user perspective since there may be unintended consequences for citizens and not everyone may benefit equally due to a digital divide (Yigitcanlar, 2020). The impact contains both positive and negative aspects, which will be analyzed separately by identifying the advantages and challenges.

One of the major existing challenges for citizen-centric management models is the lack of transparency and trust among the network infrastructure, since most smart city projects still heavily rely on centralized resources or proprietary networks that only a few major companies can finance and support. The blockchain, however, brings the possibility of a different approach for this challenge. Ibanez et al. (2017) suggest that blockchain has the potential to decentralize the Internet, opening more opportunities for citizen-centric approaches in smart city governance.

Furthermore, there are concerns among policymakers and citizens toward the lack of explainability of smart city solutions based on artificial intelligence (Thakker et al., 2020). Therefore, in this new era of Explainable Artificial Intelligence, blockchain solutions need greater transparency. Although this chapter looks at user acceptance in the smart city context in general, it also relates to why new applications need to be better understood by humans.

16.1.2 Research Questions

The first research question is: What are the factors that impact users' acceptance of blockchain technology in smart city governance?

The second research question is: How (positively or negatively) do these factors impact the acceptance of blockchain technology in smart city governance?

16.1.3 Chapter Organization

After the Introduction, Sections 16.2 and 16.3 present the selected base model and the review of related work. Section 16.4 presents a new conceptual model developed by extending the base model. Section 16.5 introduces a proposed methodology for the future justification of the conceptual model, the questionnaire, and the proposed techniques to analyze the results of the questionnaires. Section 16.6 discusses the findings and implications. Section 16.7 summarizes and concludes the study.

16.2 RELATED WORK

The systematic literature review will focus on peer-reviewed articles from recent years to extend and refine the selected model, aiming at developing a more detailed model to help understand the impact of blockchain in smart city governance specifically from the end-user perspective.

As mentioned in Dong et al. (2018), blockchain possesses four typical features which make it an attractive and beneficial technology: decentralization, distributed nodes and storage, consensus and smart contract, and asymmetric encryption. Meanwhile, Swan (2015) reported certain limitations of blockchain such as limited throughput capability, latency of transactions, limited blockchain size, and the question of unsustainable use of computing resources.

Blockchain is a "public ledger", a decentralized database, shared among users with a high level of transparency, immutability, and security of transactions (Pilkington, 2015). As the name suggests, the blockchain ledger database contains a chain of blocks and every node in the system keeps records of the blockchain. Each block carries the following: a unique header, a timestamp, the header of the previous block, and finally the data pertaining to the transaction. Once a transaction is completed, it will be broadcasted to the whole network to be validated through a chosen consensus mechanism. Currently there are quite a few different consensus mechanisms, each with advantages and disadvantages. Once the transaction is verified by most of the nodes in the network, a timestamp will be applied, and a block will be created with the transaction details and added to the blockchain.

Smart governance is a concept with different definitions on different levels. The first level of smart governance is the management of smart city services (Batty et al., 2012). The next level involves the prioritization of resources and the approval of expanding the smart city (Alkandari et al., 2012). The next higher level has even more impact by creating the governance model and structures, in other words, introducing smart administration in a city (Batty et al., 2012). The last level of smart governance focuses on smart collaboration between different sectors of the city, which will involve more stakeholders and information systems (Batagan, 2011).

As a unique system that offers decentralization as its key feature, blockchain can provide various potential benefits for smart city governance. Ølnes et al. (2017) argued that blockchain can bring great value to governments in many ways such as data quality and integrity, corruption reduction, and trust enhancement, as well as transparency, privacy, and security. Two recent systematic literature reviews on the application of blockchain in smart city developments (Shen & Pena-Mora, 2018; Xie et al., 2019) identified two major fields of interest in the implementation of blockchain in a smart city: (1) as part of the innovative IT transformation and (2) for creating a citizen-centric collaborative urban governance. Blockchain-based applications in the first category are designed to improve existing government records management systems in areas such as land titles, assets, contracts, vaccinations, and financial transactions. The high level of immutability and transparency can make the data auditing task much easier and increase the quality and reliability of the data.

As the main purpose of this study is to investigate the impact of blockchain in smart city governance from the citizens' perspective, looking at existing models from previous research is a good starting point. The selected conceptual model and related definitions will be introduced in the following section.

16.3 RESEARCH METHOD

16.3.1 Search Criteria

This section describes the systematic literature review of this study. The search of articles was conducted using the following databases and search engines: SpringerLink, IEEE Xplore Library, ACM Digital Library, Elsevier, ProQuest, and Google Scholar. Articles selected from the search were published since 2016, published in English, and were peer-reviewed. The search was conducted according to the following criteria in Table 16.1.

After the search for articles, data is extracted and stored in Microsoft Excel spreadsheets. An example of the data extraction table is shown in Table 16.2.

TABLE 16.1 Inclusion and exclusion criteria

INCLUSION	*EXCLUSION*
Article discusses technology acceptance and smart city	Article does not mention factors impacting user's acceptance of technology
Article discusses benefits and challenges of blockchain	Article focuses on the introduction of blockchain technology without discussing the benefits or challenges
Article proposes model/framework for either smart city or blockchain technology	

TABLE 16.2 Article summary data extraction table

Title
Author(s)
Methodology
Objective(s)
Finding(s)

16.3.2 Conceptual Model

Analyzing the impact of blockchain technology in smart city governance requires understanding of influential factors on the acceptance of blockchain technology. One classic model for the measurement of technology acceptance in any area is the technology acceptance model (TAM). Specifically designed for smart cities, Sepasgozar et al. (2019) combined TAM with the social cognitive theory (SCT) and created an Urban Service Technology Acceptance Model (USTAM). Based on this model, features of blockchain can be analyzed accordingly to identify the potential impact of blockchain on smart city governance.

16.3.2.1 Social Cognitive Theory (SCT)

SCT is an important theory in human behaviour studies (Bandura, 2001) and introduces various concepts for understanding citizens' acceptance of technologies. The main goal of SCT is to measure the social influence on human behaviour. In relation to technology, this theory focuses on factors of self-efficacy, anxiety, and outcome expectation. Additional factors were identified by Sepasgozar et al. (2019) including: work facilitating, cost reduction, energy saving, and time saving. Work facilitation measures the users' satisfaction on the technology to support their daily tasks when they use it. Cost reduction, energy saving, and time saving measures the economic advantage of the technology in terms of saving money, reducing loss of energy, and preventing loss of time (Sepasgozar et al., 2019).

16.3.2.2 Technology Acceptance Model (TAM)

TAM is recognized as one of the most commonly used theories in describing the acceptance of any information technology. This model by Davis (1989) identified two major factors affecting an individual's acceptance of technologies: Perceived Usefulness (PU) and Perceived Ease of Use (PEOU). PU measures people's belief on how useful this technology can be in improving their task performance. PEOU measures their belief on whether the technology can be used easily and can make their job easier (Venkatesh et al., 2003) (Figure 16.1).

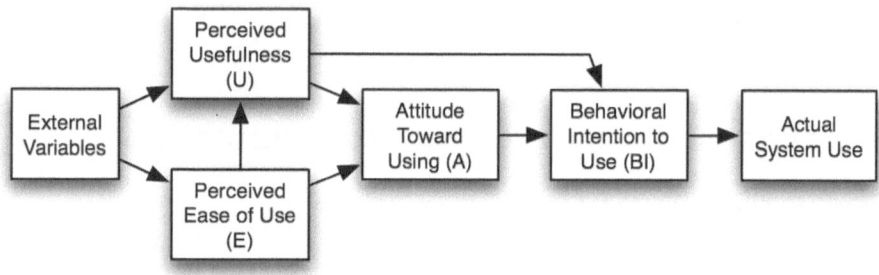

FIGURE 16.1 Technology Acceptance Model (Davis et al., 1989).

16.3.2.3 *Urban Service Technology Acceptance Model (USTAM)*

USTAM expanded the contents in TAM and SCT, created a new model that specifically focuses on urban services, and can be applied in evaluating the acceptance of blockchain in smart city development. In the USTAM, a total number of 12 factors were identified as factors that influence people's acceptance of technology in the case of urban development. Besides the original TAM model, there are other important factors including Perceived Security, Relative Advantage, Compatibility, and Reliability. Table 16.3 presents the definitions of these factors.

16.4 LITERATURE REVIEW

This review evaluates the USTAM model against other models, addressing the research question and extending the model. Many studies have been conducted on technology acceptance factors in smart

TABLE 16.3 Definitions of factors

NAME OF FACTOR	DEFINITION
Perceived security	Perception that technology is secure and privacy is ensured (Carter & Bélanger, 2005)
Relative advantage	Perception that technology is superior over its competitors (Rogers, 2010)
Perceived ease of use	'The degree to which a person believes that using a particular system would be free of effort' (Davis, 1989, p. 2)
Perceived usefulness	'The degree to which a person believes that using a particular system would enhance his or her job performance' (Davis, 1989, p. 2)
Compatibility	Perception that technology is compatible with existing hardware and software (Moore & Benbasat, 1991)
Reliability	Perception that technology is capable of providing reliable service (Carter & Bélanger, 2005)
Service quality	Perception that technology provides precise service as it should (Bandura, 2001)
Self-efficacy	Confidence among the users on their knowledge of the technology (Sepasgozar et al., 2019)
Work facilitating	Belief that technology will increase task efficiency (Sepasgozar et al., 2019)
Cost reduction	Belief that technology will reduce cost (Chiu et al., 2006)
Energy saving	Belief that technology will save energy (Chiu et al., 2006)
Time saving	Belief that technology will save time (Chiu et al., 2006)

cities, impact of blockchain, and blockchain initiatives in smart cities. Twenty out of 65 articles were identified to be highly relevant to this study.

16.4.1 Prior Research on Technology Acceptance

In 2017, a research study was conducted in Taiwan investigating the citizens' perspectives of how ICT-based smart city government services can improve the quality of life (Yeh, 2017). A theoretical framework was created and evaluated through a set of questionnaires randomly distributed among adult Taiwan citizens living in smart cities. Out of 1189 responses, 1091 valid responses were analyzed, and the results suggested that all the factors in the framework are influential in a positive way. The six factors that positively influenced the acceptance and usage of a technology included innovation concept, personal innovativeness, city engagement, service quality, perceived privacy, and trust. In addition to USTAM, the study suggested that the factors of perceived security, service quality, and self-efficacy (personal innovativeness) are critical factors with a positive impact on the acceptance of technologies. Yeh (2017) also stated that innovation and city engagement are important factors to be taken into consideration in measuring citizens' acceptance of technology. Lastly, government trust was also identified as a factor that plays a part in technology adoption.

A research study conducted in India had a similar objective, attempting to investigate the factors influencing the adoption of ICT-based smart city services from the user's perspective (Chatterjee et al., 2018). A framework was proposed and then tested through a questionnaire randomly distributed to adults of different ages and genders. This framework introduced seven factors that might have an impact on people's adoption of technologies in smart cities: overall innovation and creativity, personal creativity and innovativeness, resident engagement, IT-enabled service quality, perceived privacy, perceived security, and trust level. Unlike USTAM, perceived security was separated into two different factors by Chatterjee et al. as perceived privacy and perceived security. Service quality and self-efficacy (personal creativity and innovativeness) again proved to be important in terms of technology adoption. Innovation, citizen engagement, and government trust were all recognized to be critical factors under the situation of people's perception of technology adoption in smart cities.

A different approach was carried out in 2018, based on a secondary model derived from TAM. Buyle et al. (2018) conducted a research project to analyze the factors that influence the adoption of data standards in smart cities. The model was called Technology Readiness and Acceptance Model (TRAM), which was designed to broaden the original TAM by adding technological readiness into the picture (Lin et al., 2007). A questionnaire was designed based on the feedback from a preliminary qualitative study. Their findings pointed out that innovativeness is a factor that positively impacts both perceived usefulness and perceived ease of use. The study also found that the perceived ease of use positively affects the intention of use, and has a positive impact on the perceived usefulness of the technology. The importance of self-efficacy was also emphasized, as the research revealed that the discomfort of lacking knowledge could cause a negative impact on the adoption of the technology.

Another study investigated the factors that influenced end user's acceptance of technology in smart cities in the United Arab Emirates (Almuraqab & Jasimuddin, 2017). This research, however, was fully based on a literature review. This research also utilized the well-known TAM as a basis. This study analyzed 29 articles from 2005 to 2015 and eventually identified eight additional factors that could have an impact on end user's acceptance of technologies in smart cities besides the TAM factors. These eight factors are awareness, facilitating conditions, social influence, perceived cost, perceived trust in government, perceived trust in technology, perceived risk, and perceived compatibility. There appears to be an overlap between the factors identified in this research and those proposed in USTAM. The USTAM factor of work facilitating was recognized also in this study as facilitating conditions, and self-efficacy was represented as awareness. Social influence, compatibility, and cost saving appeared in both studies as potential influence factors with positive impacts. The result of this study also included several factors that were not part of USTAM, including a negative factor, perceived risk, and a positive factor, perceived trust in government and technology.

A Malaysian research study investigated the determinants of the behavioural intentions of adopting the Internet of Things (IoT) in smart cities (Leong et al., 2017). This study proposed a framework using Unified Theory of Acceptance and Use of Technology (UTAUT) as the base models, and tested it through an online questionnaire. The researchers received 289 responses and the result of the analysis indicated that 5 factors appear to have a significant positive impact on the adaptation of IoT in smart cities in Malaysia. The five factors were performance expectancy, effort expectancy, hedonic motivation, price value, and smart perceived trust. Two of the factors from UTAUT (performance expectancy and effort expectancy) are equivalent to perceived usefulness and perceived ease of use in TAM, measuring similar variables. Hedonic motivation, price value, and smart perceived trust were three new factors not mentioned previously in USTAM. Hedonic motivation measures the enjoyment users receive from using a technology. Price value refers to the cost efficiency of the technology, and measures a trade-off between the perceived usefulness of the technology and the perceived cost of it. While USTAM did not include perceived trust, it was a factor commonly recognized by many other studies, indicating the amount of trust citizens possess toward the technology.

Chaiyasoonthorn et al. (2019) investigated the acceptance of new technologies in smart cities in Thailand. This study started with the classic TAM and added new factors that influence the community's acceptance of technology. The authors proposed a conceptual model with nine factors: perceived usefulness, perceived ease of use, social influence, perceived trust, compatibility, facilitating condition, perceived security, and two new factors – perceived prosperity and perceived sustainability. Perceived prosperity measures the overall financial and resourcing capabilities of the cities and their citizens and is expected to have a positive impact on citizens' acceptance of technology. Perceived sustainability is also a positive impact factor, measuring how well the technology may support the smart city in achieving long-run sustainability and resilience.

Slightly different from the previous studies, Marzooqi et al. (2017) introduced a different approach focusing on citizen's acceptance and engagement in e-Governance. The research proposed a G2C (government to citizen) SF (success factor) Framework by combining other existing frameworks. The G2C-SF Framework identified four major groups of factors that affect citizens' acceptance of e-Government system: environmental, organizational, behavioural, and technological. The environmental factor measures the stability of political environment of the cities and the current degree of government-citizen interaction. The organizational factor is close to the perceived usefulness in TAM, fulfilling the needs of the citizens and organizations. The behaviour factor includes the trust factor between citizens and the government, and the factor of ease of use based on the system's design. The technological factor encompasses security, accessibility, efficiency, and compatibility.

Much of the research mentioned above focused on the smart city side of the question, other studies started from the blockchain side. One such study specifically looked at the adoption of blockchain in supply chain (Francisco & Swanson, 2018). This research began with UTAUT, and then included four more factors with influence on the users' behaviour and intentions. The four factors are performance expectancy, effort expectancy, social influence, and facilitating conditions. Compared to USTAM, while social influence and facilitating conditions appear similar, the factors perceived ease of use and perceived usefulness are formulated differently as effort expectancy and performance expectancy. The limitation of this research was the lack of practical tests and questionnaires to test the reliability of the model.

Considering the critical success factor (CSF) of blockchain acceptance, recent research aimed at identifying what influences the successful implementation of blockchain-based cloud services (Prasad et al., 2018). There are some relevant factors here as cloud services are recognized as important resources in a smart city. This research identified 19 CSFs through a literature review and focused on the end users' perspective of blockchain: user engagement, trust in government, trust in technology, cost efficiency, energy efficiency, talent pool availability, leadership readiness, compatibility, reliability, security, and privacy. In terms of technology in smart cities, most of the factors overlap with USTAM. The new factors identified are talent pool availability and leadership readiness.

In 2016, a systematic literature review was carried out to understand the state-of-art of blockchain technology at that time (Yli-Huumo et al., 2016). The study identified seven challenges of blockchain: throughput, latency, size and bandwidth, security, wasted resources, usability, and versioning, and privacy as an additional challenge. Of the eight challenges, throughput, latency, versioning, and size and bandwidth are technical. The other four challenges are those that concern a non-technical person when making decisions about accepting blockchain technologies. The four factors are security, wasted resource, usability, and privacy. Security and privacy are common factors mentioned in most articles so far. Wasted resource refers to energy and cost concerns. The study also highlighted usability, that is, both the ease of use and the perceived usefulness of the technology.

Another study in 2018 intended to identify the advantages and disadvantages of blockchain technologies in education (Chen et al., 2018). While education is a part of a smart city, it is expected that there may not be as many factors involved as other research which covered the wider smart city. Nevertheless, the items recognized in this research have value in understanding the impact of blockchain in smart cities. The conclusions pointed out four major advantages of blockchain: reliability, security, trust, and efficiency. On the downside, they also recognized a few potential issues that came with blockchain. Firstly, the authors believed that in an educational scenario, blockchain might not function properly without sufficient human intervention. Secondly, the immutable nature of blockchain may prove as a double-edged sword, as it denies the possibility of necessary modifications. Thirdly, technical challenges remain as obstacles in adopting blockchain technology. Putting the findings as factors that could impact the acceptance of blockchain in smart city, we namely have perceived security, perceived trust, reliability, perceive usefulness (efficiency, immutability), and perceived ease to use (human intervention and technical challenges).

The systematic literature review conducted by Casino et al. (2019) investigated the current status of blockchain-based applications and issues that were revealed until then. The study identified several limitations of blockchain mainly from the technical side. The first one is the suitability of blockchain. As an alternative to a traditional relational database, a blockchain database cannot be adopted blindly for every case. It is critical for organizations to carefully analyze their own requirements. The second issue has to do with latency and scalability, as the relatively slow transaction rate limits the efficiency and scale of the blockchain system. The third issue is the sustainability of blockchain, due to the high energy consumption in many cases. Fourth, the concerns around security and privacy still remain. Although blockchain uses encryption-based mechanisms and is pseudonymous, it might not be enough against certain advanced attacks. The last issue has to do with the standardizing and interoperability of blockchains. Interaction and compatibility among different blockchain systems could potentially become an issue.

Another empirical study on e-Governance looked at the existing challenges in adopting blockchain technology (Batubara et al., 2018). The result of this research categorized challenges in adopting blockchain technology in e-Governance in three major categories which are environmental, organizational, and technological. Environmental challenges were the least mentioned group, where the main issue is about regulation and standard issues. Organizational challenges were the second most mentioned group, including the main issue of cost efficiency, requirement on governance model, and trust in government. Most identified challenges are still the technological challenges, from security, scalability, and reliability to compatibility, flexibility, and usability. These challenges would act as factors that bring negative effects to the user's acceptance of blockchain technology.

There were also researchers studying blockchain from the technology acceptance perspective. A 2018 study investigated blockchain in the sharing-economy from a technology adoption perspective, to find the factors that influence adoption (Tumasjan & Beutel, 2018). The authors constructed a conceptual model using the unified theory of acceptance and use of technology (UTAUT) as a basis with several other factors that they believed to be important. Their final conceptual model introduced four main factors. Performance expectancy measures the usefulness and effectiveness of the technology, similar to perceived usefulness in TAM, which the authors believed to have the strongest influence on the adoption of blockchain. Effort expectancy is the second factor that measures the ease of use of the technology and the amount of training/education required for users. Attitude is the third factor, measuring the users' intention to adopt the technology. This incorporates other factors such as social influence and facilitating conditions. The last factor proposed in their model is the pervasiveness and it measures how widely the technology is distributed across businesses.

While much research covers the technical side of blockchain technology, studies investigating the user's perspectives of this new technology also exist. A 2019 study looked at the potential of blockchain from the users' perspective. The study discussed the advantages and disadvantages of blockchain in building trust between humans and machines in different applications (Sawal et al., 2019). The authors concluded that, because of its decentralized nature, blockchain has the advantages of being reliable, resilient, secure, and transparent. They also pointed out the difficulty in transitioning from traditional systems.

Other research on blockchain from the user's perspective witnessed the challenges of adopting blockchain (Schlegel et al., 2018). The authors discovered that there is a limitation of an institutional nature, due to political and legal environments in different countries. Another challenge identified by the authors is the potential lack of willingness of users to change the way they work. The authors also mentioned another problem 'the blockchain never forgets' (Schlegel et al., 2018, p. 9). Once an error is made, it is not possible to completely remove it from the database.

16.4.2 Factors and Hypotheses

This section summarizes all the factors identified in the literature to answer the research question. In addition, the basic conceptual model USTAM is extended to propose a new conceptual model. Two other factors in USTAM, related advantage and time, were not supported in the reviewed literature and will not be considered in this study.

Perceived usefulness and perceived ease of use are the two factors retrieved from the original TAM. As part of one of the most widely used models on technology acceptance, these two factors have proven their importance. The first two hypotheses are related to these:

H1: Perceived usefulness has a positive impact on user's acceptance of blockchain technology in smart city governance.

H2: Perceived ease of use has a positive impact on user's acceptance of blockchain technology in smart city governance.

The first factor proposed by USTAM is perceived security, which measures people's perception of the security level of the technology and privacy (Sepasgozar et al., 2019), and is one of the most common factors. In a majority of the literature, security was mentioned as one of the concerns of blockchain and smart city development (Batubara et al., 2018; Casino et al., 2019; Chaiyasoonthorn et al., 2019; Chatterjee et al., 2018; Chen et al., 2018; Marzooqi et al., 2017; Prasad et al., 2018; Sawal et al., 2019; Schlegel et al., 2018; Sun et al., 2016; Yli-Huumo et al., 2016). Unlike USTAM where security and privacy were considered as one factor, many other studies identified security as a separate factor, including both blockchain-related concerns (Batubara et al., 2018; Casino et al., 2019; Sawal et al., 2019) and the challenge of ensuring security in smart city applications (Chaiyasoonthorn et al., 2019; Malchenko, 2020; Marzooqi et al., 2017).

H3: Lack of perceived security has a negative impact on user's acceptance of blockchain technology in smart city governance.

The next critical factor agreed by different studies is the compatibility of blockchain technology. Compatibility measures how easily blockchain applications can function with existing hardware and software environments, and how much extra effort may be needed (Sepasgozar et al., 2019). Nine studies believed that this factor has a significant impact on a user's intention of accepting the technology (Almuraqab & Jasimuddin, 2017; Batubara et al., 2018; Chaiyasoonthorn et al., 2019; Chen et al., 2018; Marzooqi et al., 2017; Prasad et al., 2018; Sawal et al., 2019; Tumasjan & Beutel, 2018; Yli-Huumo et al., 2016). Therefore, the hypothesis is:

> **H4**: Compatibility has a positive impact on users' acceptance of blockchain technology in smart city governance.

Another important factor is the reliability of blockchain, that is, after being implemented in real-world scenarios. Six recent studies identified reliability as a major challenge of the implementation of blockchain technology and an important factor in smart city development (Almuraqab & Jasimuddin, 2017; Chen et al., 2018; Francisco & Swanson, 2018; Leong et al., 2017; Prasad et al., 2018; Sawal et al., 2019). The hypothesis on reliability is:

> **H5**: Reliability has a positive impact on users' acceptance of blockchain technology in smart city governance.

Moving from the technology level to the service level, service quality has been identified by USTAM as a critical factor. Again, this factor was mentioned by six studies, pointing out that the better the quality of service a technology provides, the more people intend to adopt it (Chatterjee et al., 2018; Chen et al., 2018; Francisco & Swanson, 2018; Leong et al., 2017; Tumasjan & Beutel, 2018; Yeh, 2017). The hypothesis on service quality is:

> **H6**: Service quality has a positive impact on users' acceptance of blockchain technology in smart city governance.

Self-efficacy was a factor introduced in USTAM, and five studies reviewed here support it. Almuraqab and Jasimuddin (2017) used the word awareness to represent this concept, while attitude was used by Tumasjan and Beutel (2018). Prasad et al. (2018) pointed out that leadership readiness and talent pool are challenges faced by blockchain applications, with agreement from Glomann et al. (2019) who believed that there is an onboarding challenge. Taking into consideration the limitations of institutions (Schlegel et al., 2018), self-efficacy is a critical factor for the likelihood of end-user's adoption of blockchain. The hypothesis on self-efficacy is:

> **H7**: Self-efficacy has a positive impact on users' acceptance of blockchain technology in smart city governance.

With regard to work facilitating, in many articles, the name facilitating condition was used instead, but still measuring how people believe the infrastructures were available to support the system (Almuraqab & Jasimuddin, 2017; Chaiyasoonthorn et al., 2019; Francisco & Swanson, 2018; Prasad et al., 2018). The hypothesis on facilitating conditions is:

> **H8**: Facilitating conditions have a positive impact on users' acceptance of blockchain technology in smart city governance.

Cost saving and energy saving are two other factors for technology acceptance. Sometimes they appear to be combined. Chaiyasoonthorn et al. (2019) and Casino et al. (2019) used the factor sustainability for measuring both. Yli-Huumo et al. (2016) only focused on the energy consumption of blockchain, while Prasad et al. (2018) separately discussed cost efficiency and energy efficiency. Here they are presented as two hypotheses:

H9: Cost saving has a positive impact on users' acceptance of blockchain technology in smart city governance.

H10: Waste of energy has a negative impact on users' acceptance of blockchain technology in smart city governance.

16.4.3 Factors Additional to USTAM

The first factor that was proposed by several other studies but was not included in USTAM is innovativeness. Three other studies on smart city found that innovativeness has a positive impact on citizens' acceptance of new ICT services in smart cities (Buyle et al., 2018; Chatterjee et al., 2018; Yeh, 2017). The hypothesis on innovativeness is:

H11: Innovativeness has a positive impact on users' acceptance of blockchain technology in smart city governance.

The next factor mentioned by studies on both smart city and blockchain is citizen engagement. Prasad et al. (2018) listed user engagement as a critical factor for successful blockchain initiatives, while Chatterjee et al. (2018) and Yeh (2017) found it important for technology acceptance in smart cities. The hypothesis on citizen engagement is:

H12: Citizen engagement has a positive impact on users' acceptance of blockchain technology in smart city governance.

In the literature, there is one common factor not included in USTAM, that is, perceived trust. Nine articles believed that trust is critical for a technology to be adopted (Almuraqab & Jasimuddin, 2017; Chaiyasoonthorn et al., 2019; Chatterjee et al., 2018; Leong et al., 2017; Prasad et al., 2018; Yeh, 2017). Additionally, the potential of blockchain to increase trust levels across the system is mentioned (Batubara et al., 2018; Chen et al., 2018; Marzooqi et al., 2017; Sun et al., 2016), in addition to citizens' trust in government (Almuraqab & Jasimuddin, 2017; Malchenko, 2020; Prasad et al., 2018).

H13: Perceived trust has a positive impact on users' acceptance of blockchain technology in smart city governance.

Social influence describes how the community may influence people's intention of using a new technology (Almuraqab & Jasimuddin, 2017; Chaiyasoonthorn et al., 2019; Francisco & Swanson, 2018; Glomann et al., 2019). The related hypothesis is:

H14: Social influence has a positive impact on users' acceptance of blockchain technology in smart city governance

Because blockchain is a recent technology, this brings up the challenge of finding qualified human resources. Five articles have discussed this, pointing out that more educated experts are required, as well as readiness for the administration and maintenance after setup (Chen et al., 2018; Duy et al., 2018; Glomann et al., 2019; Prasad et al., 2018; Schlegel et al., 2018). The hypothesis on human resources is:

H15: Lack of human resource has a negative impact on users' acceptance of blockchain technology in smart city governance.

As mentioned with perceived security, many studies considered perceived security and perceived privacy as two unique factors. Eleven articles mentioned both opportunities and challenges for privacy. Perceived privacy will be considered on its own here (Batubara et al., 2018; Casino et al., 2019; Chatterjee et al., 2018; Chen et al., 2018; Duy et al., 2018; Marzooqi et al., 2017; Prasad et al., 2018; Schlegel et al., 2018; Sun et al., 2016; Yeh, 2017; Yli-Huumo et al., 2016). The hypothesis on perceived privacy is:

H16: Lack of perceived privacy has a negative impact on users' acceptance of blockchain technology in smart city governance.

16.4.4 Summary

While studying the users' acceptance to e-Government systems in Dubai, Marzooqi et al. (2017) placed the success factors of e-Government into four categories: environmental factors from social and political environment level, organizational factors from organizational level, behavioural factors from individual level, and technological factors from the technology itself. These four categories are used in the proposed conceptual model here, which consists of sixteen factors that impact the users' acceptance of blockchain technology for smart city governance.

The factors under the behavioural category are self-efficacy, perceived trust, perceived usefulness, and perceived ease of use, all of which have a positive impact on users' acceptance of blockchain. The environmental factors are citizen engagement, social influence, and energy saving. Citizen engagement and social influences are positive factors while energy saving is a negative factor since energy costs are a challenge for blockchain. Factors categorized as technological are: perceived security, reliability, and service quality. While reliability and service quality positively impact people's acceptance, perceived security has a slightly negative impact, due to uncertainty. Organizational factors are: facilitating conditions, compatibility, perceived privacy, human resource, innovativeness, and cost reduction. Facilitating conditions, compatibility, innovativeness, and cost reduction are positive factors. Perceived privacy and human resources may limit blockchain use in smart city initiatives (Figure 16.2).

The new conceptual model provides a wider view of users' acceptance of blockchain in smart city governance. Among the 16 factors, 12 of them support adoption: citizen engagement, social influence, compatibility, facilitating conditions, innovativeness, cost reduction, self-efficacy, perceived trust, perceived usefulness, perceived ease of use, reliability, and service quality. The other four factors may inhibit user's acceptance of blockchain: energy saving, perceived privacy, human resource, and perceived security.

16.5 PROPOSED FUTURE METHODOLOGY

16.5.1 Approach

In Section 16.3, conceptual framework has been presented with factors that have a potential impact on user's acceptance of blockchain in smart city governance. This section introduces the proposed methodology that will be used to verify the conceptual model in a future study.

While qualitative research excels at understanding concepts and thoughts, quantitative research may better assist with the evaluation and verification of theories and assumptions (Streefkerk, 2019). In this

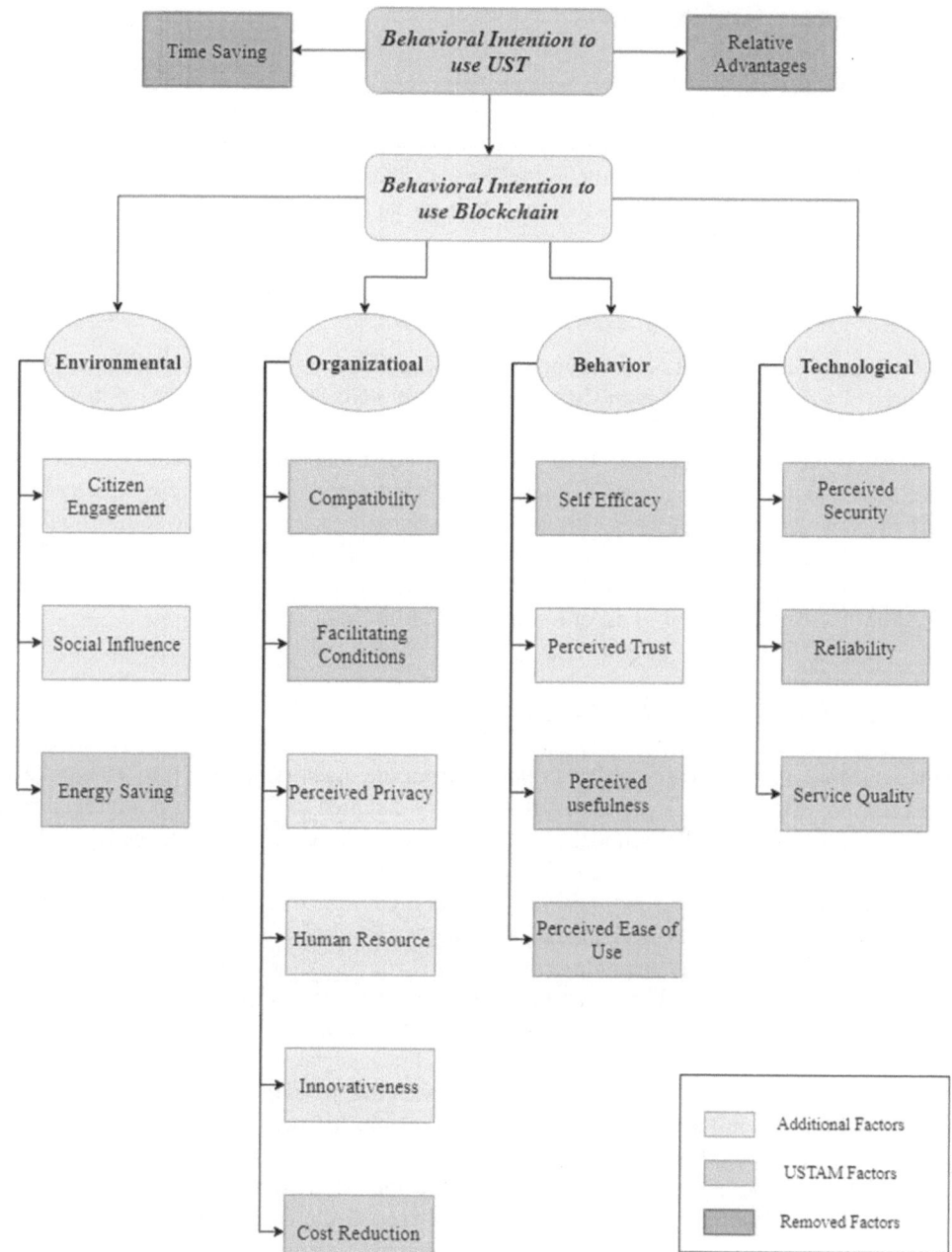

FIGURE 16.2   Conceptual Model.

study, a quantitative research will be conducted to analyze the reliability and accuracy of the conceptual model.

The research questions in this study are: What are the factors that impact user's acceptance of blockchain technology in smart city governance? How do these factors impact the acceptance of blockchain technology in smart city governance?

16.5.2 Research Instrument

To verify the proposed conceptual model, a quantitative method will be employed, and a structured questionnaire will be used as the instrument to collect data. The questionnaire will cover all 16 factors included in the conceptual model with questions adopted from previous studies with relevant objectives and adjusted to fit this study. The questionnaire will use a Likert scale from 1 for strongly disagree to 5 for strongly agree. Below is the list of factors and their measurement (Table 16.4).

16.5.3 Data Sample

In New Zealand, Land Information New Zealand (LINZ) has been working with Auckland, Wellington, and Christchurch city councils on a Smart City Programme to increase the 'smartness' of the cities (Smart Cities, 2020). As the objective of this study is to investigate the factors impacting user's acceptance of blockchain in smart city government and their effect, the target respondents would be citizens in these three major cities. According to statistics, the combined resident population of these three cities equates to about half of New Zealand (Stats NZ, 2020). Using a sample size calculation tool, to achieve 95% confidence interval with 5% margin of error, the minimum sample size required would be 385 ('Sample Size Calculator | SurveyMonkey', 2020). The data collection will be conducted through online communities (Reddit, Twitter, Facebook, Discord, and other blockchain forums).

16.5.4 Data Analysis Methodology

To verify the conceptual model with the data collected from the questionnaire, a frequency analysis will be performed. Based on the Likert scale with 1 for strongly disagree to 5 for strongly agree, a frequency analysis can reflect the respondents' tendency of the questions. An example of frequency analysis is as follows: the numbers are not actual results (Figure 16.3; Table 16.5).

Based on the result of the frequency analysis we can recognize the tendency of the participants. A flat line means participants are neutral on the statements, otherwise the participants may either be agreeing or disagreeing with the statements.

A linear regression analysis will be used to determine the relationship between each factor and the users' intention (UI) toward blockchain-based systems. The linear regression analysis will utilize the Analysis ToolPak add-in for Microsoft Excel. The type of linear regression model here is a simple linear regression model, which measures the outcome (UI) and each independent variable. The main goal is not to find out a model that quantitatively describes the relationship between the factors and user intention. Instead, the model is introduced to validate whether the hypothesis on the nature of each relationship is correct. We will look at the coefficient of correlation and the coefficient of determination. These two values will be used to decide the reliability of the hypotheses in the conceptual model.

Cronbach's Alpha is a commonly used measurement for reliability and internal consistency (Colquitt, 2001). For a questionnaire to be considered "acceptable" the Cronbach's alpha value would be above 0.7, the higher the better (Cronbach, 1951). The equation to calculate Cronbach's Alpha is as follows (Mondal & Mondal, 2017):

$$\alpha = \left(\frac{k}{(k-1)} \right) * \left(1 - \left(\left(\sum s_i^2 \right) / s_t^2 \right) \right)$$

In this equation, k is the number of items (questions) in the questionnaire, S_i is the standard deviation of the ith item, S_T is the standard deviation of the sum score. With the Likert scale ranging from 1 to 5, an Excel spreadsheet can be used to calculate the Cronbach's Alpha.

TABLE 16.4 Constructs of the framework

CONSTRUCT	MEASURE
Citizen engagement (CE)	• CE1: Citizens are well connected with the government through different blockchain-based services. • CE2: The government uses blockchain-based technologies to deliver service to the citizens. • CE3: Citizens participate in different community services using blockchain-based technologies. • CE4: Citizens are socially engaged with blockchain-based services. • CE5: Citizens are well educated to use blockchain-based technologies.
Social influence (SI)	• SI1: Other people in the community are intending to use blockchain-based services.
Energy saving (ES)	• ES1: Using blockchain technology reduces inter- and intra-city travel and thus reduces fuel cost. • ES2: Blockchain has high energy costs.
Compatibility (CM)	• CM1: Blockchain-based service can be used from anywhere remotely. • CM2: Blockchain-based services are available at any time.
Facilitating conditions (FC)	• FC1: Blockchain-based services provide good customer service. • FC2: Use blockchain-based technology makes users' tasks easier.
Perceived privacy (PP)	• PP1: Citizen information is not disclosed to unauthorized organizations or personnel. • PP2: Confidentialities of information remain protected. • PP3: Information is not provided to any government department without authority for related entities.
Human resource (HR)	• HR1: Human resource for the blockchain-based system is lacking.
Innovativeness (IN)	• IN1: Citizen's use of service is of their free will. • IN2: Systems are creatively designed for people from different backgrounds to use. • IN3: The blockchain-based services are visible to citizens and the citizens are aware of it. • IN4: Blockchain-based services are available on multiple platforms. • IN5: Users keep up with new technologies within their area of interest.
Cost reduction (CR)	• CR1: Blockchain reduces costs on paperwork. • CR2: Blockchain reduces costs on administration. • CR3: Blockchain reduces resource cost.
Self-efficacy (SE)	• SE1: Users believe they can use blockchain-based technologies efficiently. • SE2: Users believe they can use blockchain-based technologies successfully. • SE3: Users believe they can use blockchain-based technologies by themselves.
Perceived trust (PT)	• PT1: Government can protect citizens from security and privacy risks. • PT2: Citizens trust blockchain-based systems and believe in them. • PT3: The integrity of the systems can remain intact.
Perceived ease of use (PEU)	• PEU1: Learning to interact with blockchain-based systems will be easy for users. • PEU2: It will be easy for users to become skillful with blockchain-based technologies. • PEU3: Users will find it easy to use blockchain-based systems. • PEU4: Users will find it easy to work with blockchain-based technologies and do what they want to do.
Perceived usefulness (PU)	• PU1: Users would find blockchain-based technology useful.

(Continued)

TABLE 16.4 (Continued) Constructs of the framework

CONSTRUCT	MEASURE
	• PU2: Using blockchain-based systems gives me better control.
	• PU3: The use of blockchain-based systems in my job increases my productivity and efficiency.
	• PU4: Users think blockchain-based systems would provide valuable services to them.
	• PU5: The use of the blockchain-based system makes users' jobs easier.
Perceived security (PS)	• PS1: The security aspect of blockchain technology will not be compromised.
	• PS2: People working on the security of blockchain-based systems are well-trained.
	• PS3: Users are well trained to use blockchain-based services safely and securely.
	• PS4: Blockchain-based system provided to the users with high security features to protect the systems.
Reliability (RL)	• RL1: The systems can provide reliable service to citizens.
Service quality (SQ)	• SQ1: Using blockchain-based technologies prevents human errors.
	• SQ2: Blockchain-based technologies are efficient and user-friendly.
	• SQ3: The blockchain-based systems are regularly maintained to provide service with high quality.
	• SQ4: The functionalities of the blockchain-based systems are designed to meet the requirement of the users.
User Intension (UI)	• Users will use blockchain-based technologies in the future.

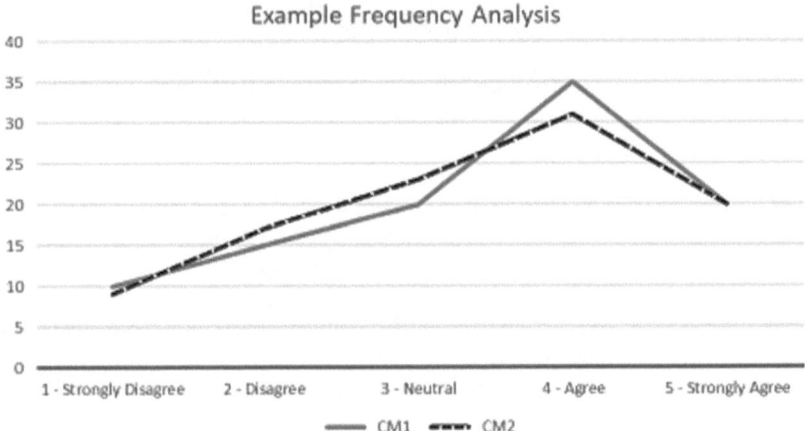

FIGURE 16.3 Example Frequency Analysis Graph.

TABLE 16.5 Example frequency analysis chart

	COMPATIBILITY 1 (CM1) (%)	COMPATIBILITY 2 (CM2) (%)
1 – Strongly Disagree	10	9
2 – Disagree	15	17
3 – Neutral	20	23
4 – Agree	35	31
5 – Strongly Agree	20	20

16.6 DISCUSSION

This study proposed a conceptual model based on USTAM and a systematic literature review. The conceptual model identified 16 factors that could have an impact on users' acceptance of blockchain technology in smart cities. This research may provide some motivation and useful information for governments and organizations to implement blockchain-based projects and services, especially in smart city sectors.

In the literature review, perceived security and perceived privacy appear to be the most mentioned factors (in 11 articles each), as challenges. It will be critical for organizations and government departments to evaluate the security and privacy aspects as they implement any blockchain-based services. Compatibility and perceived trust are the second most mentioned factors, both within nine articles. Compatibility is important to consider during the design of blockchain-based systems from a technological perspective. On the other hand, organizations and government departments need to carefully evaluate their public credit before boldly introducing blockchain-based systems, as the public may not easily adopt due to a lack of trust in them.

Among the other factors, based on how many times they were mentioned by the articles it appears that factors that related more to the technology itself have more attention than those related more to the user experience. On the technological side, reliability, service quality, cost reduction, and energy saving all received six citations out of the twenty articles, and human resource had a total of five mentions. On the other hand, self-efficacy was the only factor that appeared in five articles; other factors like facilitating conditions and social influence had four mentions. Innovativeness and citizen engagement received the least mention from the literatures, each has only three articles that recognized them as critical factors.

Based on the findings from the literature review, it appears that at the current stage, factors that could potentially impact users' acceptance are related more to the technological side rather than the user experience. This can be considered a limitation of this conceptual model because the use of blockchain is still in an early phase with a relatively small number of blockchain services currently in use. This study also has not covered general broadband access and penetration issues, and we cannot always presume that every citizen has access to or can afford high-speed broadband (Reddick et al., 2020). Two other limitations exist in this study. First, this work does not involve implementation. Second, there may be bias in the existing literature as most articles are application focused and may highlight the positive factor to justify their use case.

A questionnaire has been designed to collect data, and a set of techniques have been chosen to analyze the results and test the reliability of the proposed conceptual model. In order for participants to better understand smart city services, it can be interesting to show them a brief computer simulation so that they can better visualize the context. For this purpose, mock graphics of the following applications can be helpful: land title registrations, large asset inventory, contract management, special health records, financial incentives, and private records. All of these involve data that may be needed by multiple stakeholders but should at the same time be protected. The creation of the simulation requires further research and information gathering with experts and users. Furthermore, the simulation can be informed by the new model here, and the researchers can see to it that the simulation will help participants give more informed and fair responses. The simulation can also include notes to briefly explain the logic and algorithm of the application that is being illustrated.

16.7 CONCLUSION

This study investigated and listed factors that impact users' acceptance of blockchain in smart city governance, and understanding whether those factors are positive or negative. Starting with USTAM, a new conceptual model was proposed based on a literature review of related studies in recent years. The new conceptual model includes 12 positive impact factors: citizen engagement, social influence, compatibility, facilitating conditions, innovativeness, cost reduction, self-efficacy, perceived trust, perceived usefulness, perceived ease of use, reliability, and service quality, and four potential negative impact factors: energy saving, perceived privacy, human resource, and perceived security.

Perceived trust is one of the commonly cited factors, when the applications (such as blockchain) may be difficult for non-technical citizens to understand. It is important to be able to describe these applications in ways that make sense to human end users. White-box and explainable learning approaches can help decision makers and users better keep track of the development of complex smart city systems and trust them in the future.

Future research can identify and focus on one area of smart city that is most suitable for integrating blockchain. Then a future study can run a pilot study implementation, use the new conceptual model and questionnaire, and measure and evaluate the strengths and benefits of the new smart city application.

REFERENCES

Alkandari, A, Alnasheet, M, & Alshekhly, I (2012). Smart cities: Survey. *Journal of Advanced Computer Science and Technology Research, 2*(2), 79–90.

Almuraqab, N, & Jasimuddin, S (2017). Factors that influence end-user's adoption of smart government services in the UAE: A conceptual framework. *Electronic Journal of Information Systems Evaluation, 20*(1), 11–23.

Bandura, A (2001). Social Cognitive Theory: An agentic perspective. *Annual Review of Psychology, 52*(1), 1–26. doi: 10.1146/annurev.psych.52.1.1.

Batagan, L (2011). Smart cities and sustainability models. *Informatica Economica, 15*(3), 80–87.

Batty, M, Axhausen, K, Giannotti, F, Pozdnoukhov, A, Bazzani, A, Wachowicz, M, Ouzounis, G, & Portugali, Y (2012). Smart cities of the future. *European Physical Journal, 214*, 481–518.

Batubara, F, Ubacht, J, & Janssen, M (2018). Challenges of blockchain technology adoption for e-government. In *Proceedings of the 19th Annual International Conference on Digital Government Research Governance in The Data Age*. doi: 10.1145/3209281.3209317.

Buyle, R, Van Compernolle, M, Vlassenroot, E, Vanlishout, Z, Mechant, P, & Mannens, E (2018). Technology Readiness and Acceptance Model as a predictor for the use intention of data standards in smart cities. *Media and Communication, 6*(4), 127. doi: 10.17645/mac.v6i4.1679.

Carter, L, & Bélanger, F (2005). The utilization of e-government services: Citizen trust, innovation and acceptance factors. *Information Systems Journal, 15*(1), 5–25. doi: 10.1111/j.1365-2575.2005.00183.x.

Casino, F, Dasaklis, T, & Patsakis, C (2019). A systematic literature review of blockchain-based applications: Current status, classification and open issues. *Telematics and Informatics, 36*, 55–81. doi: 10.1016/j.tele.2018.11.006.

Chaiyasoonthorn, W, Khalid, B, & Chaveesuk, S (2019). Success of smart cities development with community's acceptance of new technologies. In *Proceedings of the 9th International Conference on Information Communication and Management, ICICM 2019*. doi: 10.1145/3357419.3357440.

Chatterjee, S, Kar, A, & Gupta, M. 2018. Success of IoT in Smart Cities of India: an empirical analysis. *Government Information Quarterly, 35*(3), 349–361. doi: 10.1016/j.giq.2018.05.002.

Chen, G, Xu, B, Lu, M, & Chen, N (2018). Exploring blockchain technology and its potential applications for education. *Smart Learning Environments, 5*(1). doi: 10.1186/s40561-017-0050-x.

Chiu, C, Hsu, M, & Wang, E (2006). Understanding knowledge sharing in virtual communities: An integration of social capital and social cognitive theories. *Decision Support Systems, 42*(3), 1872–1888. doi: 10.1016/j.dss.2006.04.001

Colquitt, J (2001). On the dimensionality of organizational justice: A construct validation of a measure. *Journal of Applied Psychology, 86*(3), 386–400. doi: 10.1037/0021-9010.86.3.386.

Cronbach, L (1951). Coefficient alpha and the internal structure of tests. *Psychometrika, 16*(3), 297–334. doi: 10.1007/bf02310555.

Davis, F (1989). Perceived usefulness, perceived ease of use, and user acceptance of information technology. *MIS Quarterly, 13*(3), 319. doi: 10.2307/249008.

Davis, F, Bagozzi, R, & Warshaw, P (1989). User acceptance of computer technology: A comparison of two theoretical models. *Management Science*, *35*(8), 982–1003. doi: 10.1287/mnsc.35.8.982.

Dong, Z, Luo, F, & Liang, G (2018). Blockchain: A secure, decentralized, trusted cyber infrastructure solution for future energy systems. *Journal of Modern Power Systems and Clean Energy*, *6*(5), 958–967. doi: 10.1007/s4 0565-018-0418-0.

Duy, P, Hien, D, Hien, D, & Pham, V (2018). A survey on opportunities and challenges of Blockchain technology adoption for revolutionary innovation. In *Proceedings of the Ninth International Symposium on Information and Communication Technology*. doi: 10.1145/3287921.3287978.

Francisco, K, & Swanson, D (2018). The supply chain has no clothes: Technology adoption of Blockchain for supply chain transparency. *Logistics*, *2*(1), 2. doi: 10.3390/logistics2010002.

Glomann, L, Schmid, M, & Kitajewa, N (2019). Improving the Blockchain user experience: An approach to address Blockchain mass adoption issues from a human-centred perspective. *Proceedings of AHFE 2019: Advances in Intelligent Systems and Computing*, 608–616. doi: 10.1007/978-3-030-20454-9_60

Ibanez, L, Simperl, E, Gandon, F, & Story, H (2017). Redecentralizing the web with distributed ledgers. *IEEE Intelligent Systems*, *32*(1), 92–95. doi: 10.1109/mis.2017.18.

Joshi, S, Saxena, S, Godbole, T, & Shreya, D (2016). Developing smart cities: An integrated framework. *Procedia Computer Science*, *93*, 902–909. doi: 10.1016/j.procs.2016.07.258.

Lee, Y, Kozar, K, & Larsen, K (2003). The Technology Acceptance Model: Past, present, and future. *Communications of the Association for Information Systems*, *12*. doi: 10.17705/1CAIS.01250.

Leong, GW, Ping, TA, & Muthuveloo, R (2017). Antecedents of behavioural intention to adopt internet of things in the context of smart city in Malaysia. *Global Business and Management Research*, *9*(4), 442–456.

Lin, C, Shih, H, & Sher, PJ (2007). Integrating technology readiness into technology acceptance: The TRAM model. *Psychology & Marketing*, *24*(7), 641–657.

Malchenko, Y (2020). From digital divide to consumer adoption of smart city solutions: A systematic literature review and bibliometric analysis. *Vestnik of Saint Petersburg University, Management*, *19*(3), 316–335. doi: 10.21638/11701/spbu08.2020.302.

Marzooqi, S, Nuaimi, E, & Qirim, N (2017). E-governance (G2C) in the public sector. In *Proceedings of the Second International Conference on Internet of Things, Data and Cloud Computing*. doi: 10.1145/3018896.3025160.

Mondal, H, & Mondal, S (2017). Calculation of Cronbach's alpha in spreadsheet: An alternative to costly statistics software. *Journal of the Scientific Society*, *44*(2), 117. doi: 10.4103/jss.jss_18_17.

Moore, G, & Benbasat, I (1991). Development of an instrument to measure the perceptions of adopting an information technology innovation. *Information Systems Research*, *2*(3), 192–222. doi: 10.1287/isre.2.3.192.

Ølnes, S, Ubacht, J, & Janssen, M (2017). Blockchain in government: Benefits and implications of distributed ledger technology for information sharing. *Government Information Quarterly*, *34*(3), 355–364. doi: 10.1016/j.giq.2 017.09.007.

Pilkington, M (2015). Blockchain technology: Principles and applications. *Research handbook on digital transformations* (pp. 225–253). Edward Edgar Publishing. doi: 10.4337/9781784717766.00019

Prasad, S, Shankar, R, Gupta, R, & Roy, S (2018). A TISM modeling of critical success factors of blockchain based cloud services. *Journal of Advances in Management Research*, *15*(4), 434–456. doi: 10.1108/jamr-03-2018-0027.

Reddick, C, Enriquez, R, Harris, R, & Sharma, B (2020). Determinants of broadband access and affordability: An analysis of a community survey on the digital divide. *Cities*, *106*, 1–12.

Rogers, EM (2010). *Diffusion of innovations*. Simon and Schuster.

Sample Size Calculator: Understanding Sample Sizes | SurveyMonkey (2020). www.surveymonkey.com/mp/sample-size-calculator/

Sawal, N, Yadav, A, Tyagi, D, Sreenath, N, & G, R (2019). Necessity of Blockchain for building trust in today's applications: A useful explanation from user's perspective. doi: 10.2139/ssrn.3388558

Schlegel, M, Zavolokina, L, & Schwabe, G (2018). Blockchain technologies from the consumers' perspective: What is there and why should who care? In *Proceedings of the 51St Hawaii International Conference on System Sciences*. doi: 10.24251/hicss.2018.441.

Sepasgozar, S, Hawken, S, Sargolzaei, S, & Foroozanfa, M (2019). Implementing citizen centric technology in developing smart cities: A model for predicting the acceptance of urban technologies. *Technological Forecasting and Social Change*, *142*, 105–116. doi: 10.1016/j.techfore.2018.09.012.

Shen, C, & Pena-Mora, F (2018). Blockchain for cities: A systematic literature review. *IEEE Access*, *6*, 76787–76819. doi: 10.1109/access.2018.2880744.

Smart Cities (2020). https://www.linz.govt.nz/about-linz/what-were-doing/projects/smart-cities

Stats NZ (2020). *Subnational population estimates: As of 30 June 2020 (provisional)*.

Streefkerk, R (2019). *Qualitative vs. quantitative research: Differences & methods.* https://www.scribbr.com/methodology/qualitative-quantitative-research/

Sun, J, Yan, J, & Zhang, K (2016). Blockchain-based sharing services: What blockchain technology can contribute to smart cities. *Financial Innovation*, 2(1). doi: 10.1186/s40854-016-0040-y.

Swan, M (2015). *Blockchain.* O'Reilly Media, Inc.

Thakker, D, Mishra, BK, Abdullatif, A, Mazumdar, S, & Simpson, S (2020). Explainable Artificial Intelligence for developing smart cities solutions. *Smart Cities*, 3(4), 1353–1382. doi: 10.3390/smartcities3040065.

Tumasjan, A, & Beutel, T (2018). Blockchain-based decentralized business models in the sharing economy: A technology adoption perspective. *Business transformation through blockchain* (pp. 77–120). Palgrave Macmillan. doi: 10.1007/978-3-319-98911-2_3

Venkatesh, V, Morris, M, Davis, G, & Davis, F (2003). User acceptance of information technology: Toward a unified view. *MIS Quarterly*, 27(3), 425. doi: 10.2307/30036540.

Xie, J, Tang, H, Huang, T, Yu, F, Xie, R, Liu, J, & Liu, Y (2019). A survey of Blockchain technology applied to smart cities: Research issues and challenges. *IEEE Communications Surveys & Tutorials*, 21(3), 2794–2830. doi: 10.1109/comst.2019.2899617.

Yeh, H. (2017). The effects of successful ICT-based smart city services: from citizens' perspectives. *Government Information Quarterly,* 34(3), 556-565. doi: 10.1016/j.giq.2017.05.001.

Yigitcanlar, T (2020). Smart city beyond efficiency: Technology-policy-community at play for sustainable urban futures. *Housing Policy Debate*, 31(1), 88–92. doi: 10.1080/10511482.2020.1846885.

Yli-Huumo, J, Ko, D, Choi, S, Park, S, & Smolander, K (2016). Where is current research on Blockchain technology? A systematic review. *PLOS One*, 11(10), e0163477. doi: 10.1371/journal.pone.0163477.

Explainable AI in Machine/Deep Learning for Intrusion Detection in Intelligent Transportation Systems for Smart Cities

17

Andria Procopiou and Thomas M. Chen
City, University of London, London, UK

Contents

DOI: 10.1201/9781003172772-17

17.1 INTRODUCTION

Overpopulation and urbanization comprise two critical problems in our society. As a result, numerous issues are caused, including environmental pollution, emissions exhaustion, global warming, energy waste, health problems, public transport congestion, and overcrowding. To overcome them, the concept of smart cities has emerged (Deakin & Waer, 2011). Smart cities aim to improve their citizens' overall quality of life, protect the environment by reducing the overall pollution produced and promote sustainability. One of the most important and vital transformations of the smart city's networks is its transportation system. In modern cities the dependence on road transportation is more than evident for the citizens, especially during the last decades due to overpopulation and urbanization. According to a United Nations report, the world's total urban population is expected to reach 2.5 billion by 2050 with more than 2.9 billion vehicles (United Nations, 2014). However, road transportation is responsible for more than one-fifth of the global CO_2 emissions, the main greenhouse gas, placing itself as a major contributor to global climate change. Hence, the need for smart and green transportation systems which are eco-friendly and cause low or no environmental impact is more essential than ever in smart cities. As a result, the concept of intelligent transportation systems (ITS) (Guerreiro et al., 2016) has emerged for a greener alternative (Xiao et al., 2015).

Road ITS consists of IoT, wireless communications, and the cloud, providing services and applications such as traffic congestion calls, notifications, emergency vehicle notification systems mapping and location technologies, automatic road enforcement, speed limits and collision monitoring, and dynamic traffic light sequence. Road ITS consists of various types of vehicles such as cars, buses, trucks, taxis, and bikes (Chauley, 2016). Road ITS also communicates with other important infrastructures in the context of smart city such as smart homes, smart airports, smart ports, and telecom networks to provide numerous benefits to the citizens and improve their life. However, cyber security is a primary concern to maintaining the operations of these networks. Road ITS is a type of cyber-physical system, a new category of systems which encompasses computational and physical capabilities interacting with the environment and humans. In cyber-physical systems, fast and precise monitoring of assets is vital. Therefore, the deployment of sensors, actuators, and controlling IoT devices is necessary. These devices constantly exchange data and relevant servers through the Internet. Hence, they can be compromised by adversaries and participate in cyberattacks. Their compromise is easier since security countermeasures are minimal to non-existent due to the constraints of such devices in power and memory. In contrast to traditional ICT networks where confidentiality is the most secure principle, availability takes the lead (Liu et al., 2012). The most popular attacks that affect availability include various types of network threats. Network attacks can target critical nodes to affect the ITS's normal functioning. Additionally, the interconnection of the ITS domains with each other, as well as with multiple and heterogeneous domains in a smart city context, gives the opportunity to attackers to conduct Distributed Denial of Service (DDoS) attacks from one network affecting other network or networks that are critical to the normal functioning of the smart city. This is particularly dangerous as it can put multiple infrastructures of the smart city at risk as there is no isolation of the networks. An attack against cyber-physical systems of the smart city is particularly dangerous since it can lead to human hazards and even result in loss of lives due to their constant interaction with them.

Therefore, the identification of vulnerabilities and their impact when exploited is more than vital. In this chapter, we review the most important security issues and challenges regarding Road ITS in smart cities networks and the different countermeasures proposed in the literature using ML and DL techniques. Based on the available literature we discuss on how explainable artificial intelligence (XAI) could assist in the improvement of such systems. The first section contains background knowledge regarding Road ITS networks, specifically the most relevant technologies and communication architectures deployed. In the second section, the most notable and critical security vulnerabilities, threats, and attacks are presented. In the next section, notable studies in the literature that focus on detecting malicious traffic in Road ITS networks using machine learning and other intelligent techniques are described. Proceeding, an extended discussion based on the presented related studies is conducted, identifying the relevant trends and the open issues that should be addressed in the future. Based on a subset of the identified open issues we were motivated to conduct an experimental case study on application-layer DDoS attacks and their impact on the Road ITS network, other interconnected networks; the concept of smart city is discussed to a considerable extent. Furthermore, we review on how XAI can be applied in IDS systems, conducting an in-depth discussion. Finally, concluding remarks are made.

17.2 ROAD ITS TECHNOLOGIES IN SMART CITIES

17.2.1 Vehicular Area Network (VANET)

For successful communication between intelligent vehicles in the Road ITS and the usage of information and telecommunication technologies sophisticated, smart wireless communication technologies had to be designed. Hence, Vehicular Area Networks (VANETs) were created. The smart vehicles deployed in VANETs are equipped with a plethora of sensing and communication technologies. The sensors gather relevant information about the vehicle and transmit it to relevant authorities through either another vehicle or Road Side Units (RSU). RSU is a static entity deployed on the road and can take the form of a traffic light, speed camera, or simply a base station. Also, information from other smart networks deployed in a smart city can be sent to vehicles through them, such as messages from a smart home's smart appliances to the owner's vehicle. A detailed description of all types of communication technologies deployed in vehicles in Road ITS is presented below.

17.2.2 VANET Communications

Vehicle-to-Vehicle (V2V) Communication: V2V forms the wireless communication between two vehicles. Through it, vehicles can communicate and exchange information such as vehicle tracking, speed, basic safety messages (Harrington et al., 2017).

Vehicle-to-Infrastructure (V2I) Communication: V2I forms the wireless communication between vehicles and an infrastructure. Mainly, there are two types of infrastructures: central entities and third parties. Central entities are servers which provide important and vital information and applications for the effective and safe journey on the roads such as weather/traffic information and collision avoidance applications. Third parties form various authorities and organizations that provide applications and services to the owners of the vehicles such as remote controlling of smart appliances in their homes, entertainment applications to the devices in the vehicle and so on (Ahmad et al., 2016). RSU connects the vehicles to other infrastructures.

Vehicle-to-Pedestrian (V2P) Communication: V2P forms the wireless communication between vehicles and mainly pedestrians, but also cyclists and individuals with disabilities and remote mobility. Through it, potential collisions can be detected (Merdrignac et al., 2017).

Vehicle-to-Grid (V2G) Communication: V2G forms the communication between Plug-In Electric Vehicles (PEVs) and the smart grid. V2G aims to provide the vehicles with all the relevant information regarding when and where to charge their batteries as well as giving the opportunity to the smart grid and gain power from them to avoid load shedding events (Nasrallah et al., 2015).

Vehicle-to-Device (V2D) Communication: V2D forms the communication between a vehicle and an electronic device connected to the vehicle itself. Through it, multiple benefits can be gained such as car sharing, vehicle unlocking with an app, and better driver experience through mobile apps.

17.3 ROAD ITS IN SMART CITY ARCHITECTURE

Given the participation of a large number of users and devices, security mechanism plays a vital role in the IoT applications. The IoT system requires a two-fold security scheme – physical device and software level security. Unlike other security mechanisms in any other system, the IoT environment also has three basic properties to maintain security: confidentiality, integrity, and availability. Security concern in the IoT environment has an impact on providing an effective IoT solution. Besides, IoT security needs to add an extra layer of security as the IoT system is involved with vastly distributed autonomous physical devices and software applications.

IoT devices may have different security issues because of the diverse nature of functionalities and services offered. The security challenges in IoT can be divided into two categories: technological challenges and security challenges. The technological challenges arise when the hardware and device architecture are not well-equipped with the security mechanism. The security challenges, on the other hand, arise when cryptographic mechanisms are not up to the mark to prevent the attacker. Despite many existing security mechanisms proposed to address the security issues in the IoT network, there are still many security issues in the network. Thus, ensuring a high level of security assurance in IoT networks is still an open challenge.

The Road ITS architecture and its communication with other critical infrastructures in smart city is shown in Figure 17.1. The architecture consists of different types of road vehicles such as trucks, cars, trams, buses, taxis, motorbikes, and bicycles (Karagiannis et al., 2011). Vehicles are from both public and private sectors (Lefvy-Benchetton & Darra, 2015). In this architecture, all the vehicles, regardless of the sector they belong to, can communicate with each other in V2V Communications to exchange important messages such as speed, basic safety messages, sensing, and vehicle-related information. Furthermore, vehicles can communicate with pedestrians and cyclists through V2P communication so passengers' safety is ensured. V2D is used inside the vehicles so passengers can connect to the Internet and use entertainment services.

Finally, the motorway vehicles (cars, buses, taxis, trucks, and motorbikes) use V2G communications to either charge their electric battery or 'sell' power to the smart grid. All of the vehicles can

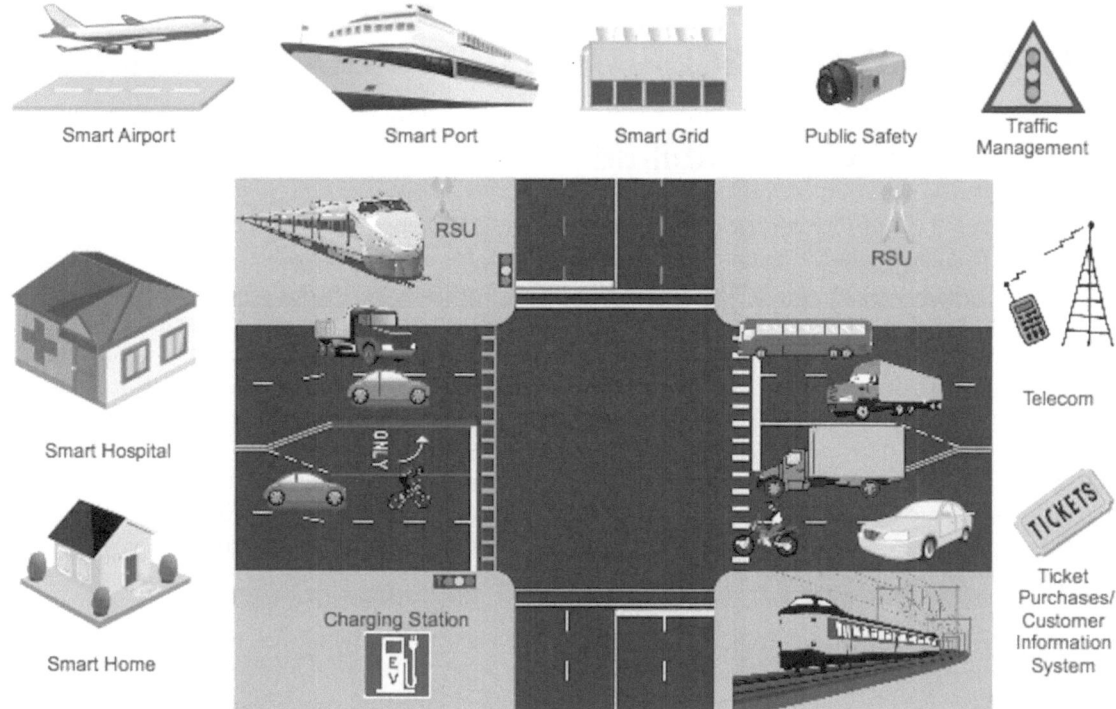

FIGURE 17.1 Road ITS in Smart City Architecture.

communicate with infrastructures using V2I communication, through the RSU. The private sector vehicles (mainly the cars) communicate with the owners smart homes to control the smart appliances and services in their homes, telecommunication networks to make phone calls, hospitals and/or police stations in case of emergency, road safety services such as forecasting and traffic applications and airports. The protocols used for these communications are GPS for location-based services, WiFi for intra-vehicular information dissemination and Internet services, 802.11p (WAVE) for communicating with other vehicles and IEC 61851 for communicating with the infrastructure. All of the vehicles from both private and public sectors communicate with traffic regulation systems which are in charge of organizing and coordinating the traffic lights on the road. The public transport vehicles also communicate with ticket purchase servers, passenger information systems, transport industry associations, and government bodies and services through LTE base stations (Menouar et al., 2017) (mainly 5G and SATCOM).

Road ITS also communicates with other ITS operators. One example is the smart airport. Road ITS communicates with the smart airport Infrastructure in the following ways. Smart Vehicles, mainly private cars, use the smart parking operations of the airport for parking purposes. Smart transportation vehicles (trains, buses, taxis, coaches, etc.) wait outside the airport to take the passengers to their destinations and vice versa. Finally, in the context of easy and comfortable passenger experience, dedicated cars wait outside to take the luggage of the passengers directly to the airport. Another smart infrastructure the Road ITS communicates with is the smart maritime transportation port. The logistics/freights trucks transmit goods and cargos from and to the smart port through maritime transportation vessels so they can be delivered to their destination. Additionally, if the smart port is used for tourism purposes the same operations described for smart airports above are carried out through Road ITS.

17.4 ROAD ITS SECURITY ISSUES AND CHALLENGES

17.4.1 Road ITS Vulnerabilities

Physical Access: All of the entities involved in the Road ITS networks from the smart vehicles to the sensors and from communication base stations to RSUs are physically accessible. There is no Therefore, they can be jeopardized by adversaries physically in an attempt to perform a cyber-physical attack.

(Wireless) Insecure Communications: In IoT and cyber-physical systems in general, the majority of communications between devices to exchange significant information is weak in terms of security. Some examples of weak security are weak or no passwords used at all and the absence of hashing and cryptographic algorithms makes the sessions easy to be hijacked by attackers. Also, since they are most of the time wireless they are under high risk.

Resource Constraints: The majority of IoT devices are low in resources such as memory, CPU power, and storage space. Therefore, successful but high in complexity, cryptographic and hashing algorithms, authentication protocols, and intrusion detection systems are unable to be deployed.

Software Bugs: IoT in Road ITS networks have their own software bugs and crushes. Since IoT specs are publicly available any software vulnerabilities can be identified and exploited.

Mobility: The majority of Road ITS entities deployed in the network are mobile. Since they are mobile they are harder to be effectively monitored.

Heterogeneity: The variety of different devices, technologies, communications, protocols and data exchanged make Road ITS a heterogeneous network which makes its distributed and effective monitoring hard to perform.

Interconnection of Networks: Road ITS communicates with multiple networks exchanging information to both the drivers but also to the infrastructure itself. Therefore, Road ITS is vulnerable to attacks being initiated from other networks and infrastructures in the context of the smart city.

17.4.2 Road ITS Threats and Attacks

Impersonation/Spoofing/Masquerading: In impersonation attacks, the adversary pretends to be a legitimate user either to make usage of the network resources or to disrupt the normal functioning of the network. This type of attack can be conducted in multiple layers, including the datalink, network/transport, and application layer. Examples of such an attack in the context of Road ITS are the impersonation of sensors or an RSU or a smart vehicle so it can communicate with the drivers themselves and carry on executing a more serious attack such as transmitting false information and threaten the drivers' lives (Ahmad et al., 2016).

Man in the Middle (MITM): MITM attacks involve the adversary acting as an intermediate agent between two legitimate ones such as a driver and an RSU or a fellow driver so messages exchanged between them can be intercepted. Also, an attacker can perform an MITM attack between an RSU and other entities from external networks in an attempt to disrupt the network's overall normal functioning. An example is an attacker dropping or alternating messages from services vital to the Road ITS such as weather forecasting, traffic congestion, or public securities to cause issues on the road (Ahmad et al., 2016).

Session Hijacking: In session hijacking, the adversary exploits a legitimate user session to gain unauthorized access to services. Examples of session hijacking attacks are the illegal capture of a driver's ID or an RSU's ID to gain unauthorized access so other more serious attacks are executed (Ahmad et al., 2016).

Information Disclosure: In such an attack, the adversary exploits private and confidential information about the owner of the vehicle and/or the services it might provide depending on the vehicle's type. The attacker can also exploit private information from external networks such as the drivers' destinations.

Data/Command Illegal Modification/Injection Attack: This attack involves the illegal modification of messages and/or commands to nodes in the Road ITS (Ahmad et al., 2016).

Replay Attack: In this attack the attacker is re-sending (replaying) that was already sent for the attacker's own benefit. Examples of such attacks can be sensors re-transmitting messages to the drivers so wrong decisions are taken or a driver/RSU re-transmitting misleading messages to other drivers to cause road disruption. Furthermore, an attacker can replay beacons frames so he can manipulate the location or the nodes routing tables (Ahmad et al., 2016).

Malware Spreading: Due to the high interconnectivity of the unauthorized installment, spreading of malware amongst nodes is a highly likely scenario. Malware can be spread from an infected RSU to the vehicles being connected to it or from one vehicle to the other (Ahmad et al., 2016).

Eavesdropping: In eavesdropping, an adversary actively intercepts a private communication by either sniffing data related to the communication or directly listening to the digital/analogue voice communication. Examples of such an attack in Road ITS are when the attacker illegally "listens" to the communications between vehicles, between the RSU and the vehicles, or between the RSU and external networks (Ahmad et al., 2016).

Distributed/Denial of Service: Distributed/Denial of Service (DDoS) attacks involve either one or more adversaries aiming to compromise the availability of a legitimate entity, by flooding it or by exploiting certain vulnerabilities. DDoS attacks can be conducted on multiple layers with some examples being Jamming, TCP SYN flooding, UDP flooding wormhole, blackhole, greyhole, sybil, and so on. Some examples are the conduction of DDoS attacks from vehicles to the RSU and vice versa, from one vehicle to others or from vehicles and RSU to external networks. DDoS attacks can prove to cause catastrophic impact on any critical infrastructure as the authors in (Procopiou & Komninos, 2015) effectively argued (Ahmad et al., 2016).

17.5 INTRUSION DETECTION AGAINST ROAD ITS ATTACKS

The most popular and effective countermeasure against network and malware attacks proposed is the usage of intrusion detection systems (IDS). IDS denotes a software or device which monitors a network or individual machines to detect any malicious activity or violation of rules and policies that might take place. An IDS consists of three main components:

Data Collection: In this component, all the relevant traffic (network data/host behaviour) is collected. Next, the data is analyzed according to a set of relevant features that potentially differentiate malicious from legitimate behaviour. In more detail, based on the set of features selected, incoming traffic is analyzed so values for each feature are calculated. These extracted values for the features chosen are forwarded to the detection engine.

Detection Engine: This component receives the extracted values for the features chosen and forwards them as input to the detection algorithm constructed. The chosen detection algorithm is constructed through a training phase. Various machine learning algorithms are used including supervised (e.g. support vector machines (SVM), random forest, k-nearest neighbours (KNN), Bayesian methods), unsupervised (e.g. clustering, outliers, isolation forest), reinforcement learning (e.g. Monte Carlo), or deep learning methods (e.g. deep belief, recurrent, convolutional and long–short-term-memories neural networks).

Alert: This component ultimately generates an alert if there is malicious behaviour present. In addition, IDS systems may encompass mitigation/prevention capabilities.

17.5.1 Related Studies

For the protection of Road ITS networks, various techniques have been proposed in the literature including machine learning algorithms, network packets analysis, and behaviour baseline construction. In this section, we present an extended set of the most notable studies on VANETs' intrusion detection using machine learning and deep learning techniques, by grouping them based on the attacks to detect. Highlighting the equal importance of detecting low- and high-intensity DDoS attacks from vehicles to the RSU, the authors in (Haydari & Yilmaz, 2018) proposed a statistical anomaly detection algorithm for their accurate detection. Their proposed scheme consisted of two algorithms: cumulative sum (CUSUM) test and geometric entropy minimization (GEM). In their results, the authors report their proposed system to effectively detect both flooding and low-rate DDoS attacks. Their system is also able to detect the attack locations and proceed with preventing the attack traffic being forwarded to the target.

A novel detection algorithm for the software-defined SDN-VANET against flooding DDoS attacks was proposed by (de Biasi et al., 2018). The authors proposed a detection algorithm deployed in controllers which constructs effective rules to prevent the forwarding of attack packets. For the detection of malicious DDoS packets time series analysis was used to monitor the number of packets and the number of flow rules generated with the destination to the specific vehicle. In addition, the number of packets since the last message was sent is also stored. Next, the average and the variance are used to determine a threshold that differentiates between legitimate and malicious behaviour. When the relevant vehicles under attack are detected, a flow tree is constructed. Based on the built flow tree the certain gateway vehicles are chosen to be sent a specific flow rule according to their last number of packets processed to the target. If the gateway is higher than the threshold, then the connection is dropped, and no packets are received. Based on the experimental results reported, the proposed solution achieves a maximum TP rate of approximately 95% and an FP rate of 30% but is managed to nearly be eliminated.

There are numerous DDoS attacks that can be conducted in network and transport layers. All of them must be considered equally critical and effective mechanisms should be developed to detect them. Hence, the authors in (Gao et al., 2019) aimed to detect a large set of D/DoS attacks in the NSL-KDD and UNSW-NB15 datasets. Their proposed solution consisted of two components: the collection component, which collects the relevant traffic, and the detection component. For detecting the malicious traffic, random forest algorithm was adopted. In their experimental results, it was reported that the constructed system achieved a 99.95% and 98.75% detection rate, and a 0.05% and 1.08% FP rate, respectively, for the two datasets.

Due to the impact of black hole attacks, the authors in (Khatoun et al., 2015) proposed a reputation system between vehicles communicating with each other by monitoring their packet transmission. When a communication takes place between two vehicles, with a third one acting as the intermediate forwarding node, the sending node receives feedback on the intermediate node correctly forwarding the data and whether the data was modified. The authors proposed a detection algorithm, which resides in the RSU and monitors these two issues using a reputation score system. The reputation score for each vehicle is calculated and stored in a score table. Proceeding, this score table is shared and synchronized between the different RSUs periodically. For each vehicle which acts as an intermediate node, its reputation score is directly dependant on the reports sent by its neighbours, and it is calculated using the weight of its historical score and reputation score and the weight of each slot. This score is updated frequently using the neighbours' observations. Furthermore, the detection algorithm recommends an efficient way for choosing the next hop that should forward the packets by using the reputation score for emergency, temporary, and web messages and their weight of integrity. In their results, they report a maximum of 90% TP rate and an FP rate ranging from 10 to 26%.

Similarly, the authors in (Baiad et al., 2014) highlighted VANETs' mobility and the constant exchange of information between vehicles as very important issues which assist in the conduction of blackhole attacks. To defend against such attacks, they proposed a cluster-based cross-layer detection algorithm on the MAC and network layers. Their proposed approach consisted of the following components: a monitoring component deployed on the network layer that used the watchdog technique, a monitoring component on the MAC layer which was responsible for counting the number of sent RTS packets and received CTS packets and comparing their numbers to check whether packet loss was due to collision. Next, the watchdogs deployed in the same clusters cooperated with each other to ultimately decide on the potential presence of a blackhole attack. In their results, they report the maximum detection rate of blackhole attacks to be 89.28% and the lowest FP rate to be 1.28%.

In VANETs, DDoS attacks in the physical layer, jamming in particular, should not be taken lightly as they can be proved particularly dangerous. Hence, the authors in (Karagiannis & Argyriou, 2018) proposed a system against wireless radio (RF) frequency jamming using K-means unsupervised machine learning. They used a new metric, the variations of the relative speed generated between the adversary who conducts the jamming attack and their target. Other features were also used such as a series of cross-layer metrics such as signal to noise and interference ratio (SINR), the packet delivery ratio (PDR), and the received signal strength and interference (RSSI). To evaluate their proposed system, they conducted three different simulated scenarios: interference, smart attack, and constant attack scenarios, the former with one interference and the latter two with a moving jammer adversary. For each scenario used different combinations of features, clusters constructed, and speed. In their results, they highlighted the impact relative speed has in successfully detecting jamming attacks.

Instead of focusing on network behaviour, the authors in (Li et al., 2015) argued that context-aware features can be used to detect various network malicious behaviour in VANETs including RTS flooding, packet dropping, and packet modification attacks. Such contextual features include velocity, temperature, wind speed, GPS coordinates, altitude, and channel status. To detect malicious nodes the authors used the SVM supervised algorithm and simulated the attack scenarios in the GloMoSim simulator. Their evaluation consisted of comparing their proposed solution to two other similar frameworks, where SVM has outperformed both of them. Regardless of the node density, number of misbehaviour nodes, and node motion speed, SVM has achieved a minimum detection rate of 95%.

In the context of machine learning, a new subarea has gained considerable popularity over the recent years is deep learning. Deep learning denotes the set of mainly neural networks that consist of multiple layers in their networks. Previous research demonstrated that a single-layer neural network is unlikely to produce substantial results. Therefore, there was need for the number of hidden layers to be increased effectively. This would allow for the implementation to be optimized in further but at the same time maintaining its theoretical universality. The authors in (Shu et al., 2020) effectively pointed out that the majority of IDS proposed systems for VANET networks considered only sub-networks of it, hence deploying their proposed system either in the vehicle, the RSU, or the cluster-head. However, such an approach is likely to fall short on distributed attacks at different parts of the network. Therefore, they followed an SDN (software-defined network) approach and treated VANET as such an entity. Hence, they proposed a collaborative IDS solution using deep learning against various types of cyberattacks, including DDoS attacks in multiple layers, from the KDD99 and NSL-KDD datasets. In brief, their proposed solution operates in the following way: each SDN controller is responsible for collecting and analyzing the flow information from vehicles and RSUs of the specific sub-network it is deployed in. Based on the information collected from all the controllers, an intrusion detection model is trained in the cloud server. The controllers perform their own detection using rule-based mechanism and send feedback to the intrusion detection model. The model merges the collected evidence and ultimately decides on whether there is an intrusion. Their proposed intrusion detection model makes usage of generative adversarial networks. In their experimental evaluation, the authors proved their system's correctness to both independent identical distribution and non-independent identical distribution cases. Proceeding, they evaluated its performance, reporting an overall 98.37% accuracy in the KDD99 dataset and 96.75% for the NSL-KDD dataset.

Highlighting the need to consider V2V, V2I, and in-vehicle communications for autonomous vehicles, the authors in (Ashraf et al., 2020) proposed a deep learning architecture based on long–short-term memory (LSTM) autoencoder algorithm to detect various types of cyberattacks, including denial of service, sniffing, distributed denial of service, spoofing and replay attacks. Over a three-second time window the following features are monitored: traffic initiating from this packet's source IP, traffic sent between this packet's source and destination IPs, time to live from this packet's source to destination and destination to source, bits per second for this packet's source to destination and destination to source, packets count sent from this packet's source to destination and destination to source. Their proposed architecture was composed of two types of layers, namely: LSTM and a fully connected layer. For evaluating their proposed system, the authors made use of two datasets: CAN car hacking dataset for in-vehicle communications and UNSW-NB15 dataset for external network communications. In their results, the authors report a 99% and 96% overall accuracy, respectively, with the TP rates being 100% and 97%, respectively.

17.5.2 Discussion and Open Issues

Without doubt, extended work was conducted in detecting malicious traffic in Road ITS networks which is considered remarkable. A plethora of different machine learning and statistical techniques were adopted to detect various types of network attacks. Analyzing the studies' reported results, we observe machine learning was proved successful as the detection rate is at least approximately 89% or more regardless of the attack. Furthermore, various types of attacks were investigated from the numerous studies provided, all of them detected successfully by machine learning. Based on the justifications provided, it is proved further that machine learning algorithms can be an ideal solution to cybersecurity attacks regardless of the type, TCP/IP layer they are conducted, and evasion techniques that might be used.

Despite the remarkable work conducted in the security, safety, and overall protection of Road ITS networks there are security challenges and open issues yet to be addressed. Firstly, the number of nodes participating in the networks changes rapidly along with their mobility which clearly affects the routing protocols and the overall network functioning. Therefore, research on routing attacks such as blackhole, greyhole, wormhole, and sybil attacks can be of particular interest. Secondly, since the literature puts great focus on detecting DDoS attacks, their proposed countermeasures could be expanded further by including effective mechanisms in zombie devices which are part of a botnet. Proposed countermeasures could focus on preventing the infection of other nodes in the network during the propagation phase. This could be proved particularly effective against DDoS attacks that involve spoofing or dynamic IP address changing evasion mechanisms.

Furthermore, the majority of studies in the literature deploy their proposed solutions to either the vehicles or the RSU. We believe that other assets in Road ITS networks could host the IDS or an agent of the IDS solution proposed such as the traffic lights, CCTVs, and environmental and safety sensors monitoring the environment and ensuring public safety. However, the majority of such sensors and devices are low in resources such as CPU power, memory, and storage. Therefore, any security countermeasures proposed should be lightweight and take into consideration these devices' lightweight nature and limited computational capabilities.

Another important limitation involves the possible exploitation of IDS systems. Distributed IDSs involve the cooperation and collaboration of multiple agents that can be responsible for collecting traffic, performing feature extraction, or even first-pass detection of incoming traffic before they forward the results. In addition, beyond detection capabilities, distributed agents can participate in the decision process of incoming traffic being classified as malicious, with a notable example being majority voting mechanisms. However, if one or more of them are compromised then the impact can be critical. Some examples include the insufficient forward of traffic to the central authority, false detection results, as well as unreliable voting behaviour. Hence, we believe these are some important arguments and such scenarios are worth being explored. By extending our previous argument, it is more than evident that the majority of literature is in favour of machine learning algorithms and approaches as they are proved to be particularly accurate and effective. However, adversarial machine learning is becoming a critical issue to various networks and infrastructures and can, therefore, cause major disruptions to Road ITS networks and the smart city where the most catastrophic scenario would consist of the cost of human lives. Hence, effective and robust workarounds must be designed that protect machine learning security models and minimize the impact any adversarial models aim to cause.

Similarly, another important issue from the extended usage of machine learning techniques and algorithms in intrusion detection denotes their blackbox approach regarding their results and decisions outputs. It is generally known that despite their accurate decision and/or predictions, the majority of ML and AI algorithms lack explainability, interpretability, and transparency. Despite ML and DL and AI experiencing major advancements in both speed and performance in intrusion detection (Nisioti et al., 2018), there was little work done providing substantial explanation and details regarding the attack itself. This can be proved a major drawback in critical systems and infrastructures such as the Road ITS. A cyber-attack in such a system can be a real catalyst to the drivers and their lives. Hence, it is more important than ever for the machine and deep learning and A.I. algorithms to share with human operators important explanations and provide assistance to them regarding possible intrusions so quick and safe decisions can be made (Buczakand & Guven, 2015). Hence, we briefly describe what explainable artificial intelligence (XAI) denotes and how these concepts could be applied to Road ITS.

Finally, two major open issues are the following: to our knowledge, none of the available related studies has performed any investigation on application-layer DDoS attacks and/or have considered an attack being initiated from the Road ITS network targeting other interconnected networks or vice versa. Although the literature in malicious traffic detection in Road ITS networks is rich in studies, they tend to focus more on DDoS attacks in lower layers. However, application-layer DDoS attacks can be an ideal attack meant for adversaries as application-layer DDoS attacks need fewer resources to be conducted and are stealthier than their lower layer perspectives as they can mimic legitimate behaviour to a great extent. The second and, to our belief, the most major issue is that to our knowledge no studies have considered cross-network attacks to and from Road ITS networks. Road ITS is one of the most important cyber-physical systems of the smart city and therefore, constantly communicates with other important networks and cyber-physical systems of the smart city, including smart hospitals, ports and airports, residential homes, ticketing systems, traffic management services, and the smart grid. This interconnection brings multiple benefits to the citizens of the smart city, provides real-time data to the services and authorities, and improves the overall functionality of the networks themselves as well as the smart city as a whole entity comes with a price. Due to this tight interconnection and constant information exchange, cyber-attacks can target Road ITS but be initiated from another network or the other way around. If this interconnection is exploited, it can cause a great impact, from the disruption of services to the threat of human lives being put at risk. Based on the final two open issues described, we were motivated to conduct a case study where we simulated a Road ITS which communicates with other important networks and systems in the smart city. Proceeding, we simulated application-layer DDoS attacks, specifically flooding and slow-rate to illustrate the impact such scenarios can have on both the Road ITS and the smart city.

17.6 EXPLAINABLE ARTIFICIAL INTELLIGENCE IN CYBER SECURITY

Evidently, there is little work conducted in the literature on how artificial intelligence can become more explainable, transparent, and Interpretable. This is of great concern in critical infrastructures, such as

Road ITS systems, as a cyberattack can lead to human lives being put into risk. This need for more evidence on the decisions made from such systems is so important that is has become part of the legislation from countries (European Union, 2016; Goodman & Flaxman, 2017). Therefore, it is important for all the security countermeasures deployed to be able to give as much information as possible for any detected malicious or suspicious behaviour. In that way, correct decisions will be taken quickly and precisely, and the safety of people is likely to be ensured. In this section, we begin by explaining the background knowledge behind XAI, giving appropriate explanations to the new concepts introduced and discussed. Proceeding, we present notable studies from the literature which focus on applying XAI to intrusion detection. Finally, we briefly discuss the work done so far and give appropriate future directions research can take.

17.6.1 Explainable Artificial Intelligence Background Knowledge

Explainable artificial intelligence (XAI) was firstly introduced in (Van Lent et al., 2004). However, the requirement for artificial intelligence to provide more information regarding their constructed models was raised since the 1970s, specifically focused on expert systems (Goodman & Flaxman, 2017), but also became more popular in the 1990s with artificial neural networks (Andrews et al., 1995). Evidently, machine and deep learning have become increasingly popular due to their high-rate success in various areas, including intrusion detection. This has made the need for interpretability on the models' outputs more important than ever. Hence, according to (Barredo Arrieta et al., 2020), the main characteristics an XAI model should encompass the following:

Understandability (also known as intelligibility) includes all the properties an algorithmic model consists that make its functioning more understandable to humans. Examples include on how it works on a high level, leaving out particular details regarding its structure and algorithmic training procedure (Montavon et al., 2018).

Comprehensibility is inspired by Michalski's (Michalski, 1983) the results of computer induction should be symbolic descriptions of given entities, semantically and structurally similar to those a human expert might produce observing the same entities'; it denotes the algorithmic model being able to explain its knowledge to humans using natural language. In practice, it is also linked to the direct evaluation of the algorithm's complexity (Guidotti et al., 2018).

Interpretability denotes the algorithmic model's ability to interpret the results in a way meaningful to humans.

Explainability denotes an algorithmic model's ability to provide humans with meaningful explanations behind its results so human operatives can take the best decision possible (Guidotti et al., 2018).

Transparency denotes an algorithmic model's ability to be fully understandable by humans without leaving any sort of details or knowledge hidden or vague.

A plethora of different proposed works attempted to define a set of goals for XAI systems in an attempt to measure its relative success in delivering the characteristics described above. As authors (Barredo Arrieta et al., 2020) the fundamental goals of an XAI model should consist of the following:

Trustworthiness denotes the confidence level of a model functioning in a logical and meaningful way in case of a problem or a set of unforeseen circumstances rising. An XAI model being explainable is not explicitly trustworthy. Furthermore, trustworthiness is not easy to be measured and/or quantified hence why it should be the primary objective of an XAI model (Kim et al., 2015; Ribeiro et al., 2016).

Causality denotes the model's ability to identify causal links between the different variables. For causality to be realized, a great amount of prior knowledge is required so effects that observed are proved as casual (Pearl, 2009).

Transferability denotes the model's ability to transfer its acquired knowledge to other problems correctly and not let humans misinterpret its results and make incorrect assumptions (Caruana et al., 2015; Szegedy et al., 2013). Furthermore, transferability also deals with a model's ability to set boundaries, allowing for a more-in depth comprehension and understanding of its implementation.

Informativeness primarily denotes a model's ability to positively impact decision making (Huysmans et al., 2011). Furthermore, the model should prove humans with enough valuable information so any misconception can be minimized and/or avoided.

Confidence denotes the model's reliability, robustness, and stability (Basu et al., 2018; Ruppert, 1987; Yu et al., 2013). A model should provide a significant level of confidence through relevant information provided to humans.

Fairness: A model should be fair and ethical in all its functioning. In literature (Chouldechova, 2017; Goodman & Flaxman, 2017) it is suggested that a model should provide a transparent visualization of which relations and variables directly impact a result. In that way, fairness can be provided on the results, analysis, and interpretation. Furthermore, bias could be identified as early as possible so the model is not incorrectly trained (Bennetot et al., 2019; Burns et al., 2018).

Accessibility: Studies have effectively pointed out that there should be a limit to the extent humans are involved in the construction of the ML model (Chander et al., 2018; Olden & Jackson, 2002). It is highlighted that in an attempt to better understand and eventually improve the model, they might unintentionally negatively impact its overall construction process.

Interactivity: Studies (Harbers et al., 2010; Langley et al., 2017) suggested that a model should give the opportunity to humans for direct interaction with them such that more explainability is once again provided.

Privacy awareness: One of the most important aspects, that is often not given the proper importance, is for the model to ensure privacy. If human operators are not aware of the data stored in the model's internal database then privacy violation issues can potentially rise (Castelvecchi, 2016).

17.6.2 Related Studies in Explainable Artificial Intelligence for Cyber Security

Highlighting the importance of transparency, lack of explainability in results, and overall blackbox nature in IDS systems, the authors in (Wang et al., 2020) proposed a framework based on SHapley Additive exPlanations (SHAP) (Lundberg & Lee, 2017). Using it any IDS can demonstrate more interpretation through local and global explanations. Local explanations are responsible for providing details for every feature and appropriate descriptions on how each affects the system's accuracy. Regarding the global explanations there are two types of them. The first one deals with identifying the most important features of an IDS. The second aims to investigate in-depth the relationships between the various attacks and the features. For the experimental analysis, the authors used the NSL-KDD dataset.

Similarly, focusing on the drawback of ML and DL techniques consisting of a blackbox approach and the difficulty of explaining the results, the authors in (Ahn et al., 2021) supported that machine learning IDS systems can be deceived by adversarial learning mechanisms. Hence, they proposed an XAI framework that identifies the most dominant features in a given dataset such as a detailed explanation of how a deep learning classifier (specifically ResNets) operates. In detail, they proposed a fitting score that quantifies a feature's importance. Using it, they can identify the trade-off between the accuracy in correctly classifying incoming traffic and the reduced number of features. For the reduction of features, the genetic algorithm was used. For providing more explainability on their deep learning classifier they defined a dominance rate that justifies to what extent each feature refers to which deep learning model.

Focusing on the lack of robust explainable methods in realistic threat models, the authors in (Kuppa & Le-Khac, 2020) conducted an in-depth security analysis regarding gradient-based XAI proposed methodologies. The authors aimed to improve an IDS's robustness against adversarial ML that aims to mislead the security countermeasures in place in an attempt to cause malfunction. After suggesting a set of different properties, a threat model should consist of XAI methods they proposed in a multi-dimensional XAI taxonomy. Specifically, the taxonomy contained three layers: XPLAIN: relevant explanations and predictions regarding the data. XSP-PLAIN: relevant explanations regarding the security and privacy characteristics of both the predictions and the data. XT-PLAIN: relevant explanations regarding the threat model of the data and predictions taken into consideration. To evaluate their taxonomy, the authors used three different algorithmic models: Mimicus, a multilayer perceptron neural network that detects malicious PDF documents (Guo et al., 2018; Smutz & Stavrou, 2012), another multilayer perceptron neural network that detects Android malware (Grosse et al., 2017) and finally an IDS system using Adversarial Auto Encoder (AAE to detect attacks from the UGR16 dataset (Fernández et al., 2018). In these three models they performed adversarial ML techniques to evaluate their model's robustness. In their experimental analysis, they demonstrate that the adversarial ML techniques used have a direct impact on the XAI methods. More specifically, it evidently shows the gaps that exist between the model's internal functioning and structure and the relevant explanations. Hence, they list future directions being the study of defence countermeasures for the relevant attacks investigated, the inclusion of properties regarding privacy and confidentiality in their XAI methodology, and the evaluation of robustness regarding other XAI frameworks with other types of neural network architectures.

Once again, highlighting the machine and deep learning models' blackbox nature, the authors in (Marino et al., 2018) used the decision tree in an attempt to give more explainability and interpretability to the results achieved against the KDD99 dataset. Specifically, they used feature engineering and rule-based methods to investigate the trained decision tree algorithm's internal structure. They proceeded with conducting an in-depth analysis of each feature's importance using entropy. Hence, they proposed minimum modifications for improving the model's accuracy. These modifications were used to find the features which incorrectly assisted in the misclassification of samples. Their proposed framework has provided satisfactory results in giving appropriate explanations through visualization.

17.6.3 Discussion and Future Directions

Without any doubt, XAI is certainly going to be essential in IDS systems in the Road ITS domain. Due to the domain's critical nature, XAI is likely to bring a plethora of different benefits. As expected, the studies aimed to make the IDS's accuracy results more explainable to human operators. All of the studies presented identified features as an important element that should assist in explaining results in a meaningful and more human-readable way. Furthermore, they performed an in-depth analysis on how features can influence the classification of malicious traffic to an attack and how influential a feature is in the model's training.

Despite the remarkable efforts conducted so far, there is need for more and extended work to be done. Future work should extend to other important parts of IDS systems, both during the training and testing phases. In detail, more focus should be placed on how a model's training parameters directly impact its accuracy. Examples of such parameters for neural networks (regardless of the architecture used) include learning rate, number of epochs, weight initialization methods, and momentum. Through their possible visualization, human operators are likely to understand how each variable affects the model's construction and accuracy and proceeds accordingly. Another possible approach, with regard to training, would be to investigate how different parts of the training set influence the model's internal construction. Imbalanced datasets and overfitting/underfitting were two long-lasting problems in machine learning. These issues can have a great impact on IDS systems. Therefore, appropriate explanations and visualizations regarding the training procedure and these potential issues can assist in minimizing these problems after human intervention.

For the testing phase, the quantification of results is a promising approach. Instead of providing a single classification label that characterizes an instance, a more numerical approach would be beneficial to the human operators. Each instance would consist of a confidence score, providing appropriate explanations on how it was labelled with the specific attack. In addition, the majority of training data can be classified as private and confidential. Therefore, it is essential that such data can remain secured from any intentional or unintentional leakage. XAI could assist in highlighting this requirement. Finally, in large-scale systems more than one ML algorithm is used so that the classification of incoming traffic could be deployed. The different algorithms collaborate with each other, sharing knowledge, passing evidence from collected data, and ultimately cooperating in deciding whether there is an intrusion or not. Once again, a blackbox approach leaves human operators completely in the dark on how incoming traffic data was handled by the different systems and agents and how they cooperated to classify it accordingly.

17.7 CASE STUDY: DDOS ATTACKS IN ROAD ITS AND THEIR IMPACT ON SMART CITIES

For the simulation of the Road ITS network and the networks having direct communication with it, the Network Simulator 3 (NS3) was used. A summary of the networks' parameters and protocols used is given in Table 17.1, along with a general representation of the overall simulation in Figure 17.1.

17.7.1 Road ITS and Smart City Simulation Environment in NS3

17.7.1.1 Road ITS Simulation

For the Road ITS network, 100 nodes were simulated. Ninety-nine consisted vehicles, such as trucks, cars, trams, buses, taxis, motorbikes, and bicycles (Cunha et al., 2016; Jiang & Du, 2015). Vehicles are from both the public and private sectors (ENISA, 2015). The remaining one (RSU) was static. Examples of RSU include a traffic light, speed camera, or simply a base station and acts as a gateway to exchange information with their infrastructures and networks in the smart city (Jiang & Du, 2015). All 99 vehicles communicate with each other and the RSU wireless to simulate the V2V and V2I communications (Abboud et al., 2016). The protocols simulated on each TCP/IP layer are summarized in Table 17.1.

TABLE 17.1 NS3 road ITS-smart city simulation parameters (TPC/IP layers protocols)

NETWORK	PHYSICAL/MAC LAYER	NETWORK/TRANSPORT LAYER	APPLICATION LAYER
Road ITS	WAVE, AODV	IPv6, TCP, UDP	HTTP, CoAP, MQTT, AMQP
Smart Home	ZigBee, AODV	IPv6, TCP, UDP	HTTP, CoAP, MQTT, AMQP
Smart Hospital	ZigBee, Ethernet, WiFi, AODV	IPv4, IPv6, TCP, UDP	HTTP, CoAP, MQTT, AMQP
Telecom Network	LTE PHY/MAC	IPv4, TCP	HTTP, VoIP
Smart Grid	Ethernet	IPv4, TCP	HTTP, CoAP, MQTT, AMQP
Smart Airport	Ethernet	IPv4, TCP	HTTP, CoAP, MQTT, AMQP
Smart Port	Ethernet	IPv4, TCP	HTTP, CoAP, MQTT, AMQP
Public Safety	Ethernet	IPv4, TCP	HTTP, CoAP, MQTT, AMQP
Traffic Management	Ethernet	IPv4, TCP	HTTP, CoAP, MQTT, AMQP
Ticket Purchases	Ethernet	IPv4, TCP	HTTP, CoAP, MQTT, AMQP

17.7.1.2 Smart Home Simulation

The private sector vehicles (mainly cars) communicate with the owners' smart homes to control the smart appliances and services in their homes. For the smart home network, 20 nodes were simulated. Nineteen were IoT devices and smart appliances and the final one was the gateway. The nodes were static, with no mobility.

17.7.1.3 Smart Hospital Simulation

The vehicles communicate with hospitals in case of accidents and emergencies. For the smart hospital, 50 nodes were simulated. Forty-eight of them were IoT medical devices and medical equipment nodes. Twenty of them were mobile and the remaining 28 were static. The remaining nodes were the gateway and the external server communicating with the Road ITS receiving and sending emergency messages when needed. The communications between the medical devices and equipment were wireless.

17.7.1.4 Telecom Network Simulation

The private sector vehicles communicate with telecommunication networks to make phone calls. For simulating the mobile telecommunications network the LTE module in NS3 was used. In the telecom network, 53 nodes were simulated. Fifty of them simulated UEs, one was eNB, and one was PGW, and the final node acted as the gateway. The 50 nodes were mobile and the remaining 3 were static. The communications were all wireless.

17.7.1.5 Other Infrastructures Simulations

The Road ITS motorway vehicles (cars, buses, taxis, trucks, and motorbikes) can receive demand/response signals from the smart grid in an attempt to reduce the functioning of the grid. The Road ITS communicates with other ITS operators. One example is the smart airport. Road ITS communicates with the smart airport Infrastructure in the following ways. Smart Vehicles, mainly private cars, use the smart parking operations of the airport for parking purposes. Smart transportation vehicles (trains, buses, taxis, coaches, etc.) wait outside the airport to take the passengers to their destinations and vice versa. Finally, in the context of easy and comfortable passenger experience, dedicated cars wait outside to take the luggage of the passengers directly to the airport. Another smart infrastructure Road ITS communicates with is the Smart Maritime Transportation Port. The logistics/freights trucks transmit goods and cargos from and to the smart port such as maritime transportation vessels so they can be delivered to their destination. Additionally, if the smart port is used for tourism purposes the same operations described for smart airports above are carried out through Road ITS. All of the vehicles communicate with hospitals in case of emergency and road accidents. All of the vehicles from both private and public sectors communicate with traffic regulation systems which are in charge of organizing and coordinating the traffic lights on the road. The public transport vehicles also communicate with ticket purchase servers to exchange updated information regarding the tickets. Since these infrastructures, although critical and important, are not the main focus of our research, for each infrastructure only one node was simulated acting as the main server communicating with the Road ITS. The communications for all the infrastructures were wired.

17.7.2 DDoS Attacks Modelling Scenarios

A summary of the attacks parameters is presented in Table 17.2.

TABLE 17.2 DDoS attack scenarios parameters (source, destination, and protocols)

ATTACK TYPE	SOURCE (ATTACKERS NUMBER)	DESTINATION	PROTOCOLS	ATTACK DURATION (SECONDS)	APPLICATIONS NUMBER (PER NODE)	INTERVAL (SECONDS)
Flood	Road ITS (100)	Traffic Management	HTTP, CoAP, MQTT, XMPP, AMQP	240	10	1.0
Flood	Road ITS (70)	Smart Airport	HTTP, CoAP, MQTT, XMPP, AMQP	240	15	1.0
Flood	Road ITS (50)	Smart Port	HTTP, CoAP, MQTT, XMPP, AMQP	240	15	1.5
Slow-Rate	Road ITS (100)	Smart Grid	OpenADR	240	10	1.0
Flood	Road ITS (100)	Public Safety	HTTP, CoAP, MQTT, XMPP, AMQP	240	15	1.0
Flood	Road ITS (100)	Smart Hospital	HTTP, CoAP, MQTT, XMPP, AMQP	240	15	1.0
Flood	Road ITS (100)	Ticket Purchases	HTTP, CoAP, MQTT, XMPP, AMQP	240	10	1.0
Slow-Rate	Telecom (50)	Road ITS	HTTP	240	10	1.5
Slow-Rate	Smart Hospital (25)	Road ITS	HTTP, CoAP, MQTT, XMPP, AMQP	240	15	1.5
Flood	Smart Home (20)	Road ITS	HTTP, CoAP, MQTT, XMPP, AMQP	240	10	1.0

17.7.2.1 Scenarios Targeting Road ITS Scenarios

Smart Home to Road ITS: In this scenario the DDoS attack initiated from the smart home has the RSU as a target. A possible disruption of the RSU prevents other networks from sending information to the Road ITS vehicles, therefore threatening its normal functioning.

Telecom Network to Road ITS: In this scenario, the DDoS attack initiated from the telecom network has the RSU as a target. A possible disruption of the RSU prevents other networks from sending information to the Road ITS vehicles and therefore threatening its normal functioning.

Smart Hospital to Road ITS: In this scenario, the DDoS attack initiated from a smart hospital network has the RSU as a target. A possible disruption of the RSU prevents other networks from sending information to the Road ITS vehicles, therefore threatening its normal functioning.

17.7.2.2 Scenarios Initiated from Road ITS

Road ITS to Traffic Management: The DDoS attack initiated from the Road ITS involves the vehicles and the RSU flooding the Traffic Coordination Service. Traffic Coordination Service is critical for the normal and safe functioning of Road ITS and their unavailability can cause accidents on the road.

Road ITS to Smart Airport: The DDoS attack initiated from the Road ITS involved the vehicles and the RSU flooding the smart airport's server in charge of the available transportation timetables to and from the airport. A possible disruption to this airport service can cause issues in the normal functioning of the Airport, Public Transportation Services, and the smart city.

Road ITS to Smart Port: The DDoS attack initiated from the Road ITS involved the vehicles and the RSU flooding the smart port's cargo and logistics server which is in charge of the transportation of goods and cargo to and from the port. A possible disruption to this port service can cause issues in the normal functioning of the Port, the Public Transportation Services logistics and the smart city.

Road ITS to Smart Grid: The DDoS attack initiated from the Road ITS involved the vehicles and the RSU flooding the smart grid's Demand/Response Server. A possible disruption of the Demand/Response Server will cause issues to the smart grid's normal functioning and threaten its stability.

Road ITS to Public Safety Services: The DDoS attack initiated from the Road ITS involved the vehicles and the RSU flooding the Public Security Services server. A disruption of this server leaves the Road ITS vehicles in a rather unsafe situation in case an accident happens on the road.

Road ITS to Smart Hospital: The DDoS attack initiated from the Road ITS involved the vehicles and the RSU flooding the Hospital Emergency server. A disruption of this server leaves the Road ITS vehicles in a rather unsafe situation in case an accident happens on the road the server will not be able to send an ambulance. Furthermore, if ambulances located on the road are needed to go somewhere in case of emergency the server will be unable to notify them.

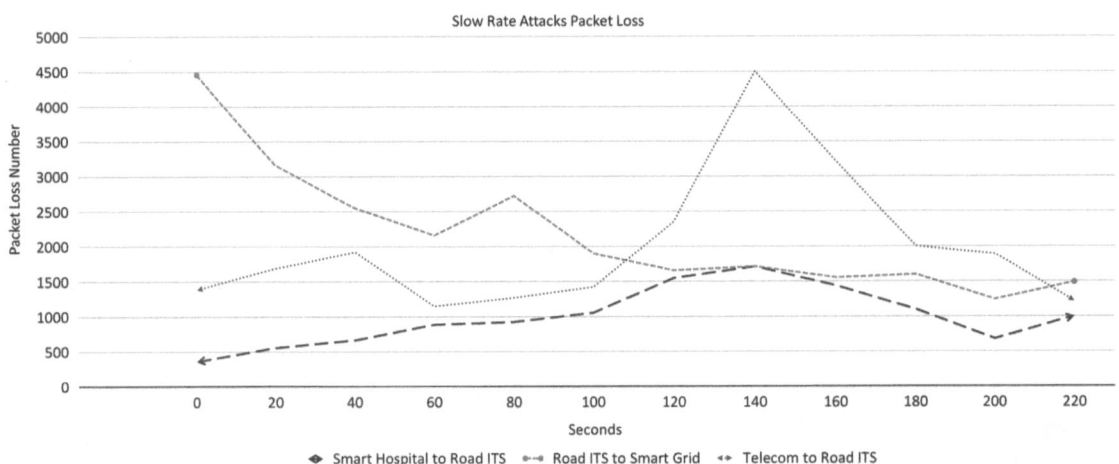

FIGURE 17.2 Slow-Rate DDoS Scenarios Packet Loss.

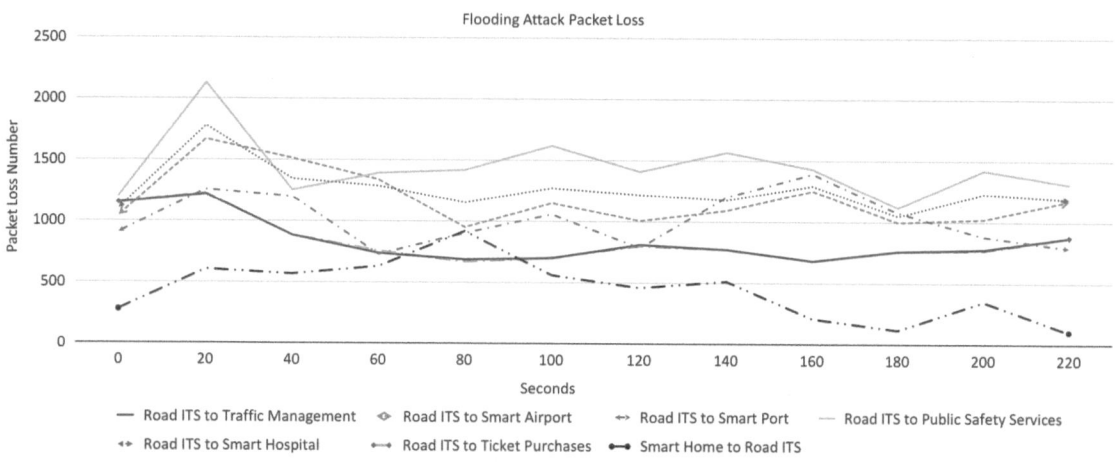

FIGURE 17.3   Flooding DDoS Scenarios Packet Loss.

Road ITS to Ticket Purchases: The DDoS was initiated from the Road ITS involved the vehicles and the RSU flooding the ticket purchases Server. A disruption of this server leaves the smart city citizens unable to make any ticket purchases for the public transport ITS.

17.7.3 Results, Analysis, and Discussion

The results of the dropped packets from the three slow-rate scenarios are shown in Figure 17.2. In all three of the scenarios the target starts losing packet around the 2nd–3rd second of the attack. The results of the dropped packets from the seven flooding scenarios are shown in Figure 17.3. When comparing the two types of attacks, it is evident that slow-rate attack causes a larger loss of packet with less resources used than the flooding scenarios. The slow rate scenarios only used either less number of attackers, less applications, and/or less number of protocols. As opposed to the most successful flooding scenario which achieved an average of 1441 dropped packets per 20 seconds, the most successful slow-rate scenario achieved 2183 dropped packets per 20 seconds. Furthermore, the slow rate scenarios achieve packet loss from the 2nd–3rd second as opposed to the flooding scenarios that achieve this from the 4th–5th second. Also, the flooding scenarios send a new packet/ request every 1.0 seconds from each node while the slow-rate sends a new subsequent packet or initiates a new incomplete request every 1.5 seconds. In conclusion, the slow-rate attacks can achieve impact with less resources and potentially without being detected from security countermeasures due to them being relatively silent as opposed to flooding attacks which are relatively 'noisy'.

17.7.4 Road ITS-Smart City Security Impact Framework Evaluation

In the security industry, for the assessment of critical and sensitive assets under threat, cybersecurity impact and risk-assessment frameworks are used. Many frameworks were proposed such as FIPS 199 by NIST (NIST, 2004). However, Road ITS and the smart city pose differences from traditional ICT networks. Hence, they have different security impacts due to different attack vectors that can be adopted, the introduction of cyber-physical attacks as well as the impact, both cyber and physical, after the execution of such attacks and the impact that an attack can cause to other networks of the smart city that communicate with the affected network as well as the overall functioning of the smart city. Based on these arguments, we constructed a cybersecurity impact assessment framework tailored for the purpose of the study. Although our case study focuses on Road ITS and how a DDoS attack to and from it can affect its normal functioning and the smart city's in general, the framework could be applied to any cyber-physical systems of the smart city. It could also be extended to consider other types of cyber-attacks that threaten confidentiality and/or integrity. The framework has two dimensions to quantify the impact. The first one is the scale level. Scale level is defined as how assets are affected under a cyber-attack. The term node represents the target of the DDoS attack, the Road ITS is how the Road ITS network is affected, and the smart city is how other networks in the smart city are affected by the DDoS attack. The second is the attack consequence from a cyber and a cyber-physical point of view. The attack consequence across these three domains can be classified as minor, moderate, and critical depending on its severity in terms of asset impact and the attack intensity and is summarized in Table 17.3.

Minor: A minor impact from a cyber point of view indicates that the attack incident has a limited and minor adverse effect on the target's (whether it is a single node or a network/infrastructure) operations. In detail, this is indicated as minor unavailability to some services the target provides, minor damages to the assets, minor physical damage to equipment, and minor financial losses. A minor consequence from a cyber-physical point of view indicates that the attack has a limited and minor adverse effect on the human individuals without putting them physically at risk, just causing frustration to them. Also, the minor consequence causes limited physical malfunction and/or damage to the equipment involved.

Moderate: A moderate impact from a cyber point of view indicates that the attack incident has a significant adverse effect on the target's (whether it is a single node or a network/infrastructure) operations and assets. In detail, significant adverse indicates that the target fails to provide its primary functions to the individuals, significant cyber damage to assets occurs, significant physical damage to equipment is done and significant financial causes occur. A moderate consequence from a cyber-physical point of view indicates that a significant effect on people that could potentially harm them physically is possible, but not causing loss of life. Also, the moderate consequence causes significant physical malfunction and/or damage to the equipment involved.

Critical: A critical impact from a cyber point of view indicates that the attack incident has a severe and catastrophic adverse effect on the target's (whether it is a single node or a network/infrastructure) operations and assets. Severe/catastrophic could mean the target completely failing to provide the services it is supposed to provide and that there are major financial losses. A critical consequence from a cyber-physical point of view indicates that catastrophic consequences occur to humans and loss of life is more than likely to occur. Also, catastrophic consequences cause physical destruction to the equipment and infrastructures involved.

17.7.5 Results and Discussion

All three DDoS attacks from the smart home, the telecom network, and the smart hospital to the Road ITS/RSU cause great impact to the target RSU as well as the Road ITS. From a cyber point of view, the target RSU becomes unavailable to requests from outside infrastructures and networks. This can put the

TABLE 17.3 Road ITS-smart city security impact framework

	CYBER IMPACT			CYBER-PHYSICAL IMPACT		
	LOW	*MEDIUM*	*HIGH*	*LOW*	*MEDIUM*	*HIGH*
Node	Minor	Minor	Moderate	Moderate	Critical	Critical
Road ITS	Moderate	Critical	Critical	Critical	Critical	Critical
Smart City	Critical	Critical	Critical	Critical	Critical	Critical

vehicles in frustrating to life-threatening situations. For example, traffic congestion services will not be able to communicate with the drivers to assist them into taking alternative routes to avoid traffic. This can cause frustration to the drivers and traffic congestion to the roads but it is certainly not a life-threatening situation. However, other more important services will not be able to communicate with the vehicles to assist them. For example, if there was an accident on the road or if the weather is dangerous to the drivers there is no communication between the two networks to warn them fast. This can cause serious issues on the road and put drivers' lives at risk. Next, the security impact from the attacks initiated from the Road ITS to other networks and infrastructures of the smart city is discussed.

Road ITS to Traffic Coordination: A DDoS attack from the Road ITS to the Traffic Coordination server causes an overall critical impact from both a cyber as well as a cyber-physical point of view. From a cyber point of view, the target (Traffic Coordination) becomes unavailable to legitimate requests from other parts of the Road ITS. Also, it is unable to send legitimate commands and coordinate the traffic lights to other parts of the Road ITS in the smart city. From a cyber-physical point of view, the un-availability of the traffic coordination server can cause issues to the traffic lights in the motorway thus resulting in accidents between vehicles.

Road ITS to Smart Airport: A DDoS attack from the Road ITS to the smart airport Public Transportation server can cause a moderate overall impact. From a cyber point of view, the server responsible for the Public Transportation for the smart airport becomes unavailable to handle requests from the public transportation and responds to changes in the timetable. From a cyber-physical point of view, it will be unable to give fast and accurate information to airport passengers requesting what public transportation means are available to and from the airport. However, such an attack is unlikely to threaten any human lives, only causing frustration among passengers; however, the financial impact is likely to be great.

Road ITS to Smart Port: A DDoS attack from the Road ITS to the smart port Freight/Logistics server can cause a moderate overall impact. From a cyber point of view, the server responsible for managing the Freights/Logistics and communicating with the vehicles transporting various goods and cargo will be unable to legitimate requests as well as with the relevant companies and organizations providing the goods. From a cyber-physical point of view, the cargo vehicles will be unable to communicate with the smart port Server. Thus, the cargo and goods will have delays in being delivered to the port and the transportation of them in the vessels will be severely delayed. This will result in great financial impact but no human lives will be threatened.

Road ITS to Smart Grid: A DDoS attack from the Road ITS to the smart grid serve can cause moderate to critical impact. Smart vehicles can send and receive signals from the smart grid to sell electricity to the infrastructure when more energy is needed in an attempt to prevent its load shedding. However, the compromised vehicles can attack the demand/response server. From a cyber point of view, the demand/response server in the utility centre of the smart grid becomes unavailable to legitimate demand/response requests from either normal vehicles or other entities such as smart homes. From a cyber-physical point of view, this attack can cause moderate to critical impact. The unavailability of the server can leave smart home residents frustrated while they attempt to request or sell energy to the smart grid. However, the server's unavailability can leave the smart grid in an unstable state when in need of energy. The server will be unable to request energy from microgrids and distributed energy resources or perform selective load shedding. As a result, its normal functioning is likely to be put at risk and massive outages can be experienced.

Road ITS to Public Safety Services: A DDoS attack from the Road ITS to the public safety services can have a drastic impact. From a cyber point of view, the server responsible for managing the Public Safety of the City becomes unavailable to legitimate requests. From a cyber-physical point of view, the unavailability of the server can cause major issues to the smart city. Since it will become unavailable it can no longer monitor the safety on the road. Therefore, in case of abnormal behaviour, whether these are caused by criminals or by road accidents, the server will be unable to detect it through the cameras and communicate with relevant authorities such as the smart hospitals and the Smart Police Stations.

Road ITS to Smart Hospital: A DDoS attack from the Road ITS to the smart hospital External Server can cause a critical impact. From a cyber point of view, the server responsible for managing the external medical requests becomes unavailable. This causes a serious and critical cyber-physical impact as this can pose a great threat to individuals who need to communicate with the Hospital as soon as possible in case of a road accident or individuals with medical problems and are in need of immediate medical assistance. Also, due to its unavailability, the server cannot communicate with Smart Ambulances in case they are needed for work thus resulting in great threat to human lives.

Road ITS to Ticket Purchases: A DDoS attack from the Road ITS to the ticket purchases server can cause moderate impact. From a cyber point of view, the Ticket Purchase server becomes unavailable to legitimate requests from tickets purchase actions from customers. It is also unable to communicate with public transportation means to provide accurate information to the customers interested in booking tickets. Although human lives are not threatened in this scenario the financial impact is great.

All attacks cause significant impact to the target as well as the smart city both from a cyber as well as cyber-physical point of view. Moreover, since the DDoS attacks are initiated from the Road ITS, the infrastructure itself is negatively impacted. From a cyber point of view, the RSU and any compromised vehicles are going to cause issues to any other passing-by vehicles that will try to communicate with either of them by delaying the legitimate exchange of information or even dropping this communication. Furthermore, these attacks can have cyber-physical impacts on the Road ITS normal functioning. Legitimate vehicles will be unable to communicate with any of the infected vehicles thus putting their drivers' lives at risk. Also, drivers of the infected vehicles are likely to experience problems with controlling their cars since they participate in the DDoS attack and do not have a clear view of what is happening to their vehicles. Additionally, the drivers of the area the DDoS was initiated, whether they have participated in the attack or not, are likely to experience problems if they want to communicate with external infrastructures and networks. One example is in the event of a road accident the emergency signal will not be able to be delivered to the nearest smart hospital due to the unavailability of the relevant RSU or nearby vehicles for the emergency message to be transmitted. Also, the driver will be unable to communicate with other networks such as smart homes and telecom networks. Road ITS's interconnectivity with other infrastructures makes it a complex entity for it to be managed and monitored. Hence, it makes an ideal target for adversaries who not only want to disrupt the Road ITS's normal functioning but also the smart city's in general. A DDoS attack will be spread fast due to the interconnectivity and the lack of security solutions and cause a serious consequences on multiple domains. In addition, the lack of standardization of the IoT spectrum, the interconnection of the various systems and domains, and the deployment of so many new and different technologies bring additional vulnerabilities to the complex Road ITS.

17.8 CONCLUSION

Road ITS is already deployed in multiple cities of the world as a starting point towards the realization of the smart city vision. However, the Road ITS brings its own vulnerabilities to its own network, other networks/infrastructures, and most importantly to the smart city and its citizens. Since these kinds of infrastructures adopt technologies such as IoT, cyber-physical systems, and wireless communications which, although pioneering, have limited resources, they are therefore easier to be exploited by the attackers due to the weak security countermeasures deployed. ML and DL proved to be particularly effective in detecting malicious network behaviour in various networks and against multiple attacks.

In this chapter, we discussed how ML or DL is used to protect the Road ITS against network attacks. We provided important background knowledge regarding the technologies used in Road ITS networks, including the communications and the architectures used. In the next section, the most critical security issues and challenges were described, including the most relevant vulnerabilities and threats Road ITS networks encompass. Based on these threats, a set of relevant and notable studies detecting various types of network attacks using machine learning were presented. A detailed discussion was made about relevant trends and important gaps.

Next, we outlined the importance of XAI approaches in IDS systems for the Road ITS systems. We provided relevant background knowledge the XAI model should have as well as a list of the most important objectives it should ensure. Next, we reviewed some of the most recent studies conducted in proposing XAI systems for IDS algorithmic models. Finally, we gave future directions on XAI methods that could be adopted to improve the IDS systems' explainability and transparency. Proceeding, we have conducted a case study where we illustrate how cross-network application layer D/DoS attacks can occur to and from the Road ITS network and evaluate their impact on the target system itself as well as the smart city. Our results indicate that DDoS attacks conducted in the Road ITS network can easily disrupt the critical nodes of the target infrastructure by making them unavailable to accept legitimate requests and potentially threatening human lives.

REFERENCES

Abboud, K, Omar, HA, & Zhuang, W (Dec. 2016). Interworking of DSRC and cellular network technologies for V2X Communications: A survey. *IEEE Transactions on Vehicular Technology*, *65*(12), 9457–9470, doi: 10.1109/TVT.2016.2591558.

Ahmad, F, Adnane, A, & Franqueira, V (2016). A systematic approach for cyber security in vehicular networks. *Journal of Computer and Communications*, *4*, 38–62. doi: 10.4236/jcc.2016.416004.

Ahn, S, Kim, J, Park, SY, & Cho, S (2021). Explaining deep learning-based traffic classification using a genetic algorithm. *IEEE Access*, *9*, 4738–4751, doi: 10.1109/ACCESS.2020.3048348.

Andrews, R, Diederich, J, & Tickle, AB (1995). Survey and critique of techniques for extracting rules from trained artificial neural networks. *Knowledge-Based Systems*, *8*. Elsevier.

Ashraf, J, Bakhshi, AD, Moustafa, N, Khurshid, H, Javed, A, & Beheshti, A (2020). Novel deep learning-enabled LSTM autoencoder architecture for discovering anomalous events from intelligent transportation systems. *IEEE Transactions on Intelligent Transportation Systems*, pp. 1–12. doi: 10.1109/TITS.2020.3017882.

Baiad, R, Otrok, H, Muhaidat, S, & Bentahar, J (2014). Cooperative cross layer detection for blackhole attack in VANET-OLSR. In 2014 International Wireless Communications and Mobile Computing Conference (IWCMC), Nicosia, 863–868, doi: 10.1109/IWCMC.2014.6906469.

Barredo Arrieta, A , Díaz-Rodríguez, N , Del Ser, J , Bennetot, A , Tabik, S, Barbado, A, Garcia, S , Gil-Lopez, S , Molina, D , Benjamins, R, Chatila, R, & Herrera, F (2020). Explainable Artificial Intelligence (XAI): Concepts, taxonomies, opportunities and challenges toward responsible AI. *Information Fusion*, *58*, 82–115.

Basu, S, Kumbier, K, Brown, JB, & Yu, B (2018). Iterative random forests to discover pre- dictive and stable high-order interactions. *Proceedings of the National Academy of Sciences*, *115*(8), 1943–1948.

Bennetot, A, Laurent, JL, Chatila, R, & Díaz-Rodríguez, N (2019). Towards explainable neural-symbolic visual reasoning. In NeSy Workshop IJCAI 2019, Macau, China.

Buczak, AL, & Guven, E (Second quarter 2016). A survey of data mining and machine learning methods for cyber security intrusion detection. *IEEE Communications Surveys & Tutorials*, *18*(2), 1153–1176, doi: 10.1109/COMST.2015.2494502.

Burns, K, Hendricks, LA, Saenko, K, Darrell, T, & Rohrbach, A (2018). Women also snow-board: Overcoming bias in captioning models.

Caruana, R, Lou, Y, Gehrke, J, Koch, P, Sturm, M, & Elhadad, N (2015). Intelligible models for healthcare: Predicting pneumonia risk and hospital 30-day readmission. In Proceedings of the 21th ACM SIGKDD International Conference on Knowledge Discovery and Data Mining, in KDD '15, 1721–1730.

Castelvecchi, D (2016). Can we open the black box of AI? *Nature News*, *538*(7623), 20.

Chander, A, Srinivasan, R, Chelian, S, Wang, J, & Uchino, K (2018). Working with beliefs: AI transparency in the enterprise. In Workshops of the ACM Conference on Intelligent User Interfaces.

Chauley, NK (2016). Security analysis of vehicular ad hoc networks (VANETs): A comprehensive study. *International Journal of Network Security and Its Applications*, *10*(5), 261–274.

Chouldechova, A (2017). Fair prediction with disparate impact: A study of bias in recidivism prediction instruments. *BigData*, *5*(2), 153–163.

Cunha, F, Villas, L, Boukerche, A, Maia, G, Viana, A, Mini, RAF, & Loureisro, AAF (2016). Data communication in VANETs: Protocols, applications and challenges. *Ad Hoc Networks*, *44*, 90–103.

de Biasi, G, Vieira, LFM, & Loureiro, AAF (2018). Sentinel: Defense mechanism against DDoS flooding attack in software defined vehicular network. In 2018 IEEE International Conference on Communications (ICC), Kansas City, MO, 1–6, doi: 10.1109/ICC.2018.8422303.

Deakin, M, & Waer, HA (2011). From intelligent to smart cities. *Intelligent Buildings International*, *3*, 140–152.

ENISA (2015). Cyber security and resilience of intelligent public transport good practices and recommendations, 1–68.

Fernández, GM, et al. (2018). UGR'16: A new dataset for the evaluation of cyclo stationarity-based network IDSs. *Computers & Security*, *73*, 411–424.

Gao, Y, Wu, H, Song, B, Jin, Y, Luo, X, & Zeng, X (2019). A distributed network intrusion detection system for distributed denial of service attacks in vehicular ad hoc network. *IEEE Access*, *7*, 154560–154571, doi: 10.1109/ACCESS.2019.2948382.

Goodman, B, & Flaxman, S (2017). European union regulations on algorithmic decision-making and a "right to explanation". *AI Magazine*, *38*(3), 50–57.

Grosse, K, et al. (2017). Adversarial examples for malware detection. In Proceedings of the European Symposium on Research in Computer Security (ESORICS), 62–79.

Guerreiro, G, Figueiras, P, Silva, R, Costa, R, & Jardim-Goncalves, R (2016). An architecture for big data processing on intelligent transportation systems. An application scenario on highway traffic flows. In 2016 IEEE 8th International Conference on Intelligent Systems (IS), Sofia, 65–72, doi: 10.1109/IS.2016.7737393.

Guidotti, R, Monreale, A, Ruggieri, S, Turini, F, Giannotti, F, & Pedreschi, D (2018). A survey of methods for explaining black box models. *ACM Computing Surveys*, *51*(5), 93:1–93:42.

Guo, W, et al. (2018). LEMNA: Explaining deep learning based security applications. In Proceedings of the ACM Conference on Computer and Communications Security (CCS), 364–379.

Harbers, M, Van Den Bosch, K, & Meyer, JJ (2010). Design and evaluation of explainable BDI agents. In IEEE/WIC/ACM International Conference on Web Intelligence and Intelligent Agent Technology, 2, IEEE, 125–132.

Harrington, J, Lacroix, J, El-Khatib, K, Lobo, FL, & ABF Oliveira, H (2017). Proactive Certificate Distribution for PKI in VANET. In Proceedings of the 13th ACM Symposium on QoS and Security for Wireless and Mobile Networks, 9–13.

Haydari, A, & Yilmaz, Y (2018). Real-time detection and mitigation of DDoS attacks in intelligent transportation systems. In 2018 21st International Conference on Intelligent Transportation Systems (ITSC), Maui, HI, 157–163, doi: 10.1109/ITSC.2018.8569698.

Huysmans, J, Dejaeger, K, Mues, C, Vanthienen, J, & Baesens, B (2011). An empirical eval- uation of the comprehensibility of decision table, tree and rule based predictive models. *Decision Support Systems*, *51*(1), 141–154.

Karagiannis, D, & Argyriou, A (2018). Jamming attack detection in a pair of RF communicating vehicles using unsupervised machine learning. *Vehicular Communications*, *13*, 56–63.

Karagiannis, G, et al. (Fourth Quarter 2011). Vehicular networking: A survey and tutorial on requirements, architectures, challenges, standards and solutions. *IEEE Communications Surveys & Tutorials*, *13*(4), 584–616, doi: 10.1109/SURV.2011.061411.00019.

Khatoun, R, Gut, P, Doulami, R, Khoukhi, L, & Serhrouchni, A (2015). A reputation system for detection of black hole attack in vehicular networking. In 2015 International Conference on Cyber Security of Smart Cities, Industrial Control System and Communications (SSIC), Shanghai, 1–5, doi: 10.1109/SSIC.2015.7245328.

Kim, B, Glassman, E, Johnson, B, & Shah, J (2015). iBCM: Interactive Bayesian case model empowering humans via intuitive interaction, Technical Report, MIT-C-SAIL-TR-2015-010.

Kuppa, A, & Le-Khac, N-A (2020). Black Box Attacks on Explainable Artificial Intelligence (XAI) methods in Cyber Security. In 2020 International Joint Conference on Neural Networks (IJCNN), Glasgow, UK, 1–8, doi: 10.1109/IJCNN48605.2020.9206780.

Langley, P, Meadows, B, Sridharan, M, & Choi, D (2017). Explainable agency for intelligent autonomous systems. In AAAI Conference on Artificial Intelligence, 4762–4763.

Lefvy-Benchetton, C, & Darra, E (2015). ENISA: Cybersecurity for smart cities: An achitecture model for public transport. *Cyber Security for Smart Cities*, 1–54.

Li, W, Joshi, A, & Finin, T (2015). SVM-CASE: An SVM-based context aware security framework for vehicular ad-hoc networks. In 2015 IEEE 82nd Vehicular Technology Conference (VTC2015-Fall), Boston, MA, 1–5, doi: 10.1109/VTCFall.2015.7391162.

Liu, J, Xiao, Y, Li, S, Liang, W, & Chen, CLP (Fourth Quarter 2012). Cyber security and privacy issues in smart grids. *IEEE Communications Surveys & Tutorials*, *14*(4), 981–997, doi: 10.1109/SURV.2011.122111.00145.

Lundberg, SM, & Lee, S-I (2017). A unified approach to interpreting model predictions. In Proceedings of the Advances in Neural Information Processing Systems, 4765–4774.

Marino, DL, Wickramasinghe, CS, & Manic, M (2018). An adversarial approach for explainable AI in intrusion detection systems. In IECON 2018 - 44th Annual Conference of the IEEE Industrial Electronics Society, Washington, DC, USA, 3237–3243, doi: 10.1109/IECON.2018.8591457.

Menouar, H, Guvenc, I, Akkaya, K, Uluagac, AS, Kadri, A, & Tuncer, A (March 2017). UAV-enabled intelligent transportation systems for the smart city: Applications and challenges. *IEEE Communications Magazine*, *55*(3), 22–28, doi: 10.1109/MCOM.2017.1600238CM.

Merdrignac, P, Shagdar, O, & Nashashibi, F (July 2017). Fusion of perception and V2P communication systems for the safety of vulnerable road users. *IEEE Transactions on Intelligent Transportation Systems*, *18*(7), 1740–1751, doi: 10.1109/TITS.2016.2627014.

Michalski, RS (1983). A theory and methodology of inductive learning. *Machine learning*. Springer, 83–134.

Montavon, G, Samek, W, & Müller, K-R (2018). Methods for interpreting and understanding deep neural networks. *Digital Signal Processing*, *73*, 1–15, doi: 10.1016/j.dsp.2017.10.011.

Nasrallah, YY, Al-Anbagi, I, & Mouftah, HT (2015). Mobility impact on the performance of electric vehicle-to-grid communications in smart grid environment. In 2015 IEEE Symposium on Computers and Communication (ISCC), Larnaca, 764–769, doi: 10.1109/ISCC.2015.7405606.

Nisioti, A, Mylonas, A, Yoo, PD, & Katos, V (4th Quart. 2018). From intrusion detection to attacker attribution: A comprehensive survey of unsupervised methods. *IEEE Communications Surveys and Tutorials*, *20*(4), 3369–3388.

NIST (2004). FIPS 199: Standards for security categorization of federal information and information systems.

Olden, JD, & Jackson, DA (2002). Illuminating the"blackbox": A randomization approach for understanding variable contributions in artificial neural networks. *Ecological Modelling*, *154*(1–2), 135–150.

Parliament and C. of the European Union (2016). General data protection regulation.

Pearl, J (2009). *Causality*. Cambridge University Press.

Procopiou, A, & Komninos, N (2015). Current and future threats framework in smart grid domain. In 2015 IEEE International Conference on Cyber Technology in Automation, Control, and Intelligent Systems (CYBER), Shenyang, China, 1852–1857, doi: 10.1109/CYBER.2015.7288228.

Ribeiro, MT, Singh, S, & Guestrin, C (2016) Why should I trust you?: Explaining the predictions of any classifier. In ACM SIGKDD International Conference on Knowledge Discovery and Data Mining, ACM, 1135–1144.

Ruppert, D (1987). *Robust statistics: The approach based on influence functions*. Taylor & Francis.

Shu, J, Zhou, L, Zhang, W, Du, X, & Guizani, M (2020). Collaborative intrusion detection for VANETs: A deep learning-based distributed SDN approach. *IEEE Transactions on Intelligent Transportation Systems*, pp. 1–12. doi: 10.1109/TITS.2020.3027390.

Smutz, C, & Stavrou, A (2012). Malicious PDF detection using metadata and structural features. In Proceedings of the Annual Computer Security Applications Conference (ACSAC), 239–248.

Szegedy, C, Zaremba, W, Sutskever, I, Bruna, J, Erhan, D, Goodfellow, I, & Fergus, R (2013). Intriguing properties of neural networks.

Van Lent, M, Fisher, W, & Mancuso, M (2004). An explainable artificial intelligence system for small-unit tactical behavior. *Proceedings of the National Conference on Artificial Intelligence*, 900–907. Cambridge.

Wang, M, Zheng, K, Yang, Y, & Wang, X (2020). An explainable machine learning framework for intrusion detection systems. *IEEE Access*, *8*, 73127–73141, doi: 10.1109/ACCESS.2020.2988359.

Xiao, Z, Xiao, Z, Wang, D, & Li, X (2015). An intelligent traffic light control approach for reducing vehicles CO2 emissions in VANET. In 2015 12th International Conference on Fuzzy Systems and Knowledge Discovery (FSKD), Zhangjiajie, 2070–2075, doi: 10.1109/FSKD.2015.7382270.

Yu, B, et al. (2013). Stability. *Bernoulli*, *19*(4), 1484–1500.

Real-Time Identity Censorship of Videos to Enable Live Telecast Using NVIDIA Jetson Nano

18

Shree Charran R.[1] and Rahul Kumar Dubey[2]

[1]*Indian Institute of Science, Bengaluru, India*
[2]*Robert Bosch Engineering and Business Solutions Private Limited, Bengaluru, India*

Contents

DOI: 10.1201/9781003172772-18

18.1 INTRODUCTION

Aggressive journalism has evolved over time as a product of globalization. The massive boost in technology has helped electronic media to live telecast news in a 24-hour format. Initially, the media remained engaged in covering the news round the clock in order to create awareness among viewers. But, with the spike in the number of private channels in the early 21st century, a new pattern of attracting the viewers towards their own channel has evolved, that is by being the first to spread the news quickly. Sensationalized stories, media paparazzi has become common on broadcasting channels. This has further led to a massive shift towards a culture of live interviews and live streaming of content. Broadcasting of live content comes with its own set of technical, ethical, and legal issues. Identity censorship is one of the issues to be handled by media houses to avoid safety and legal issues for itself and for the interviewees. The most common identity censorship we have witnessed in news telecasts are blurred faces and morphed voices of interviewees. This is done for many reasons: for privacy and security issues, avoiding display of disturbing content, etc. Figure 18.1 gives an example of identity censorship with face blurring and face pixelating which is a common industry practice.

Most editing software will allow you to both blur images and distort voices beyond recognition. During the editing process, you can also edit the sound to remove any identifying names or places mentioned in the interview. Although editing tools allow you the most options for concealing identity, the raw unedited footage will always be a liability for both the media house and the people they film. Another technical concern is the time lag involved in the whole process before it can be telecasted. We propose to tackle both these areas of concern in this chapter. Figure 18.2 shows the standard workflow for the activity of streaming edited content which a standard media house adopts before telecasting sensitive content. Smart city development has an immense need for use of such small-edge computing devices combined with deep learning and artificial intelligence applications for security and monitoring purposes. They can be directly adopted for surveillance, parking, and vehicle identification.

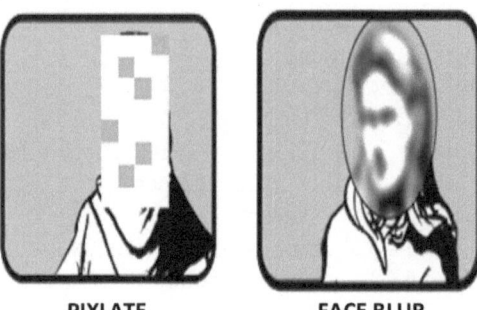

PIXLATE   FACE BLUR

FIGURE 18.1   Identity Censorship Example.

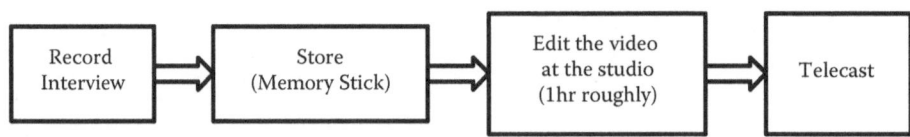

FIGURE 18.2   Identity Censorship Workflow.

18.2 BACKGROUND

Shah et al. (2016) performed face detection using Raspberry PI devices with basic image processing techniques. They used Haar cascades to detect faces and by locating the boundary box of the face by turning on the colour gradients of the range. The limitation of their experiment was that they could not detect multiple faces. Inthanon and Mungsing (2020) built a tool to detect drowsiness in vehicle drivers from facial images. They trained and tested their algorithm on the NVIDIA Jetson Nano device to obtain real-time predictions. The device was able to accurately evaluate the image by tracking closing eye motions more than 35 FPS or 1.5 seconds, and yawning or opening mouth motion more than 50 FPS equals 2 seconds. The best accuracy they could achieve on a test video was 83.31%. Salih and Gh (2020) made comparisons between the Jetson Nano, Raspberry Pi 4, and Raspberry Pi 3 for the purpose of Face detection and face identification. They concluded that the NVIDIA Jetson Nano was comparatively a superior device with 12.1 FPS and 8.9 FPS for face detection and identification tasks. The authors claim the speed per prediction to be 80 milliseconds which was faster than the Traditional CPU performance. Lindner et al. (2020) performed face detection for security and monitoring using a camera mounted on 3 different single-board computers. The authors explored Haar & MTCNN for face detection. The MTCNN was the better performing model with 2.3% false predictions. Further, they also benchmarked performance speed of various single board computers and Laptops and found the speed of the Jetson Nano device similar to those of Standard I-7 laptops. Süzen et al. (2020) performed tests with respect to deep learning applications on big data using single board computers. The authors conducted benchmark analysis on 3 single-board computers – Jetson TX2, Jetson Nano, and Raspberry PI for deep learning applications. Performance metrics included accuracy, time, memory, and power consumption during parallel processing. The model used for the experiment was a standard deep CNN algorithm with a standard fashion dataset for the training and testing. The results showed that for large datasets Jetson TX2 has more power consumption, but it performs better than the others with low processing time and higher accuracy. Lofqvist and Cano (2020) tested pre-trained Single Shot MultiBox Detector (SSD) and Region-based Fully Convolutional Network (R-FCN) on Jetso Nano to Benchmark it based on inference time, memory consumption, and accuracy. It was found that on average the processing time decreased by 2 seconds and 1050M less memory was used across both models after applying the image scaling technique. Although accuracy was not significantly improved, applying compression techniques did indeed decrease the inference time and memory consumption, the two aspects that are highly important to consider for implementing deep learning applications on constrained devices in space. Davis et al. (2011) performed face recognition in real time from surveillance videos. The authors used nested Haar Cascades for eyes, nostrils, and ear-tips combined with analysis of colour skin-tone information. Furthermore, they used Support vector machines to match features and identify persons.

Despite several works in the field of face detection and identification, there is limited literature on the combination of edge devices plus deep learning for the broadcasting industry. This shows there is some scope for ideation and problem solving. In this chapter we propose the use of an edge device a.k.a single-board computer for the purpose of real-time image processing and streaming

18.3 MATERIALS

18.3.1 Hardware

18.3.1.1 NVIDIA Jetson Nano

NVIDIA's Jetson Nano is a budget, single-board hybrid-computer for AI applications. Jetson Nano is equipped with a GPU with CUDA core, which is very useful for computation related to image

TABLE 18.1   Technical specification of the Jetson Nano (NVIDIA Developers, 2020)

Performance	472 GFLOPS
CPU	Quad-Core ARM Cortex-A57 64-bit @ 1.42 GHz
GPU	NVIDIA Maxwell w/ 128 CUDA cores @ 921 MHz
Memory	4 GB LPDDR4 @ 1600 MHz, 25.6 GB/s
Networking	Gigabit Ethernet
Display	HDMI 2.0 and eDP 1.4
USB	4x USB 3.0, USB 2.0 Micro-B
Video encode	H.264/H.265 (4Kp30)
Video decode	H.264/H.265 (4Kp60, 2x 4Kp30)
Storage	16 GB eMMC
Power under load	5 W–10 W
Price	$89 (NVIDIA Developers, 2020)

processing. It provides high performance despite having a very simple architecture. Artificial intelligence-enabled devices require GPU-enabled systems and usually consume low power to run algorithms; hence a device like the Jetson Nano is most suited. Similar computers are already in use in ATMs, smart home systems, robotic systems, etc. Table 18.1 provides the technical specification of the Jetson Nano Device.

18.3.1.2 Camera

The camera that was used is the XIMEA model MX200CG-CM. The camera has 20MPx and the ability to stream 4K images at 25 fps and Full HD 1080p (1920 × 1080) images at 30 fps simultaneously. The camera and the Jetson Nano setup are shown in Figure 18.3.

18.3.2 Dataset

The custom dataset used comprises 1050 annotated images of interviewers and interviewees. The dataset is prepared using videos from news clippings available online and few videos shot personally to validate

FIGURE 18.3   NVIDIA Jetson Nano + 20 Megapixel Camera MX200CG-CM Setup.

FIGURE 18.4   Sample from the Dataset.

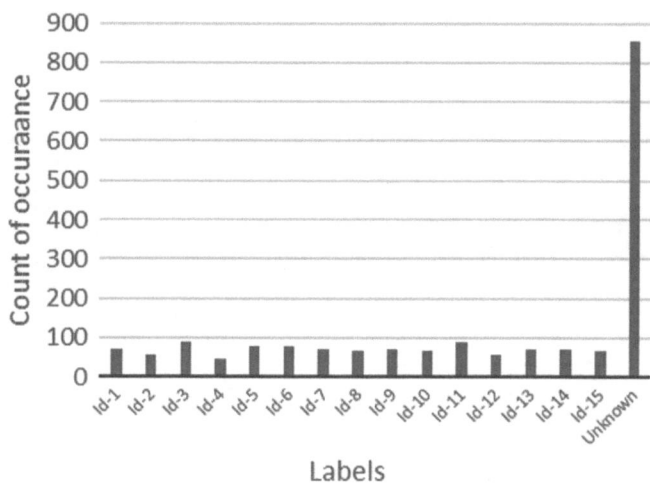

FIGURE 18.5   Classes Distribution in the Dataset.

the results. There are 15 identified interviewers labelled by their IDs and all unidentified faces are labelled as "Unknown". Figure 18.4 is a random sample from the annotated dataset where the interviewer is identified with their respective ID and the interviewee is not identified from the database and hence has no ID.

Figure 18.5 shows the distribution of the labels across the custom dataset. Figure 18.6 shows the distribution of people in each image across the whole dataset. It can be seen that generally there are 2 people in an image (1 Interviewee + 1 Interviewer) in the images labelled. In certain cases there are multiple interviewees.

18.3.3 Proposed Methodology

Face Identification forms the core component of censoring identity. We first detect and identify faces in the image, then censor out the unidentified faces. The 4 individual components of Face Identification are: (i) Detecting and locating one or more faces in the image and marking the face with a bounding box.

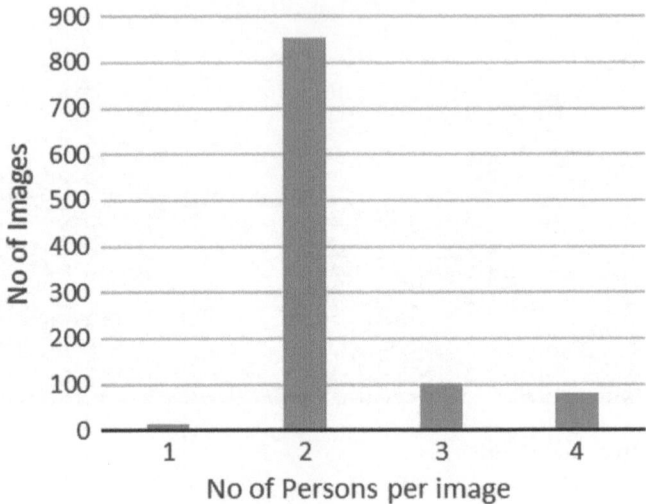

FIGURE 18.6 People Distribution in the Dataset.

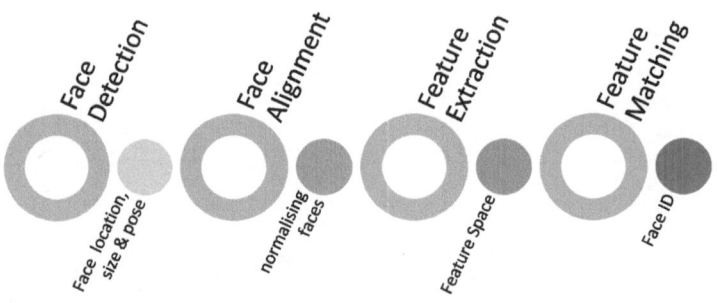

FIGURE 18.7 Face Recognition Components.

(ii) Normalizing the face with respect to geometry and photo metrics. (iii) Extracting features from the face bounded. (iv) Performing feature matching of the face against one or more known faces in the database. Figure 18.7 depicts the components of a typical face recognition algorithm.

We use Multi-task Cascaded Convolutional Neural Networks (Zhang et al., 2016) to detect and localize the face and create a bounding box around it. We used FaceNet (Schroff et al., 2015) to compare the face within the bounding with the database and identify the face. Multi-task Cascaded Convolutional Neural Networks has three blocks, that is, P-Net, R-Net, and O-Net, and is able to outperform many face-detection benchmarks while retaining real-time performance.

a. Candidate windows are produced through a fast Proposal Network (P-Net). This creates multiple scaled copies of the image and proposes candidate windows. Non-Maximum Suppression is used to merge highly overlapped candidates.

b. It then refines these candidates in the next stage through a Refinement Network (R-Net). This is done by recalibrating the bounding boxes and performing Non-Maximum Suppression.

c. The Final Output Network removes the bounding boxes that have low confidence scores, finds five facial landmarks, calibrates bounding boxes and landmarks further, and then does Non-Maximum Suppression. Output Network (O-Net).

The learning objective of MTCNN is a multi-task loss, with one loss as binomial cross-entropy loss is the probability that the box has a face, the second one is Euclidean distance loss for bounding box regression, and third one is Euclidean loss for facial landmark regression. The 3 losses are weighted and summed up in a cumulative multi-task formula. The output bounding box with the face from MTCNN is fed to FaceNet for face recognition in the bounding box. FaceNet is a face recognition model by Google. The FaceNet system can be used to extract high-quality features from faces, called face Embedding's that can then be used to train a face identification system. Finally, an SVM classifier is used to identify the face in the last stage. The face Embedding's are multidimensional numerical vector representations of a face which represent the unique identity of the face. Face net provides a 128 dimension embedding for a face. The triplet loss involves comparing face Embedding's for three images, one being an anchor (reference) image, second, a positive image (matching the anchor), and third, a negative image (not matching the anchor). Embedding's are learnt by a Deep CNN network such that the positive embedding is closer to anchor embedding compared to negative embedding distance to the anchor

$$F(A) - F(P) + margin < F(A) - F(N) \tag{18.1}$$

where

A, P, and N represent Anchor, positive, and negative.

This equation forms the basis of multi-task loss function that FaceNet uses. The triplet selection starts with triplets that violate the above constraint – this leads to faster convergence. The MTCNN and FaceNet models are pre-existing NVIDIA model architecture.

18.3.4 Model Pipeline

The input video stream captured by the camera is transformed by the inbuilt plugin for format conversion and scaling based on network requirements and passes the transformed data to MTCNN for face detection. These detected bounding boxes are then passed to the FaceNet to recognize the interviewer and interviewee. All unrecognized interviewees bounding boxes are masked with a Gaussian blur and relayed back for streaming using NVIDIA Deep Stream software development kit (Figure 18.8).

18.4 EXPERIMENTAL STUDY

In this study we have performed experimental tests on face recognition on images in the database on Jetson Nano devices. The experiments are done using the existing Tensor RT & Deep Stream pipeline existing on the Jetson Nano device in addition to OpenCV image processing library.

18.4.1 Data Augmentation

Since the custom dataset prepared is small compared to traditional datasets, we implemented data augmentation to increase the size and diversity of the existing dataset. This augmented data is acquired by performing a series of preprocessing transformations to the existing dataset. Transformations like horizontal and vertical flipping, skewing, cropping, and rotating are performed. Collectively, this augmented data is able to simulate a variety of subtly different data points, as opposed to just duplicating the same data. The augmentation settings are randomly set so as to create a wide variety of permutations and are performed as below:

```
rotation_angles=15,30,45,60,75,90
width_shift_range=0.05,0.10,0.15,0.20,0.25,0.30,0.35,0.40,
0.45,0.50
height_shift_range=0.05,0.10,0.15,0.20,0.25,0.30,0.35,0.40,
0.45,0.50
shear_range=0.05,0.10,0.15,0.20,0.25,0.30,0.35,0.40,0.45,0.50
zoom_range=0.05,0.10,0.15,0.20,0.25,0.30,0.35,0.40,0.45,0.50
horizontal_flip=True,
vertical_flip=True,
```

Training dataset of sizes 1.15K, 2K, 5K, and 10K are prepared so as to analyze performance and accuracy of the system for different dataset sizes.

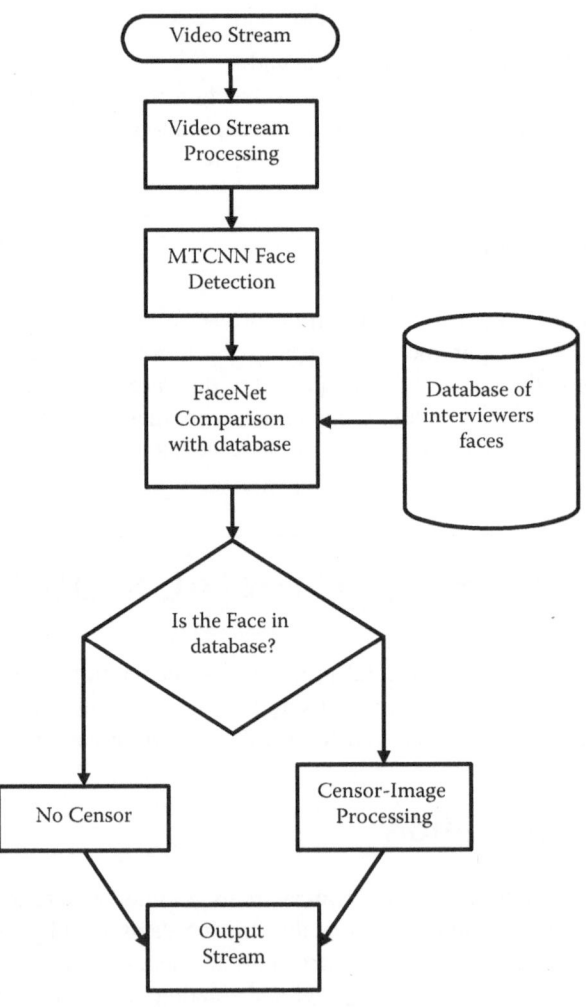

FIGURE 18.8   Overall Pipeline.

18.4.2 Experiments

a. We first determine the ideal size of the dataset to be used for training the MTCNN and FaceNet Algorithms. The algorithms are run on the dataset of sizes 1.15K, 2K, 5K, and 10K to identify the model with the lowest error rate and better run time.

b. The final model parameters are set and the setup is evaluated on 4 custom real-time test videos and the final predictions are noted.

18.4.3 Metrics

The evaluation metrics for face detection and face identification experiments are Frames/Second and the metric for the final test of the face recognition setup is accuracy.

a. **Frames/Second** = Frames/Time Period Elapsed. This indicates how many frames are processed and displayed per second. Higher frames per second indicate computationally fast models.

b. **Accuracy** = Total correct faces Masked or Not masked/Total Faces found.

18.5 RESULTS

The training of the face detector is done for 100 epochs across all the datasets. From Table 18.2, it can be observed that the augmented dataset of size 10K performs best achieving a test accuracy of 99.5%. This is intuitive as more data usually helps perform and generalize better. Figure 18.9 shows samples of faces extracted by the detector accurately by MTCNN. The individual performance of MTCNN for face detection and FaceNet for face identification on the test dataset for the optimal dataset size i.e. 10K trained on 100 epochs is shown in Table 18.3.

To further test our setup in real time we use 4 real-time videos with interviewers' faces in the database and with random interviewees whose faces are not in the database. Table 18.4 shows the performance of the setup when recording and streaming 4 live videos and performing identity masking in real time. It can be noticed that the performance is similar to that of the training with the best accuracy of 95.07%. It can be noticed that the frames per second decrease for the video with lower accuracy. This decrease is due to multiple faces being present in a single frame which requires more processing. The overall accuracy is acceptable. Lower frames per second indicate lag due to processing. Figure 18.10 gives two sample frames of the output stream with the interviewees censored as required.

TABLE 18.2 Performance of the face detectors across training dataset sizes

DATASET	1.15K	2K	5K	10K
Test Accuracy (%)	87.8	95.6	98.2	99.5
Training Time (sec)	20	26	32	58
Memory (GB)	1.5	1.9	2	2.5
CPU (Power/W)	0.47	0.99	1.5	2.32
GPU (Power/W)	0.76	1.12	2.84	2.92

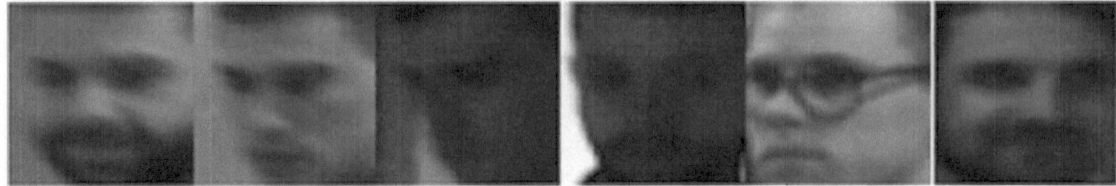

FIGURE 18.9   Random Sample of Faces Extracted by MTCNN.

TABLE 18.3   MTCNN and facenet performance

	PROCESSING TIME IN SEC	*ERROR RATE IN %*
MTCNN	0.75	2.3
FaceNet	0.16	1.8

TABLE 18.4   Technical specification of the Jetson Nano

SL NO.	*VIDEO (SEC)*	*FRAMES PER SECOND ACHIEVED*	*ACCURACY ACHIEVED IN %*
1	59.01	25	93.31
2	58.82	23	90.52
3	59.71	19	89.83
4	57.33	25	95.07

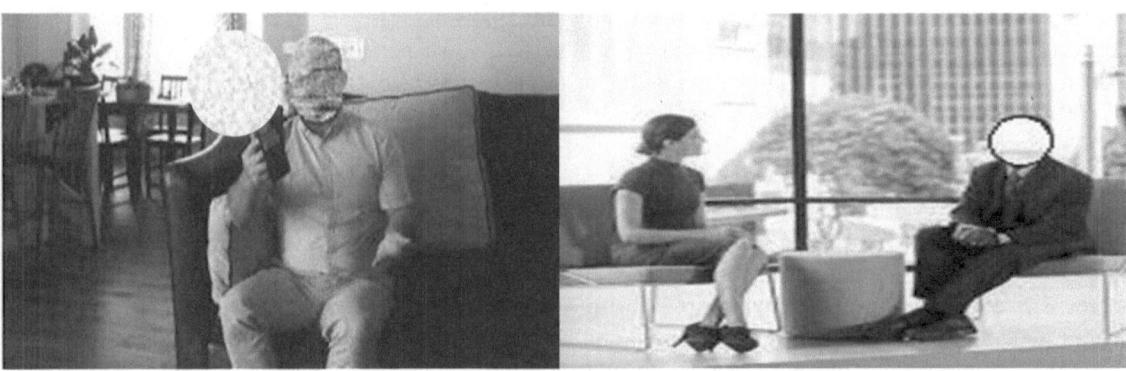

FIGURE 18.10   Final Output Sample.

18.6 CONCLUSIONS

In this chapter, our goal is to build an effective and low-cost identity censorship mechanism so as to enable instant live streams by censoring select faces to allow identity privacy. The video is captured using the Jetson Nano device mounted with a High Definition recording camera. We censor the interviewees' faces in real time using a combination of deep learning-based face detection and image processing making use of the GPU architecture of the Jetson Nano. This enables the video stream to be ready to be telecasted with no editing. The added advantage being that there will be no raw unedited footage being handled manually, thereby enhancing privacy. Our mechanism when enabled was able to detect the interviewer with 0.5% error rate in training. It was able to censor the faces of intervieews at 1.5% error rate in training. On live video, censorship of interviewees with best accuracy of 95.07% at 25 Frames per second speed was achieved. This method hence enables us to incorporate a single-board device and AI approaches to improve and ease the privacy management in media streaming services. The limitation of the approach lies with the fact that the device being a budget device, it tends to have larger error rates as the number of persons to be identified in the frame increases beyond 2 persons. Plus the high error rate is associated with lower frames per second generated. But given the fact that any other alternative is 10–15 times more expensive and also slower, this approach still seems viable for development.

Future study can involve the use of edge devices with more computational capacities and more refined algorithms. We could also try to evaluate the performance on faster moving videos. Smart city development has an immense need for use of such small-edge computing devices combined with deep learning and Artificial Intelligence applications for security and monitoring purposes among multiple other uses.

REFERENCES

Davis, M, Popa, S, & Surlea, C (February 2011). Real-time face recognition from surveillance video. doi: 10.1007/978-3-642-17554-1_8.

Inthanon, P, & Mungsing, S (2020). Detection of drowsiness from facial images in real-time video media using Nvidia Jetson Nano. In 2020 17th International Conference on Electrical Engineering/Electronics, Computer, Telecommunications and Information Technology (ECTI-CON), Phuket, Thailand, 246–249, doi: 10.1109/ECTI-CON49241.2020.9158235.

Lindner, T, Wyrwał, D, Białek, M, & Nowak, P (2020). Face recognition system based on a single-board computer. In 2020 International Conference Mechatronic Systems and Materials (MSM), Bialystok, Poland, 1–6, doi: 10.1109/MSM49833.2020.9201668.

Lofqvist, M, & Cano, J (2020). Accelerating deep learning applications in space, arXiv e-prints.

NVIDIA Developers (2020, April). https://developer.nvidia.com/embedded/faq

Salih, TA, & Gh, MB (2020). A novel face recognition system based on Jetson Nano developer kit. In 2020 IOP Conf. Ser.: Mater. Sci. Eng. 928 032051.

Schroff, F, Kalenichenko, D, & Philbin, J (2015). FaceNet: A unified embedding for face recognition and clustering. In 2015 IEEE Conference on Computer Vision and Pattern Recognition (CVPR), Boston, MA, 815–823.

Shah, AA, Zaidi, ZA, Chowdhry, BS, & Daudpoto, J (2016). Real time face detection/monitor using raspberry pi and MATLAB. In 2016 IEEE 10th International Conference on Application of Information and Communication Technologies (AICT), Baku, 1–4, doi: 10.1109/ICAICT.2016.7991743.

Süzen, AA, Duman, B, & Şen, B (2020). Benchmark analysis of Jetson TX2, Jetson Nano and Raspberry PI using Deep-CNN. In 2020 International Congress on Human-Computer Interaction, Optimization and Robotic Applications (HORA), Ankara, Turkey, 1–5, doi: 10.1109/HORA49412.2020.9152915.

Zhang, K, Zhang, Z, Li, Z, & Qiao, Y (Oct. 2016). Joint face detection and alignment using multitask cascaded convolutional networks. *IEEE Signal Processing Letters*, 23(10), 1499–1503, doi: 10.1109/LSP.2016.2603342.

Smart Cities' Information Security: Deep Learning-Based Risk Management

W. Abbass, A. Baina, and M. Bellafkih

National Institute of Posts and Telecommunication 'INPT', Madinat Al Irfane, Rabat, Morocco

Contents

19.1 INTRODUCTION

A 'smart city' is a systemic model of urban development that relies on information and communication technologies (ICT) such as the Internet of Things (IoT), Big Data applications Cloud Computing, and Machine Intelligence in order to promote smart management of institutions, businesses, and citizens economically, socially, and environmentally. Its main goal is to build up a smart government, smart people's livelihood, and smart production. Bringing changes into smart cities radically alters our daily lives. Morocco serves as the best example of urban development. It has over the past 50 years introduced a large economic growth by implementing the strategy of decentralization in order to empower provinces and municipalities in responding to the economic and environmental needs (El Alaoui, 2017). In fact, the introduction of the concept of 'smart city' in Morocco has attracted considerable attention that it seen as a vital opportunity to achieve sustainable development. It is opening up to citizens by constantly offering new forms of mobility (Song et al., 2017). It still poses challenges to global security and privacy. The potential of 'smart cities' is limitless which allows the enhancement of the existing capabilities, such as:

- Energy consumption
- Agricultural development and transportation
- Security and innovation of public spaces
- Buildings and workplaces
- Waste management

A 'smart city' is a collection of different ICT systems that commonly operate with each other. Consequently, a holistic security strategy is highly recommended. In smart cities, ICT is used to conduct real-time management of data related to the interconnected objects in order to satisfy their needs of service. For an efficient development, the most advanced ICT is used in order to fully perceive the objects and their interactions, thus leading to greater exposition to vulnerabilities. Actually, at the heart of these interconnected objects, there is data, individual or collective, accessible to a large set of citizens and systems. By linking them together, the area of exposure to risk expands. Moreover, these objects are generally not made to withstand updates which would considerably slow down their use and became mismatched with the smart city requirements (Sookhak et al., 2018). It is the main challenge of the connected city: to be sufficiently adjustable and manageable while being robust enough against attacks. The 'smart city' is still in full emergence; it is imperative to define good practices and security policies to protect it. Moreover, its polymorphism complicates the definition of global security policies considering that their implementation evolves along with the development of new technologies (Baig et al., 2017).

A security-by-design approach must consist in identifying the current needs and its potential evolutions. It must propose a specific integration of the concrete responses to the functional need and the minimum security clauses that allow deploying services with a satisfactory level of confidence (Gharaibeh et al., 2017). It must fundamentally gather data and trigger effective countermeasures that continuously assure security challenges. Nevertheless, this is likely to go against the principle of the 'smart city' concept as the risks associated with various objects threaten the proper functioning of the provided services. Thereby, the addedvalue of condudcting SRM process as it tolerates Risk Management (Assessment and Monitoring). Various approaches exist; however, their deployment is at an unprecedented pace which consequently benefits the attackers. Due to resource constraints and complexity, these approaches entail several false alarms which reduce services' availability.

The current Smart City security development cannot keep up with the perpetual adoption of the new ICT technologies. It increasingly faces security challenges as the major problem is securing all the interconnected devices and their constant flow of data. Thus, considering the emerging risks, DL-based approaches depict an efficient solution to address these issues. Actually, related to XAI, the black boxes are eliminated in a manner as how the decision is made becomes clear. It offers insights leading to better outcomes and forecasts the most preferred SRM behaviour. XAI is commonly addressed in the sense of DL as it plays a major role in solving the most common tasks including regression, prediction, and classification.

This chapter tolerates using association rules in order to better assess security risks targeting a city's objects. Indeed, association rules is a key strategy for learning behaviours for smart decision taking. It is used in order to shorten the time intervening from the risk occurrence to its definition as a rule into the monitoring phase. The main contribution of our research is that it can be used to advise SRM experts on how to manage security risks related to Smart Cities.

The layout of the chapter is as follows:

- Section 19.2 provides a literature review of the Smart Cities concept and its related security risks
- Section 19.3 introduces the proposed DL-based SRM for Smart Cities
- Section 19.4 entails a discussion summarizing the main findings and limitations

19.2 LITERATURE REVIEW

A systematic analysis has been conducted to investigate current research in the Smart Cities concept and related security risks. A Smart City is the integrated product of IoT, Big Data, Cloud Computing, and security technologies. Various sensors connect objects and their related data in order to implement real-time security controls and provide access anytime and anywhere. The main goal of this chapter is to assess risks within Smart Cities. Indeed, these infrastructures are based on a layered architecture where each of the layers include different technologies for data transmission, processing, and storage. The complexity of these infrastructures is an appealing feature raising interest for new SRM approaches. However, risks related to Smart Cities' security must imperatively be analyzed and evaluated in order to protect their design and operation.

Currently, these infrastructures are experiencing more and more attacks that result in disruptions followed by cascading failures. As a result, there are large-scale consequences that spread from one object to another (Alcaraz & Zeadally, 2015). Azevedo Guedes et al. (2018) have presented the hacking attacks targeting a city's objects transmitting data over the Internet. Generally, hackers exploit easily the existing vulnerabilities of smart technologies in order to launch misuse attacks. Most existing security strategies are developed against external eavesdroppers and attackers (Elmaghraby & Losavio, 2014). However, internal attackers can violate access to data and consequently allow access to external attackers (Ismagilova et al., 2020). Braun et al. (2018) have provided a survey of the essential DL aspects to deal with a Smart City's privacy and security. Moreover, a deep neural network-based approach has been applied for Intrusion Detection against IoT gateways (Brun et al., 2018). A simple threshold detector approach has also been able to detect intrusions. The related work does not consider unknown threats. Attacks against Smart Cities significantly reduce the availability of the provided services, their integrity, and confidentiality. DoS and Distributed Denial of Service (DDoS) attacks are the most popular risks targeting the services' availability. Accordingly, SRM's main challenge consists of accurately estimating risk occurrence and impacts. Yet, Smart Cities require specific constraints which render SRM approaches very difficult to implement: the need for real-time performance and high availability of data. Therefore, it is mandatory to sufficiently identify and value the critical devices within the ever-evolving environment. A dynamic modelling of security risks is mandatory (Shayan et al., 2020), wherefor, the need of intelligent framework assessing and mitigating the potential risks. Accordingly, we tolerate:

- Determining the SRM scope within a dynamic environment
- Classifying potential threats targeting Smart Cities' main services
- Treating identified risks and applying security policy

Considering the high level of a Smart City's objects, the overall security depends greatly on effective SRM. Actually, the main limitation of the traditional SRM lies in the fact that it does not consider the perpetual evolution of the interconnected objects. Thus, the overall security must take into consideration the continuous assessment of novel services. We therefore promote improving SRM through the application of DL. The matter is a vital technology for SRM as it proactively stamps out the potential risks and boosts security through pattern detection, real-time monitoring, and systematic penetration testing. DL depends on 2 phases: the training and the detection. On the one hand, the training phase uses a mathematical algorithm in order to extract traditional knowledge as a reference input that would discover existing associations. On the other hand, the detection stage uses the identified associations in order to acquire classification. Accordingly, DL allows solving the most common tasks including regression, prediction, undefined fiber-optic and classification

19.3 SMART CITIES' SECURITY BASED ON DL

Ept of Smart City is establishing itself in the SRM domain and is becoming a critical concern (Ismagilova et al., 2020). The ICT infrastructure of a smart city entails undefinedfiber-optic channels, wireless networks, sensors, and end-point devices that allow users to be connected to the provided services. Whilst ICT is the key component, it also introduces a number of security risks. These imply impacts to the affected city such as:

- Hacking the ICT system of a water treatment facility
- Traffic incidents
- Loss of sensitive information

The exponential increase in the number of connected objects represents a considerable new risk surface. A smart city entails various interconnected sensors and actuators that process and transmit data in order to provide vital services. The objects' interconnection raises security issues that must be mandatorily lessened, considering the thorough dependence on ICT resources. If all of the computers, networks, and other resources are no longer accessible, no further service can be provided (Krundyshev & Kalinin, 2020). Actually, the adversarial threats in a Smart City are increasingly heterogeneous. As a threat targeting a critical asset of the Smart City infrastructure occurs, appropriate countermeasures must be taken to mitigate the potential impacts.

The concept of Smart City aims to respond to urban challenges through the use of ICT. Smart Cities are riddled with censors that continuously communicate with each other through wireless technologies. Because these technologies are not protected, it is very easy to access them (Lacinák & Ristvej, 2017). But within the transition, cities often neglect the security concept. However, as cities become intelligent, they also become more vulnerable to a myriad of security risks that are consequently manifold. As a result, SRM must consider the integrity and security of data within today's dynamic environment in order to efficiently embody the engineering innovations of Smart Cities. Security must be conducted through a practical strategy that can be easily and widely adopted by the vital services which are presented in Figure 19.1.

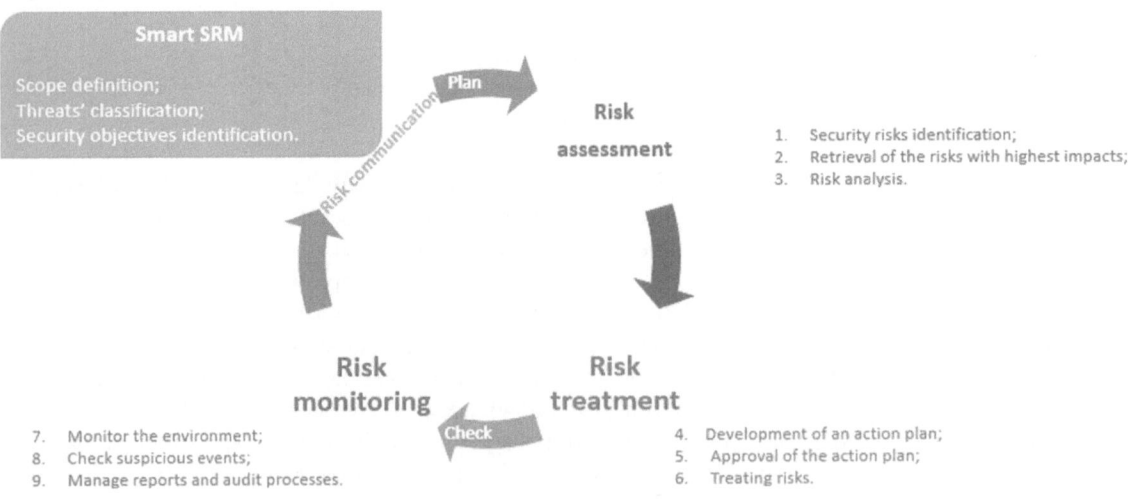

FIGURE 19.1   Smart SRM Framework.

Accordingly, we tolerate:

- A *Planning phase* in order to effectively designate the city's organizational vision. This phase primarily identifies the critical assets and the potential risks that may target the overall security;
- A *"Do"* phase formulating the needed security controls to be implemented;
- A *checking phase* that inspects the feasibility of the chosen security controls. It mainly reviews the level of security risks;
- A *monitoring* phase checking the security controls in order to be extended and integrated into corporate policies. This phase basically triggers preventative actions (ACLs, countermeasures) and maintains a display of the SRM-related information.

19.3.1 Application of Association Rules

The number and variety of security risks are continually increasing, making it more difficult to mitigate their impacts using classical approaches. Accordingly, DL tools provide a greater opportunity to withstand the potential attacks and classify it autonomously. In fact, DL allows gathering knowledge from multiple levels of abstraction. This knowledge distinguishes levels of concepts where higher concepts are defined from lower ones and these with other higher concepts. Previously, we had investigated the use of the Apriori algorithm to determine association rules between vulnerabilities and potential threats (Abbass et al., 2020). We collected data from the interconnected objects and conveyed it to a central implementation and once collected, this data can be used for smart decision making. It has promoted unintentionally bias into the various city's objects. However, due to the classical rule where every items is treated with minimum support and trust, different output rules have been generated resulting in low efficiency.

Association rules compute information from a dataset as if-then statements. It is based on two numbers expressing the degree of uncertainty: an antecedent (the "if" part of a rule) and consequent (the "then" part of a rule). Both antecedent and consequent have a set of items that do not have any items in common. The support represents the number of transactions that include all items in the antecedent and consequent parts of the rule. The confidence is the ratio of the number of transactions that contains all.

Furthermore, the minimum support threshold (min_Sup) and minimum confidence threshold (min_Conf) are generally set to satisfy the security needs. On the one hand, if support (A) ≥ min_Sup, A is then a frequent itemset. On the other hand, if support (A => B) ≥ min_Sup and confidence (A => B) ≥ min_Conf, then A => B is considered a strong association rule. Moreover, the Lift function is employed to measure the correlation between the itemsets in a rule. When the lift is larger than 1, A has a positive correlation with B; when the lift is lesser than 1, A has a negative correlation with B.

The Apriori algorithm is the best candidate mining the association rules. It uses knowledge from previous iterations in order to produce frequent itemsets. It allows finding frequent itemsets whose occurrences exceed a predefined minimum support threshold in order to derive association rules from those frequent itemsets. The application of the Apriori algorithm concerns the identification of the underlying relations between potential risks. As depicted in Figure 19.2, the chosen use case employs a knowledge-based system which is heterogeneous and encompasses sensitive data

The experimental study includes filtering assets that analyze all the potential events. Python scripts are used in order to emulate the global services. The network environment is monitored in order to capture data and build a dataset. Encryption is set with SSL and data is transferred from the GreyLog server. We carried out a bottom-up approach as we started from every single item in the audit dataset and generated the candidates by using self-joining. The length of the itemsets is extended by one item at a time. The subsets are pruned at each stage which is repeated until no more possible itemsets can be derived from the dataset. Security risks are measured through a sequence of steps, including:

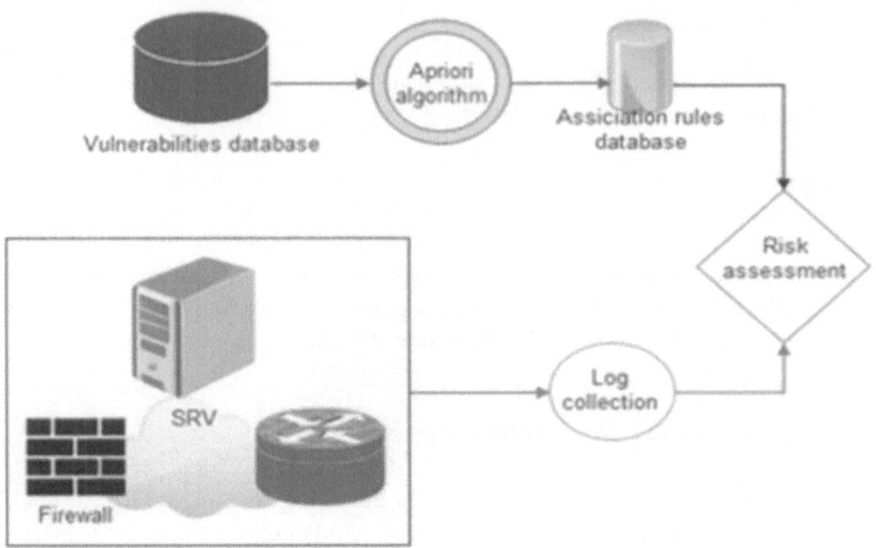

FIGURE 19.2   Experimental Study.

- *Log collection* by using the scanning tool Graylog. It is a free and open-source log management. It is used to collect, index, and analyze business services. Graylog allows real-time monitoring which captures flows from Firewall filtering equipment and export connection traces to a data container. As depicted in Figure 19.3, the logs are commonly captured which make it possible to apply different filters in order to ease their analysis;
- *Vulnerabilities database* that entails the numerous vulnerabilities and their attributes which are suitable for association rule mining. The goal is to determine matching patterns which would further extract SRM-related knowledge. The database is the key to defining the filtration rules;
- *Applying the Apriori algorithm* to the vulnerabilities database. This latter is multidimensional, so the algorithm allows determining vulnerabilities' correlation which further highlights the frequent itemsets. At first, each itemset is taken as a candidate, then their occurrence is counted. Our experimental study involves setting the min_support threshold to 50%. In fact, only the itemsets which count more than or equal to min_support are chosen for the next iteration. Our goal is to determine the correlation between risk attributes within the vulnerabilities database. An association rule, X = > Y, is formed from a set of transactions where some value of itemset X determines the values of itemset Y under 2 indicators: support and confidence. The second iteration of the Apriori algorithm allows determining 2 itemsets with the min_support threshold. Afterwards, the third iteration concludes by determining 3 itemsets falling in the min_support threshold. Subsequently, if the previous itemset is frequent then mining rules can be identified otherwise they are disqualified. The getC () function discovering the candidate data collection and the count () function measures the number of the candidate data collection and allows depicting the quantity of the raw data. The algorithm's application involves:
  - The getL () function investigating the GrayLog data as follows:

```
(1) Begin
(2) if (i>=list.size) then go to (8);
(3) q=list.get(i+j-1);
temp=q.count/ length;
```

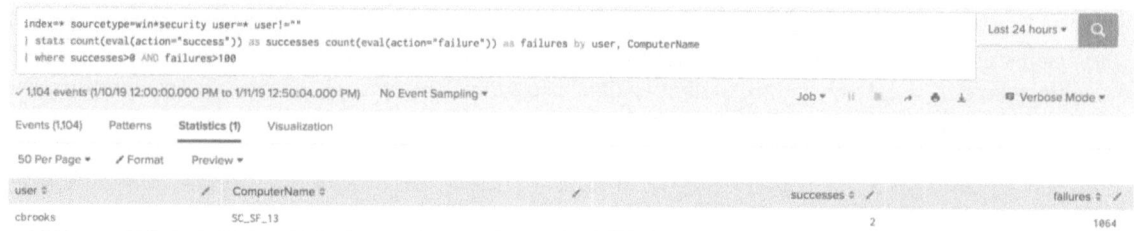

FIGURE 19.3   Log Collection Using Graylog Tool.

```
(4) if (temp>=0.3) then go to (6);
(5) i=i+j, go to (2);
(6) if (n>=j) then go to (2);
(7) list.remove(i), go to (6);
(8) End.
```

Support (X) = (The number of the transactions involving the item X)/(The total number of Transactions)

Confidence = (Transactions involving both items X and Y)/(Total Transactions involving X)

- *Building the association rules* in order to guide the independent vulnerabilities to form a holistic SRM model. Building association rules encompasses these steps:

```
(1) Begin
(2) foreach frequent itemset fi do {
generate all non-empty sub-itemsets Ij of fi
}
(3) foreach non-empty sub-itemsets Ij of fi do {
create the rule (Ij → (fi - Ij))
if Supp(fi)/Supp(Ij) ≥ Min_sup
}
(4) End.
```

The proposed approach allows mining the intrinsic correlation of the ICT environment of Smart Cities in the form of if-then statements. It perceives the potential associations between existing vulnerabilities and attacks. Association rules uncover hidden patterns in large datasets. Accordingly, we identified 5 security risks and we applied the Apriori algorithm in a quest of extracting association rules. Table 19.1 represents the set of 4 failures containing 5 risks.

For the generation of the association rules, the minimum confidence threshold is 50%. Our approach conveys two steps which are the extraction of frequent failures, and the generation of valid association rules from them: a risk is said to be frequent if its support is at least equal to a fixed minimum support; a rule is valid when its confidence exceeds or equals a fixed minimum confidence.

T = {t1, t2, … tn} is the database of transactions.

TABLE 19.1   The set of failures for generating the association rules

ATTACK	POTENTIAL RISKS				
A1	R1	–	–	R4	R5
A2	–	R2	R3	–	R5
A3	R1	R2	R3	–	R5
A4	–	R3	–	–	R5

The formula of A = {a1, a2, … an} is used to represent all attributes that has been distinguished in T (the interesting regularities between features values). Accodringly, we provide the following Tab 1 example where T = {t1, t2, t3} is a sample database with 3 records, A = {speed, acceleration, and Distance} Distance identifies all features distinguished within T. The fuzzy set of the T feature is: FDistance = {Very close, Close, Distant}.

19.4 DISCUSSION

The use of DL has become a powerful tool for SRA. It allows supporting policy makers and enforcement bodies in making strategic decisions by monitoring and improving inspections. Moreover, it results in a fairer and more selective approach and more efficient inspections. SRM of Smart Cities is vulnerable to facing new challenges compared to usual cities. The security risks are considered a compound challenge for governments and organizations. It comprises objects whether distributed or interconnected delivering critical services. These objects are continuously interconnected to the Internet which widens exposure to a myriad of security risks, mainly DoS and DDoS (Mehmood et al., 2017). In fact, assessing and managing risk effectively protects Smart Cities, reduces their vulnerabilities and the potential impacts resulting from risks. SRM is an integral process of management and corporate governance. It includes risk assessment and monitoring which further allows assessing proactively and reactively the security risks. Typically, a Smart City leverages the IoT and smart ICT to collect data from the various objects. DL has significantly contributed to the advancement of uncertainty prediction and analysis. It uses algorithms in a layered manner in order to learn and make intelligent decisions on its own.

On the one hand, DL is the emerging technology for Smart Cities (Habibzadeh et al., 2019). These are increasingly using connected objects and new ICT technologies to improve the quality of services. In fact, they collect an ever-increasing amount of data from different sensors placed in different places. This data is used as a basis for the development of security policies. However, the greater heterogeneity raises several challenges that DL responds to suitably. Indeed, it adapts models with changing parameters and enables real-time adjustments in changing environments. DL applications to Smart Cities have shown promising results, as more accurate decision-making processes are based on a large set of training data. On the other hand, applications of DL require training data with a long execution time. The association rules depend generally on the user's defined threshold values. Consequently, the user has no prior knowledge about the appropriate values. Thus, there is a large set of rules that is generated which unfortunately leads to the primary problem in decision making.

We observed that the used approach supports the rules generated in the overall association mining and discovers the strongest interrelationship with 100% confidence related to the identified vulnerabilities. With 100% confidence level and 60% support, DDoS are commonly related to the Brute Force.

19.5 CONCLUSION

SRM is one of the main concerns of smart cities as it allows protection against malicious and non-malicious threats. Indeed, an improved SRA allows:

- Better modeling data in order to discover association rules of vulnerabilities
- Applying the Apriori algorithm in order to better analyze the network's malicious activities and abnormal behaviours
- Extracting knowledge directly from the data

Smart Cities is the concept where ICT are the improving vectors for the quality and performance of the urban services. Its central interest is to build optimized communities and more livable and desirable spaces on the basis of an organizational approach guided by open virtue and transparency. Smart Cities' security should not be sacrificed for the benefit of technological advancement. The structure of the Smart City must therefore be sufficiently robust to be able to ensure continuity of service even in the event of unavailability. This protection must go beyond technical measures and must integrate a strategic reflection to overcome any eventuality, bearing in mind that risk management must be constantly conducted.

The Apriori algorithm encompasses various scans through the conduction of a scan by a transaction of cutting and hashing. The chosen audit dataset includes malicious and normal events. The improved SRA approach tolerated the analysis of the network's captured data by detecting matching patterns in order to further extract SRA-related knowledge. DL is transforming the way cities operate, deliver, and maintain services. It is expected to meaningfully support the sustainable development of future Smart Cities. Its numerous fields have transformed conventional cities into highly equipped Smart Cities. Yet, this comes with some drawbacks. Thus, there is a need to consider a smart SRM that would hold the Smart City initiatives continuing. A holistic SRA is a high priority to guarantee an organization's overall security. It has an impact on rapid decision-making due to the lack of knowledge. Apriori algorithm has indeed offered new possibilities in terms of data protection and has significantly improved risk intelligence by forecasting the potential risks and strengthening protection against them. The results from this study revealed that all organizational and project-related factors are important considerations for the adoption of SRA based on the Apriori algorithm. In order to improve the detection rate of security risks, future work would focus on a novel knowledge-based database discovery model improving Apriori association rule mining with Particle Swarm Optimization (PSO).

REFERENCES

Abbass, W, Baina, A, & Bellafkih, M (2020). Evaluation of security risks using apriori algorithm. In *Proceedings of the 13th International Conference on Intelligent Systems: Theories and Applications*, Rabat, Morocco.

Alcaraz, C, & Zeadally, S (2015). Critical infrastructure protection: Requirements and challenges for the 21st century. *International Journal of Critical Infrastructure Protection*, 8, 53–66.

Azevedo Guedes, A, Alvarenga, JC, Sgarbi Goulart, MDS, Rodriguez, MRY, & Soares, CP (2018). Smart cities: The main drivers for increasing the intelligence of cities. *Sustainability*, 10(9), 3121.

Baig, ZA, Szewczyk, P, Valli, C, Rabadia, P, Hannay, P, Chernyshev, M, Johnstone, M et al.(2017). Future challenges for smart cities: Cyber-security and digital forensics. *Digital Investigation*, 22, 3–13.

Brun, O, Yin, Y, Gelenbe, E, Kadioglu, YM, Augusto-Gonzalez, J, & Ramos, M (2018). Deep learning with dense random neural networks for detecting attacks against IoT-connected home environments. In *International ISCIS Security Workshop*.

El Alaoui, A (2017). Determinants of a smart city in Morocco. *The Journal of Quality in Education*, 7(9), 11–11.

Elmaghraby, AS, & Losavio, MM (2014). Cyber security challenges in smart cities: Safety, security and privacy. *Journal of Advanced Research*, 5(4), 491–497.

Gharaibeh, A, Salahuddin, MA, Hussini, SJ, Khreishah, A, Khalil, I, Guizani, M, & Al-Fuqaha, A (2017). Smart cities: A survey on data management, security, and enabling technologies. *IEEE Communications Surveys & Tutorials*, 19(4), 2456–2501.

Habibzadeh, H, Nussbaum, BH, Anjomshoa, F, Kantarci, B, & Soyata, T (2019). A survey on cybersecurity, data privacy, and policy issues in cyber-physical system deployments in smart cities. *Sustainable Cities and Society*, 50, 101660

Ismagilova, E, Hughes, L, Rana, NP, & Dwivedi, YK (2020). Security, privacy and risks within smart cities: Literature review and development of a smart city interaction framework. *Information Systems Frontiers*, 1–22.

Krundyshev, V, & Kalinin, M (2020). The security risk analysis methodology for smart network environments. In *International Russian Automation Conference (RusAutoCon)*, Russia.

Lacinák, M, & Ristvej, J (2017). Smart city, safety and security. *Procedia Engineering*, 192, 522–527.

Mehmood, Y, Ahmad,F, Yaqoob, I, Adnane, A, Imran, M, & Guizani, S (2017). Internet-of-things-based smart cities: Recent advances and challenges. *IEEE Communications Magazine, 55*(9), 16–24.

Shayan, S, Kim, KP, Ma, T, & Nguyen, THD (2020). The first two decades of smart city research from a risk perspective. *Sustainability, 12*(21), 9280.

Song, H, Srinivasan, R, Sookoor, T, & Jeschke, S (2017). *Smart cities: foundations, principles, and applications.* John Wiley & Sons.

Sookhak, M, Tang, H, He, Y, & Yu, FR (2018). Security and privacy of smart cities: a survey, research issues and challenges. *IEEE Communications Surveys & Tutorials,* 21(2), 1718–1743.

Syed, AS, Sierra-Sosa, D , Kumar, A, & Elmaghraby, A (2021). IoT in smart cities: A survey of technologies, practices and challenges. *Smart Cities, 4*(2), 429–475.

Index